Mathematics and Visualization

The series *Mathematics and Visualization* is intended to further the fruitful relationship between mathematics and visualization. It covers applications of visualization techniques in mathematics, as well as mathematical theory and methods that are used for visualization. In particular, it emphasizes visualization in geometry, topology, and dynamical systems; geometric algorithms; visualization algorithms; visualization environments; computer aided geometric design; computational geometry; image processing; information visualization; and scientific visualization. Three types of books will appear in the series: research monographs, graduate textbooks, and conference proceedings.

More information about this series at http://www.springer.com/series/4562

Hamish Carr • Issei Fujishiro • Filip Sadlo •
Shigeo Takahashi
Editors

Topological Methods in Data Analysis and Visualization V

Theory, Algorithms, and Applications

Springer

Editors
Hamish Carr
University of Leeds
Leeds, UK

Issei Fujishiro
Keio University
Yokohama, Kanagawa, Japan

Filip Sadlo
Heidelberg University, IWR
Heidelberg, Germany

Shigeo Takahashi
University of Aizu
Aizu-Wakamatsu City, Fukushima, Japan

ISSN 1612-3786 ISSN 2197-666X (electronic)
Mathematics and Visualization
ISBN 978-3-030-43035-1 ISBN 978-3-030-43036-8 (eBook)
https://doi.org/10.1007/978-3-030-43036-8

Mathematics Subject Classification: 76M24, 53A45, 62-07, 62H35, 65D18, 65U05, 68U10

This Springer imprint is published by the registered company Springer Nature Switzerland AG.
The registered company address is: Gewerbestrasse 11, 6330 Cham, Switzerland

Preface

Thirty years ago, the field of scientific visualization established the importance of studying systematic methods for representing data on screen, in particular for functions computed over Euclidean space. As time has passed, the data sizes have increased to the point where it is no longer feasible to present all of the data to a human interpreter, and scientific visualization has therefore depended ever more heavily on analytic techniques to extract significant information for human consideration.

One of the most successful approaches has been the application of topological analysis to data, and the past 30 years have seen a consistent expansion in theory, technique, algorithm, and application. Initially confined to vector field topology, the community developed ideas in scalar field topology, persistent homology, tensor field topology, multivariate topology, as well as specialized approaches for particular data sources.

One of the hallmarks of this work has been the application of topological analysis to scientific problems at scale and a focus on computationally tractable approaches. As a result, the TopoInVis community was developed to support visualization experts in their topological work.

Starting in 2005, biennial workshops have been held on topological visualization in Budmerice (2005), Grimma (2007), Snowbird (2009), Zürich (2011), Davis (2013), Annweiler (2015), and Tokyo (2017), where informal discussions supplement formal presentations and knit the community together. Notably, these workshops have consistently resulted in quality publications under the Springer imprint which form a significant part of the working knowledge in the area.

At the 2017 workshop at Keio University in Tokyo, scalar topology was the largest area of interest, in contrast to previous years when vector topology has often been the dominant area. Vector topology continues to be visible, as does the recent growth in multivariate topology. At the same time, tensor topology is an area of continuing work, and additional types of topology also show up from time to time, while applications continued to be of interest to all present, and the keynotes both addressed application problems.

Of the 23 papers presented at TopoInVis 2017, 16 passed a second-round review process for this volume. In grouping these papers, the largest number (7) related to scalar field topology and have been divided into three papers on persistent homology that are more theoretical plus a further four that are more applied, including one paper dealing with pathological and test cases that straddles scalar and multivariate problems.

In the first group, the first paper investigates new methods of defining hierarchies from persistence pairings, to provide a better representation of complex scalar data sets. The second paper, which received the best paper award at the workshop, gives an innovative new approach for computing merge trees in a form amenable to shared-memory parallel implementation. The third paper tackles the problem of extending existing forms of persistent topology to data sets sampled from non-manifold inputs. In all of these papers, the common concern is to extend the theoretical frameworks which support the topological investigation of data.

The second group of papers is more oriented to practical details than to the underlying theory. In the first paper, topological tools are extended to help understand search spaces from optimization problems by defining a collapsed meta landscape that represents the original space at a coarse-grained level. The second paper explores automated methods for choosing transfer functions that highlight topological features effectively in direct volume rendering (DVR). The third paper also relates to a practical problem: that persistence diagrams are expensive to compare, substituting a persistence indicator function (PIF) that is more amenable to data analysis. Finally, this group includes a paper on a part of topological computation that is rarely addressed in papers due to space limitations: effective methods and test sets for debugging topological code, including paper models for small practical Reeb spaces for instructional purposes.

In the third group of papers, the authors consider the question of topological variation over time. Here, the first paper looks at the use of merger trees to visualize dark matter halos in cosmological simulations. The second tackles a different problem, the analysis of fingers in a simulation of viscous fluid by using persistent tracking over time to identify how fingers merge, grow, and separate. Finally, the third paper tackles the problem of tracking distinct topological regions over time, and how local decision-making generates broken tracks, while global decision-making provides improved tracking of features.

All three of these groups deal primarily with the simplest case for topological analysis: that of scalar field analysis and its variants. The fourth group of papers tackles one of the recent developments in the area: the use of topological analysis for bivariate and multivariate data. The first of these papers considers the use of Joint Contour Nets, a discrete topological structure, to approximate the Pareto analysis of data. In the second paper, existing work on scalar topological user interfaces is extended to bivariate data through the use of fiber surfaces, while the final paper considers combinations of topological data structures, such as the contour tree and Morse–Smale complex.

This leaves one last group, which deals with all other forms of topology. Whereas previous years have often seen a predominance of vector field analysis, this year's workshop accepted one paper on vectors, a second on tensors, and a third on surface topology. In the first of these, Galilean invariance is applied to provide vector analysis that is independent of the frame of reference of the analysis. The second paper extends previous work on tensor field analysis by proving a maximum on the number of critical points possible in a linearly interpolated tensor field. Finally, the last paper looks at new forms of shape analysis of surfaces based on Poincaré duality.

Looking back at this collection of papers, we can see that, although the specific areas of interest ebb and flow, the concern for theory, practical techniques, and applications continues, and that as new forms of topological analysis are introduced, they stimulate a great deal of detail work. However, once established, areas continue due to their ongoing applicability to practical visual data analysis, showing the value to our community of this ongoing Springer book series.

We would therefore like to thank all of the participants of TopoInVis 2017, as well as Springer for their continued support, and anticipate future workshops will continue the process. We would also like to thank Tateishi Science and Technology Foundation and the Telecommunications Advancement Foundation for their generous financial support of the workshop.

Leeds, UK — Hamish Carr
Yokohama, Japan — Issei Fujishiro
Heidelberg, Germany — Filip Sadlo
Aizu-Wakamatsu, Japan — Shigeo Takahashi
Co-Chairs, TopoInVis 2017

Contents

Part I
Persistence

Hierarchies and Ranks for Persistence Pairs

Bastian Rieck, Filip Sadlo, and Heike Leitte

Abstract We develop a novel hierarchy for zero-dimensional persistence pairs, i.e., connected components, which is capable of capturing more fine-grained spatial relations between persistence pairs. Our work is motivated by a lack of spatial relationships between features in persistence diagrams, leading to a limited expressive power. We build upon a recently-introduced hierarchy of pairs in persistence diagrams that augments the pairing stored in persistence diagrams with information about *which* components merge. Our proposed hierarchy captures differences in branching structure. Moreover, we show how to use our hierarchy to measure the spatial stability of a pairing and we define a rank function for persistence pairs and demonstrate different applications.

1 Introduction

A wide range of application domains employ the concept of persistence, i.e., a measure of feature robustness or scale. It is particularly effective when dealing with noisy data, permitting analysts to distinguish between "signal" and "noise". Being a purely topological approach, however, the information conferred by persistence does not retain any spatial information. While this is sometimes desirable, previous work [19, 24, 28] has shown that retaining at least a minimum of geometrical information is often beneficial, as it increases the expressive power. In this paper, we develop a hierarchy that relates points in a persistence diagram. Our hierarchy makes exclusive use of topological properties of data, while still being able to distinguish between geometrically distinct data. Moreover, the hierarchy is capable of measuring stability properties of the pairing of critical points itself, yielding

B. Rieck (✉) · H. Leitte
TU Kaiserslautern, Kaiserslautern, Germany
e-mail: rieck@cs.uni-kl.de; leitte@cs.uni-kl.de

F. Sadlo
Heidelberg University, Heidelberg, Germany
e-mail: sadlo@uni-heidelberg.de

H. Carr et al. (eds.), *Topological Methods in Data Analysis and Visualization V*, Mathematics and Visualization, https://doi.org/10.1007/978-3-030-43036-8_1

additional structural stability information about input data. We demonstrate the practicality of our method by means of several datasets. Additionally, we compare it to a state-of-the-art hierarchy, point out the improvements over said hierarchy, and demonstrate how our novel approach differs from related hierarchical concepts such as Reeb graphs.

2 Related Work

We refer the reader to Edelsbrunner and Harer [17] for a detailed overview of persistence and related concepts. There are several related approaches for creating a hierarchy of persistence information. Doraiswamy et al. [16] calculate a topological saliency of critical points in a scalar field based on their spatial arrangement. Critical points with low persistence that are isolated from other critical points have a higher saliency in this concept. These calculations yield saliency curves for different smoothing radii. While these curves permit a ranking of persistence pairs, they do not afford a description of their nesting behavior. Consequently, in contrast to our approach, the saliency approach is incapable of distinguishing some spatial rearrangements that leave persistence values and relative distances largely intact, such as moving all peaks towards each other. Bauer [1] developed what we refer to in this paper as the regular persistence hierarchy. It is fully combinatorial and merely requires small changes of the pairing calculation of related critical points. This hierarchy was successfully used in determining cancellation sequences of critical points of surfaces. However, as shown in this paper, this hierarchy cannot distinguish between certain nesting relations.

In scalar field analysis, the calculation of graph structures such as the Reeb graph [15] or the contour tree [11], along with merge and split trees, has a long tradition. These graphs relate critical points with each other, but do not permit the calculation of hierarchies of persistence pairs. We will demonstrate this on a simple one-dimensional example in this paper. Recent work in this area is driven by the same motivation as our work: the merge tree, for example, turns out to be more expressive with respect to spatial differences in the domain. Thus, even if two scalar fields have the same critical pairs, their merge trees are capable of retaining differences in sublevel set merging behavior. This observation led to the development of distance measures for merge trees [3], Reeb graphs [2], and extremum graphs [22]. Since the aforementioned tree structures tend to be unwieldy, Pascucci et al. [23] proposed a hierarchical decomposition, the branch decomposition. This decomposition relates the different branches of a contour tree with each other. Recently, Saikia et al. [25] used these graphs as a similarity measure for the structural comparison of scalar data. While these works are close to our method in spirit, they rely on a different type of structural information.

3 Background and Notation

In this paper, we assume that we are working with a domain $\mathbb{D}$ and a scalar function $f\colon \mathbb{D} \to \mathbb{R}$. We make no assumptions about the connectivity of $\mathbb{D}$ or its dimension. As for f, we require it to have a finite number of critical points—a condition that is always satisfied for real-world data—and that the function values of those critical points are different—a condition that may be satisfied by, e.g., symbolic perturbation [18]. Such scalar fields commonly occur in many different applications, and their features are often described using scalar field topology. This umbrella term refers to the analysis of how certain special sets—the level sets—of the scalar function f change upon varying parameters. More precisely, given a threshold c, one typically distinguishes between, e.g., level set $\mathscr{L}_c(f)$, and sublevel set $\mathscr{L}_c^-(f)$,

$$\mathscr{L}_c(f) := \{x \in \mathbb{D} \mid f(x) = c\} \tag{1}$$

$$\mathscr{L}_c^-(f) := \{x \in \mathbb{D} \mid f(x) \leq c\} \tag{2}$$

In this paper, we also require the interlevel set $\mathscr{L}_{l,u}(f)$,

$$\mathscr{L}_{l,u}(f) := \mathscr{L}_u^-(f) \setminus \mathscr{L}_l^-(f) = \{x \in \mathbb{D} \mid l \leq f(x) \leq u\}. \tag{3}$$

Interlevel sets are commonly used to describe the topology of real-valued functions [9] or the robustness of homology classes [5, 6]. Scalar field topology refers to the investigation of changes in connectivity in these sets. Such changes are intricately connected to the critical points of f by means of Morse theory [21]. Focusing on the sublevel sets (the case for superlevel sets can be solved by duality arguments), we find that (local) minima *create* new connected components in the sublevel set, while (local) maxima—or saddles in higher dimensions—are responsible for merging two connected components, thereby *destroying* one of them. Related creators and destroyers may thus be paired with each other (using, e.g., the "elder rule" [17, p. 150] that merges the connected component with a higher—younger—function value into the one with a lower—older—function value), which permits their use in various data structures.

The persistence diagram is a generic data structure to represent such a pairing. For every creator–destroyer pair, it contains a point in $\mathbb{R}^2$ according to the corresponding function values. Persistence diagrams have many desirable stability properties [13, 14] and permit the calculation of different metrics. Unfortunately, they are sometimes too coarse to describe both the topology *and* geometry of a scalar function. Given a point (c, d) in a persistence diagram, where c is the function value of the creator and d is the function value of the paired destroyer, the absolute difference $|d - c|$ is referred to as the persistence $\mathrm{pers}(c, d)$ of the pair. Persistence permits a way to define whether certain pairs are more prominent than others. Roughly speaking, the persistence of such a pair is the magnitude of the smallest perturbation that is able to cancel it.

4 Persistence Hierarchies

The calculation of persistence always underlies the idea of a pairing, i.e., a way of relating different parts of a function with each other. Here, we shall only focus on zero-dimensional persistent homology, which describes connected components in the sublevel sets of a function, and the elder rule for pairing points. Consequently, we have a relationship between local minima and local maxima (or saddles in higher dimensions) of a function. We leave the treatment of other topological features for future work. Moreover, we only cover the case of sublevel sets; superlevel set calculations follow by analogy.

4.1 Regular Persistence Hierarchy

Bauer [1] observes that the process of merging two connected components permits the definition of a natural hierarchy between persistence pairs. More precisely, assume that we are given two connected components σ and σ', each created at a local minimum. If σ merges into σ' at, e.g., a local maximum, we consider σ' to be the parent of σ. This relation is a necessary but not sufficient condition for finding out which pairs of critical points of a Morse function cannot be canceled without affecting other points. We call this hierarchy, which is equivalent to a merge tree [11], the regular persistence hierarchy. Each of its nodes corresponds to a creator–destroyer pair. The hierarchy forms a directed acyclic graph, i.e., a tree. This is a consequence of the assumption that the function values at critical points are unique. Thus, whenever a merge of two connected components takes place, the "younger" component is uniquely defined. Moreover, a critical point cannot both create and destroy a topological feature, so there cannot be any cycles in the hierarchy. The regular persistence hierarchy has a natural root that corresponds to the global minimum, as the connected component corresponding to this value is never merged.

Example and Limitations The regular persistence hierarchy cannot distinguish some connectivity relations: for example, Fig. 1 depicts the regular persistence hierarchies for two simple functions. We can see that the hierarchy is equal for both functions even though their connectivity behavior is different. More precisely, in the red function, the two persistence pairs are connected via two different branches of the function, i.e., it is impossible to reach both minima without traversing a third minimum—the global one—as the threshold of the sublevel sets is increased. This difference in connectivity results in a different stability of the pairing. A perturbation of the critical points at ⓩ and ⓒ in the blue function, for example, is capable of changing the complete pairing: if we move the points to $f(4) = 1.9$ and $f(5) = 0.9$, respectively, the pairing of the critical point ⓐ at $x = 1$ will change, as well. The same perturbation has no effect on the red function, though. A hierarchy of persistence pairs should account for these differences in connectivity.

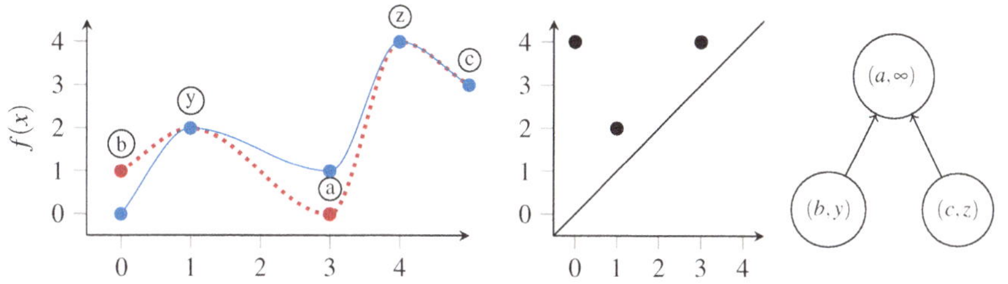

Fig. 1 Two functions with different connectivity but equal persistence diagrams. Both functions also share the same regular persistence hierarchy. (**a**) Functions. (**b**) Persistence diagram. (**c**) Hierarchy

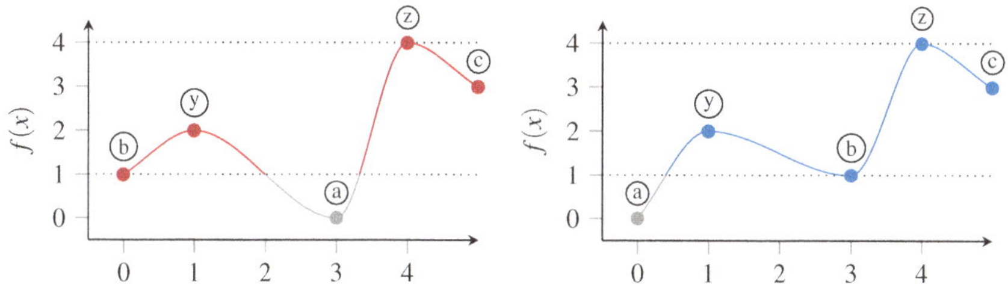

Fig. 2 The merge phase of the interlevel set persistence hierarchy (ISPH) makes use of the connectivity of the interlevel set (hatched lines): to connect the critical point pairs (b, y) and (c, z) in (**a**), a region belonging to a third critical point ⓐ needs to be traversed. This is not the case for (**b**)

4.2 *Interlevel Set Persistence Hierarchy (ISPH)*

The example depicted in Fig. 1 demonstrated a lack of discriminating information in the regular persistence hierarchy. The key observation, illustrated as running example in Fig. 2, is that not every merge of two connected components is topologically equal: a merge may either result in a different branching structure of the hierarchy or it may keep the branches intact. Figure 3 depicts our proposed interlevel set persistence hierarchy (ISPH). To measure these differences, we propose extending the traditional union–find data structure that is used to detect the merges. Instead of merely storing the parent node of a given connected component in the hierarchy, i.e., the generating critical point with lowest function value, we also store $\hat{c}$, the highest minimum—in terms of the function value—along this branch. This will permit us to decide whether an additional branch needs to be introduced, as is the case for function (a) in Fig. 2. If $\hat{c}$ of a connected component is identical to the value of the parent, we call this assignment trivial. We will use $\hat{c}$ interchangeably both for the critical point as well as for its function value. Subsequently, we distinguish between two types of merges: the first type of merge only extends a branch in

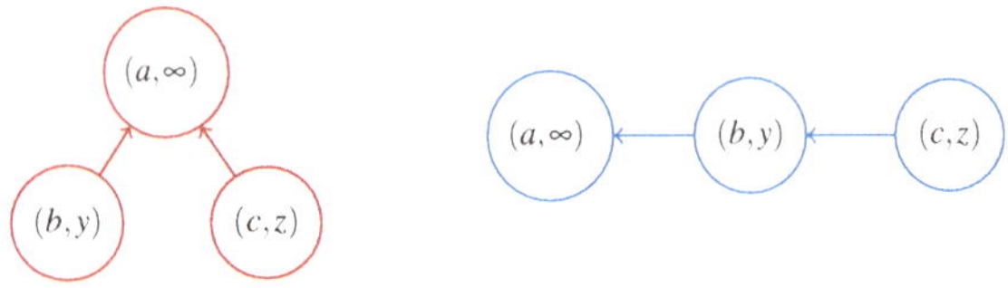

Fig. 3 The ISPHs for the example functions shown in Fig. 1. In contrast to the regular persistence hierarchy, our hierarchy is capable of discriminating between the two functions. The hierarchy on the right has been rotated for layout reasons

the hierarchy, while the second type of merge results in two branches that need to be unified at a third critical point. Figure 2 depicts the two cases for the example functions shown in Fig. 1. We can see that for function (a), the pairs of critical points are connected by an additional critical point ⓐ only. Hence, two branches of the hierarchy merge at this point. Function (b), by contrast, merely prolongs a branch in the hierarchy—both pairs of critical points of the function are already connected without the inclusion of an additional critical point.

To distinguish between these two cases, we check the stored highest critical points $\hat{c}_l$ and $\hat{c}_r$ of the two connected components that merge at a local extremum. Without loss of generality, we assume that $\hat{c}_l$ belongs to the "older" branch and $\hat{c}_r$ belongs to the "younger" branch. If both $\hat{c}_l$ and $\hat{c}_r$ are trivial, we merge their respective branches just as for the regular persistence hierarchy. Else, we have to check the induced connectivity to decide how the branches should be connected. To this end, let y_u refer to the value of the current critical point, i.e., the one at which the two connected components merge. Furthermore, let y_l refer to $\min(\hat{c}_l, \hat{c}_r)$, the oldest of the two stored critical points. We now calculate the interlevel set $\mathscr{L}_{y_l,y_u}(f)$; see the colored parts in Fig. 2 for an example. Following this, we check whether $\hat{c}_l$ and $\hat{c}_r$ are in the same connected component with respect to $\mathscr{L}_{y_l,y_u}(f)$. If so, the current branch can be prolonged and $\hat{c}_l$ is set to $\hat{c}_r$. If not, two branches meet at the current critical point and merge into one. In Fig. 2a, upon reaching ⓨ, we merge components ⓑ and ⓐ, giving rise to the pair (b, y). We have trivial critical points, i.e., $\hat{c}_l$ = ⓐ and $\hat{c}_r$ = ⓑ, so we add an edge between (b, y) and $(a, \cdot)$ in the hierarchy; we do not yet know how a will be paired, so we write "$\cdot$" for its partner. The next merge happens at ⓩ. We have $\hat{c}_l$ = ⓑ, $\hat{c}_r$ = ⓒ, and y_u = ⓩ. This gives rise to the interlevel set $\mathscr{L}_{b,z}(f)$. We now check whether $\hat{c}_l$ and $\hat{c}_r$ are connected in $\mathscr{L}_{b,z}(f)$. As this is not the case, we add an edge between (c, z) and $(a, \cdot)$, which is the parent of the older component, to the hierarchy. In Fig. 2b, by contrast, we have the same merges and the same interlevel set, but ⓒ and ⓑ are *connected* in $\mathscr{L}_{b,z}(f)$, leading to the creation of an edge between (c, z) and (b, y). The check with respect to the interlevel set connectivity is insufficient, however, for higher-dimensional domains. Instead, we need to check whether a path in the neighborhood graph of our data (or, equivalently, an integral line) connecting the two critical points does not cross any regions that are assigned to another critical point. This presumes that we classify our data according to ascending or descending regions, which can be easily integrated into standard persistent homology algorithms [12].

Alg. 1 Calculation of the ISPH

```
Require: A domain 𝔻
Require: A function f : 𝔻 → ℝ
U ← ∅
Sort the function values of f in ascending order
for function value y of f do
    if y is a local minimum then
        Create a new connected component in U
        U.critical ← y
    else if y is a local maximum or saddle then
        Use U to merge the two connected components meeting at y
        Let C′ and C be the two components meeting at y
        if both components have a trivial critical value then
            Create the edge (C′, C) in the hierarchy
        else
            Let c′ be the critical value of the older connected component
            Let c be the critical value of the younger connected component
            y_l ← min(c, c′)
            Create the interlevel set L := 𝓛_{y_l, y}(f)
            if the shortest path connecting c, c′ in L contains no other critical points then
                Create the edge (c′, y) in the hierarchy
            else
                Create the edge (C′, C) in the hierarchy (as above)
            end if
        end if
    else
        Use U to add y to the current connected component
    end if
end for
```

Algorithm and Example The ISPH requires only an additional data structure for storing information about the critical points we encounter. Moreover, we require checking the connectivity of the interlevel set—an operation that requires an additional union find data structure and a way to calculate (shortest) paths in the neighborhood graph of our data. Algorithm 1 gives the pseudocode description of our novel hierarchy based on sublevel sets. Figure 3 shows the ISPHs for the example functions in Fig. 1, demonstrating that our hierarchy represents differences in merging behavior in the sublevel sets.

Implementation We implemented our algorithm using `Aleph`,[1] a library for exploring various uses of persistent homology. Our implementation of the ISPH is publicly available and supports processing structured grids (using the VTK file format) as well as one-dimensional functions.

Comparison with Other Tree-Based Concepts The ISPH is capable of preserving more information than merge trees, split trees, and Reeb graphs. As a simple

[1] https://github.com/Submanifold/Aleph.

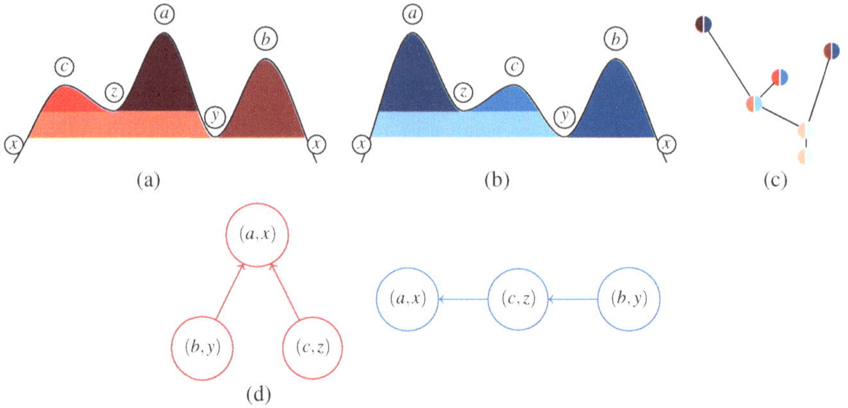

Fig. 4 Comparison with Reeb graphs. The two functions yield the same Reeb graph (or, equivalently, the merge tree of their superlevel sets), while our hierarchy is capable of telling them apart. (**a**) First function. (**b**) Second function. (**c**) Reeb graph. (**d**) Interlevel set persistence hierarchys (ISPHs)

example, consider the functions in Fig. 4. Both functions carry the same sublevel/superlevel set information; their persistence diagrams and regular persistence hierarchies coincide. Their Reeb graphs, shown in Fig. 4c with nodes whose colors indicate the corresponding contour of the function, are also equal. Of course, this does not imply that Reeb graphs (or merge trees) are generally unsuitable. During our experiments, we encountered numerous functions in which Reeb graphs (or merge trees) are able to detect differences in functions with equal persistence diagrams. At the same time, the ISPH was able to detect differences in these cases as well. Figure 4d shows the ISPHs of the functions in Fig. 4a, b.

The preceding example proves that the ISPH is unrelated to existing decompositions: since the Reeb graphs (and the merge trees) of the two functions are equal but the ISPHs differ, it is not possible to derive the ISPH from, e.g., the branch decomposition tree [23] or the extended branch decomposition graph [25].

Robustness When adding noise to a function, topological hierarchies such as the merge (split) tree and the Reeb graph are known to contain numerous short branches, which make identifying important features and comparing different trees more difficult [25]. By contrast, our novel ISPH only contains as many points as there are *pairs* in the persistence diagram. Moreover, low-persistence pairs do not result in too much clutter because they tend to only create short branches. In that sense, our hierarchy performs similarly well as the extended branch decomposition graph by Saikia et al. [25].

4.2.1 Calculating Ranks

Since the ISPH is a DAG, we can define the rank of a topological feature: given two vertices u and v in the hierarchy $\mathscr{H}$, we write $u \sim v$ if there is a directed path connecting u and v. The rank of a vertex u in the hierarchy is then calculated as the number of vertices that are reachable from it, i.e.,

$$\operatorname{rank}(u) := \operatorname{card}\{v \in \mathscr{H} \mid u{\sim}v\}, \tag{4}$$

with $\operatorname{rank}(\cdot) \in \mathbb{N}$. The minimum of the rank function is obtained for the last connected component to be destroyed, i.e., the one that merges last with another component. The rank can be easily calculated using a depth-first traversal of the tree. It may be visualized as additional information within a persistence diagram, thereby permitting datasets with similar persistence diagrams but different hierarchies to be distinguished from each other without showing the hierarchy itself. It is also invariant with respect to scaling of the function values in the data. Similar concepts, such as the rank invariant [8] in multidimensional persistence, only use existing information from the persistence diagram, whereas our ISPH goes beyond the persistence diagram by including more structural information about critical points.

4.2.2 Stability Measure

Our ISPH permits assessing the stability of the *location* of critical points in the pairing. This issue with persistence diagrams was already pointed out by Bendich and Bubenik [4], who demonstrated that small changes in the critical values of a function—while not drastically changing the persistence diagram itself—may still change the points that are responsible for creating a certain topological feature. The ISPH contains information that may be used to assess the stability of the creators of topological features. To make this more precise, consider the example in Fig. 5. Upon traversing the superlevel sets of both functions, they exhibit the same persistence pairs, namely (a, z), (b, y), and (c, z). Refer to Fig. 4d for the corresponding ISPHs. If we perturb the critical point ⓨ for both functions (indicated

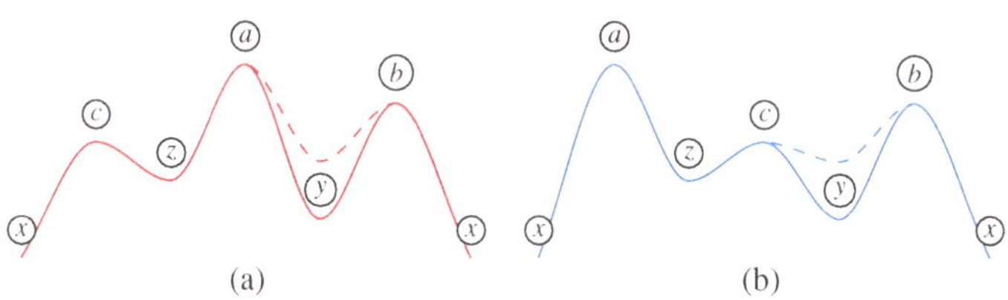

Fig. 5 Stable and unstable function whose superlevel sets yield the same persistence diagram. Since the ISPH is capable of distinguishing between the two cases, it assesses their respective stability differently. (**a**) Stable function. (**b**) Unstable function

by a dashed line), we still get the pairs (b, y) and (c, z) for the stable function. For the unstable function, however, the perturbation results in the pairs (b, z) and (c, y). In this sense, their location is less stable.

We thus define a stability measure for each critical point based on the hierarchy. First, we need to quantify the stability of an *edge*. Let $e := \{(\sigma, \tau), (\sigma', \tau')\}$ be an edge in the ISPH. We define the stability of e to be

$$\operatorname{stab}(e) := \max\left\{|f(\sigma) - f(\sigma')|, |f(\tau) - f(\tau')|\right\}, \tag{5}$$

which is the minimum amount of perturbation that is required to change the hierarchy in the sense described above. This quantity is also equal to the L_∞-distance between two points in a persistence diagram. We may now extend the stability measure to individual vertices by assigning each vertex v the minimum stability value of its outgoing edges, i.e.,

$$\operatorname{stab}(v) := \min\left\{\min\{\operatorname{stab}(e) \mid e = (v, w) \in \mathscr{H}\}, \operatorname{pers}(v)\right\}, \tag{6}$$

where w ranges over all direct children of v. Taking the second minimum ensures that we use the persistence of v if v is a leaf node.

4.2.3 Dissimilarity Measure

Since the ISPH is a directed tree, a straightforward dissimilarity measure is given by tree edit distance [7] algorithms. These algorithms attempt to transform two trees into each other by three elementary operations: relabeling a given node, deleting an existing node, and inserting a new node. Given two nodes with corresponding minima–maxima (c_1, d_1) and (c_2, d_2), respectively, we define the cost for *relabeling* to be

$$\operatorname{cost}_1 = \max\left(|c_1 - c_2|, |d_1 - d_2|\right), \tag{7}$$

i.e., the L_∞-distance between the two points. Similarly, we define the cost for *deleting* or *inserting* a node somewhere else in the hierarchy to be

$$\operatorname{cost}_2 = \operatorname{pers}(c, d) = |d - c|, \tag{8}$$

i.e., the persistence of the pair. The choice of these costs is "natural" in the sense that they are also used, e.g., when calculating the bottleneck distance [13] between persistence diagrams.

Complexity The tree edit distance has the advantage of being efficiently-solvable via standard dynamic programming techniques. It is thus scalable with respect to the size of the hierarchy—more so than the Wasserstein or bottleneck distances [20, 26] that are commonly used for comparing persistence diagrams: we have a worst case complexity of $\mathcal{O}\left(n^2 m \log m\right)$, where n is the number of pairs in the smaller hierarchy, and m is the number of pairs in the larger hierarchy. By contrast, the Wasserstein distance has a time complexity of $\mathcal{O}\left(n^3\right)$ [17, p. 196], and we observed large differences in runtime behavior (see Sect. 5).

5 Results

We exemplify some usage scenarios of the ISPH by means of several synthetic and non-synthetic datasets.

5.1 *Synthetic Data*

We created two synthetic datasets on a grid with 5000 cells. Processing each dataset takes approximately 4.5 s—regardless of whether we calculate the regular persistence hierarchy or the ISPH. Our current implementation leaves lots of room for performance improvements, though. Figure 6 shows the data together with the resulting hierarchies for the two-dimensional equivalent to the data shown in Fig. 5. We use a standard color map that uses red for large values and white for low values. Notice that the persistence diagrams of both datasets are equal, as well as their regular persistence hierarchies (which we do not show here). Figure 7a depicts the persistence diagrams of the two data sets, colored by their stability values and the ranks (mirrored along the diagonal). Even this summary information gleaned from the ISPH is capable of yielding information to distinguish different datasets.

Figure 8 depicts a more complicated example, mixing peaks and craters. The situation in Fig. 8b is less stable because a perturbation of the peak is capable of changing the complete pairing. Since this peak is connected differently in Fig. 8a, it does not decrease the stability of the global maximum. Note that the location of the peak is allowed to move; as long as it stays on the ridge, as depicted in Fig. 8b, the ISPH will not change. Likewise, as long as the peak moves along the plateau in the foreground, the ISPH will remain the same as depicted in Fig. 8a.

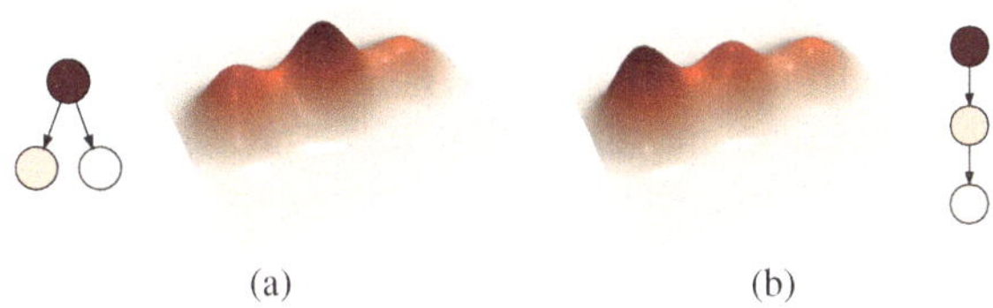

Fig. 6 In contrast to previous approaches, our hierarchy is capable of distinguishing between two peaks that are connected via a third one that is higher (**a**) and a "ridge" of peaks (**b**)

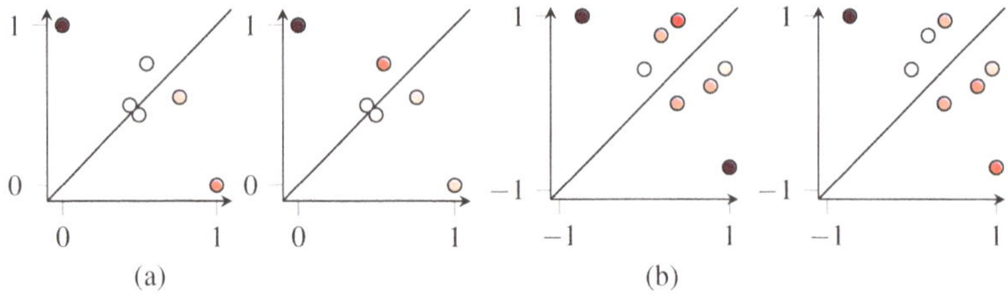

Fig. 7 Combined persistence diagrams showing the ranks (upper part) and the stability values (lower part) as colors. Both carry sufficient information to distinguish datasets. (**a**) Persistence diagrams for Fig. 6. (**b**) Persistence diagrams for Fig. 8

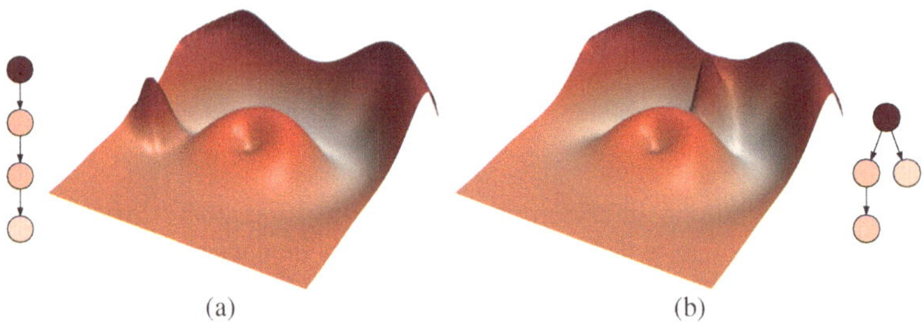

Fig. 8 Depending on the position of a single peak that moves, our hierarchies are different, because the merging behavior of critical points has been changed

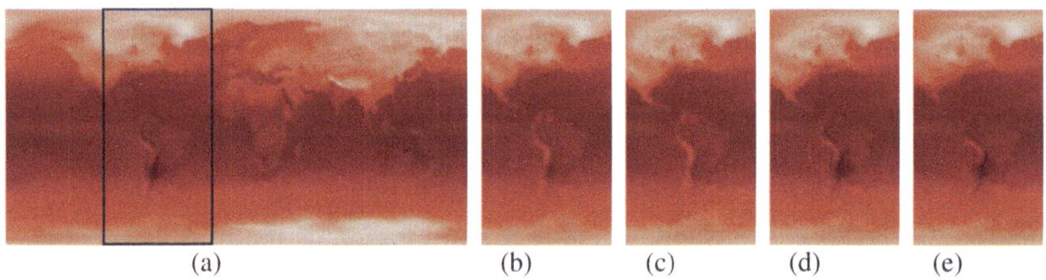

Fig. 9 Climate dataset for $t = 0$ (**a**), as well as some excerpts (box) for subsequent timesteps. We can see that the continent of Africa exhibits a temperature increase for $t = 3$ and $t = 4$. (**a**) $t = 0$. (**b**) $t = 1$. (**c**) $t = 2$. (**d**) $t = 3$. (**e**) $t = 4$

5.2 Climate Data

We used time-varying scalar field data (surface temperature) from the German Climate Computing Center (DKRZ). The large size (18,432 positions) and the large number of timesteps (1460) makes comparing these data complicated. Figure 9 shows an excerpt of the dataset. It exhibits oscillatory behavior because (global) temperature follows a day–night pattern. At the given resolution, a full day–night cycle comprises four timesteps. We would hence expect that the topological dissimilarity between corresponding timesteps is somewhat equal; or, more precisely,

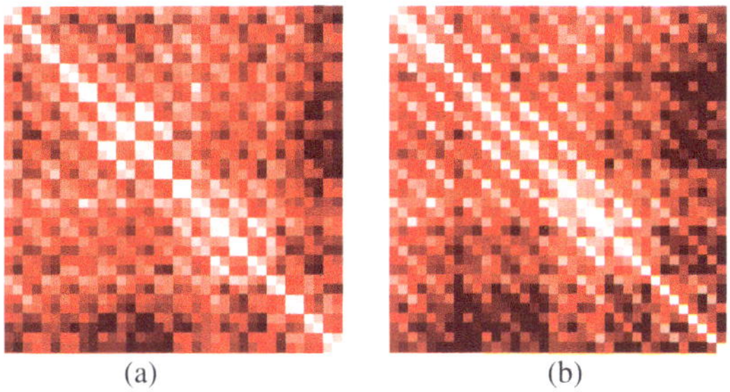

Fig. 10 Dissimilarity matrices comparing the Wasserstein distance and our proposed ISPH distance, which is capable of detecting the oscillatory behavior (minor diagonals) inherent to the data. (**a**) Wasserstein distance. (**b**) ISPH distance

we would expect that a pairwise distance matrix depicts the oscillatory behavior that occurs in the data. To assess the capabilities of the dissimilarity measure from Sect. 4.2.3, we compare it to the Wasserstein distance. We first calculate pairwise distances between the first 36 timesteps of the data. Calculating the Wasserstein distance takes about 2 min per pair, hence comparing all 36 pairs takes about 21 h. Our hierarchy-based dissimilarity measure, by contrast, takes 2.3 s for calculating each timestep and approximately 6 s to calculate the dissimilarity per pair. The full distance matrix is thus obtained in approximately 1 h. Figure 10 depicts the dissimilarity matrices of the first 36 timesteps. We can see that the ISPH distance matrix contains patterns in the minor diagonals, indicating that timesteps t_i and t_{i+4} are highly similar. These patterns appear because the corresponding ISPHs are also highly similar, even though the persistence pairs (and thus the persistence diagrams) change over time because the range of temperatures changes. The Wasserstein distance does not exhibit these patterns. We note, however, that the two matrices are highly correlated ($R^2 \approx 0.95$), showing that for *most* of the timesteps, both measures yield similar values.

6 Conclusion and Future Work

We presented a novel hierarchy that relates persistence pairs with each other. In contrast to earlier work, our hierarchy is better capable of distinguishing certain nesting behaviors, which we demonstrated by means of several example datasets. At present, it is unclear to what extent the persistence hierarchy is an invariant like the *rank invariant* of multidimensional persistence [8]. Since the hierarchy changes when the data undergoes certain transformations, it remains to be shown which operations leave it unchanged—a trivial observation is that the hierarchy is invariant with respect to uniform scaling in all critical points. Hence, the development of metrics for matching hierarchies may be beneficial. We only briefly sketched a dissimilarity measure based on labeled trees. By considering the complete connectivity structure of the tree, though, graph kernels such as the random walk kernel [27] could be employed. Furthermore, it would be interesting to analyze how the hierarchy changes when different pairing schemes for critical points are employed, such as the measures proposed by Carr et al. [10].

Acknowledgments We thank the anonymous reviewers for their helpful comments that helped us improve and clarify this manuscript.

References

1. Bauer, U.: Persistence in discrete Morse theory. Ph.D. Thesis, University of Göttingen (2011)
2. Bauer, U., Ge, X., Wang, Y.: Measuring distance between Reeb graphs. In: Proceedings of the Annual Symposium on Computational Geometry, pp. 464:464–464:473 (2014)
3. Beketayev, K., Yeliussizov, D., Morozov, D., Weber, G.H., Hamann, B.: Measuring the distance between merge trees. In: Topological Methods in Data Analysis and Visualization III: Theory, Algorithms, and Applications, pp. 151–165. Springer, Berlin (2014)
4. Bendich, P., Bubenik, P.: Stabilizing the output of persistent homology computations (2015). arXiv:1512.01700
5. Bendich, P., Edelsbrunner, H., Kerber, M.: Computing robustness and persistence for images. IEEE Trans. Vis. Comput. Graph. **16**(6), 1251–1260 (2010)
6. Bendich, P., Edelsbrunner, H., Morozov, D., Patel, A.: Homology and robustness of level and interlevel sets. Homology Homotopy Appl. **15**, 51–72 (2013)
7. Bille, P.: A survey on tree edit distance and related problems. Theor. Comput. Sci. **337**(1), 217–239 (2005)
8. Carlsson, G., Zomorodian, A.J.: The theory of multidimensional persistence. Discret. Comput. Geom. **42**(1), 71–93 (2009)
9. Carlsson, G., de Silva, V., Morozov, D.: Zigzag persistent homology and real-valued functions. In: Proceedings of the Annual Symposium on Computational Geometry, pp. 247–256 (2009)
10. Carr, H., Snoeyink, J., van de Panne, M.: Simplifying flexible isosurfaces using local geometric measures. In: IEEE Conference on Visualization, pp. 497–504 (2004)
11. Carr, H., Snoeyink, J., Axen, U.: Computing contour trees in all dimensions. Comput. Geom. **24**(2), 75–94 (2003)
12. Chazal, F., Guibas, L.J., Oudot, S.Y., Skraba, P.: Persistence-based clustering in Riemannian manifolds. J. ACM **60**(6), 41:1–41:38 (2013)
13. Cohen-Steiner, D., Edelsbrunner, H., Harer, J.: Stability of persistence diagrams. Discret. Comput. Geom. **37**(1), 103–120 (2007)
14. Cohen-Steiner, D., Edelsbrunner, H., Harer, J., Mileyko, Y.: Lipschitz functions have L_p-stable persistence. Found. Comput. Math. **10**(2), 127–139 (2010)
15. Doraiswamy, H., Natarajan, V.: Efficient algorithms for computing Reeb graphs. Comput. Geom. **42**(6–7), 606–616 (2009)
16. Doraiswamy, H., Shivashankar, N., Natarajan, V., Wang, Y.: Topological saliency. Comput. Graph. **37**(7), 787–799 (2013)
17. Edelsbrunner, H., Harer, J.: Computational Topology: An Introduction. AMS, Providence (2010)
18. Edelsbrunner, H., Mücke, E.P.: Simulation of simplicity: a technique to cope with degenerate cases in geometric algorithms. ACM Trans. Graph. **9**(1), 66–104 (1990)
19. Gerber, S., Bremer, P.T., Pascucci, V., Whitaker, R.: Visual exploration of high dimensional scalar functions. IEEE Trans. Vis. Comput. Graph. **16**(6), 1271–1280 (2010)
20. Maria, C., Boissonnat, J.D., Glisse, M., Yvinec, M.: The Gudhi library: simplicial complexes and persistent homology. In: Hong, H., Yap, C. (eds.) Mathematical Software – ICMS 2014, pp. 167–174. Springer, Heidelberg (2014)
21. Milnor, J.: Morse Theory. Princeton University Press, Princeton (1963)

22. Narayanan, V., Thomas, D.M., Natarajan, V.: Distance between extremum graphs. In: IEEE Pacific Visualization Symposium (PacificVis), pp. 263–270 (2015)
23. Pascucci, V., Cole-McLaughlin, K., Scorzelli, G.: The Toporrery: computation and presentation of multi-resolution topology. In: Mathematical Foundations of Scientific Visualization, Computer Graphics, and Massive Data Exploration, pp. 19–40. Springer, Berlin (2009)
24. Rieck, B., Leitte, H.: Structural analysis of multivariate point clouds using simplicial chains. Comput. Graph. Forum **33**(8), 28–37 (2014)
25. Saikia, H., Seidel, H.P., Weinkauf, T.: Extended branch decomposition graphs: structural comparison of scalar data. Comput. Graph. Forum **33**(3), 41–50 (2014)
26. Tierny, J., Favelier, G., Levine, J.A., Gueunet, C., Michaux, M.: The topology ToolKit. IEEE Trans. Vis. Comput. Graph. **24**(1), 832–842 (2018)
27. Vishwanathan, S.V.N., Schraudolph, N.N., Kondor, R., Borgwardt, K.M.: Graph kernels. J. Mach. Learn. Res. **11**, 1201–1242 (2010)
28. Zomorodian, A.J., Carlsson, G.: Localized homology. Comput. Geom. **41**(3), 126–148 (2008)

Triplet Merge Trees

Dmitriy Smirnov and Dmitriy Morozov

Abstract Merge trees are fundamental data structures in computational topology. They track connected components in sublevel sets of scalar functions and can be used to compute 0-dimensional persistence diagrams, to construct contour trees on simply connected domains, and to quickly query the relationship between connected components in different sublevel sets. We introduce a representation of merge trees that tracks the nesting of their branches. We present algorithms to construct and manipulate the trees in this representation directly. We show that our algorithms are not only fast, outperforming Kruskal's algorithm, but they are easy to parallelize in shared memory using double-word compare-and-swap operations. We present experiments that illustrate the scaling of our algorithms as functions of the data size and of the number of threads.

D. Smirnov
Massachusetts Institute of Technology, Cambridge, MA, USA
e-mail: smirnov@mit.edu

D. Morozov (✉)
Lawrence Berkeley National Laboratory, Berkeley, CA, USA
e-mail: dmitriy@mrzv.org

H. Carr et al. (eds.), *Topological Methods in Data Analysis and Visualization V*, Mathematics and Visualization, https://doi.org/10.1007/978-3-030-43036-8_2

1 Introduction

Merge trees are widely used in computational topology. These data structures track how components appear and merge in sublevel sets of functions, as one sweeps the threshold from negative to positive infinity. Once constructed, merge trees can be used to generate contour trees [1] or 0-dimensional persistence diagrams [2]. They are used in applications ranging from combustion to cosmology to materials science.

Most algorithms to compute merge trees are closely related to algorithms for minimum spanning tree construction. Of these, the most common is Kruskal's algorithm [3]. Recently, Carr et al. [4] introduced an algorithm that constructs merge trees by incrementally pruning extrema. It can be viewed as an adaptation of Borůvka's algorithm [5] to the merge tree problem, and, as Borůvka's, their algorithm is amenable to parallelization.

The algorithm of Bremer et al. [6] deserves special attention. It works by incrementally adding edges to the domain and updating the tree by merging paths inside it. This algorithm features several desirable properties. (1) It can process edges in a streaming fashion, without access to the full sequence. This property is most convenient when the domain is represented implicitly, and the number of edges is significantly higher than the number of vertices. (2) It can be used to combine a pair of merge trees to find the merge tree of the union of underlying domains, in sublinear time. In other words, it does not need to access those parts of the trees that do not change in the union. The need to combine merge trees comes up naturally during distributed computation [7, 8]. (3) It can be easily parallelized in shared memory, in a lock-free manner, using compare-and-swap operations for synchronization.

The algorithm of Bremer et al. would be perfect if it was not so slow. In theory, it can scale quadratically in the number of input vertices. In practice, it is orders of magnitude slower than Kruskal's algorithm. Section 3 includes this algorithm in several experiments, where it leaves a lot to be desired.

In this paper, we introduce a different representation of merge trees. Instead of recording the nesting of sublevel sets explicitly, the new *triplet representation* records the nesting of the branches of the merge tree. We present algorithms that construct merge trees in this representation directly by incrementally adding edges to the domain. They possess the same desirable properties (1)–(3) as above, and as our experiments show, the new algorithms perform better in practice than Kruskal's algorithm. The new algorithms are also sufficiently simple that they can be parallelized in shared memory using double-word compare-and-swap primitives for synchronization.

2 Background

Given a graph G and a scalar function $f : \operatorname{Vrt} G \to \mathbb{R}$ on its vertices, we assume that all vertex values are distinct, breaking ties lexicographically in practice. For $a \in \mathbb{R}$, the *sublevel graph at* a, denoted G_a, is the subgraph induced by the vertices whose function value does not exceed a. The *representative* of vertex u at level $a \geq f(u)$ is the vertex v with the minimum function value in the component of u in G_a. The *merge tree* of function f on graph G is the tree on the vertex set of G that has an edge (u, v), with $f(u) < f(v)$, if the component of u in $G_{f(u)}$ is a subset of the component of v in $G_{f(v)}$, and there is no vertex v' with $f(u) < f(v') < f(v)$ such that the component of u is a subset of the component of v' in $G_{f(v')}$. Figure 1 illustrates a function on a graph and its merge tree. (If G is disconnected, a "merge tree" is actually a forest. We abuse terminology and do not distinguish this case, continuing to call it a tree, to not clutter the language.)

Intuitively, a merge tree keeps track of the connected components in the sublevel graphs. If we sweep threshold a from $-\infty$ to ∞, as we cross values of specific vertices, they either start new components in the tree, or they are added to existing components, possibly joining multiple such components together. Such a sweep lies at the core of the most common algorithm for merge tree construction, which adapts Kruskal's algorithm, originally developed for the minimum spanning tree construction. This algorithm processes vertices in sorted order and maintains connected components in a disjoint set data structure that supports fast component identification and merging. For each vertex, it queries the representatives of the connected components of its neighbors with lower function value and unites them, recording the changes to the connected components in the merge tree.

The problem of constructing a merge tree is related to the minimum spanning tree construction. The latter can be reduced to the former by subdividing the edges of the input graph; the new vertices are assigned the edge values, while the values of the original vertices are set to $-\infty$. The merge tree of the resulting function identifies the minimum spanning tree of the original graph: degree-3 nodes of the merge tree are the vertices subdividing the edges of the minimum spanning tree.

However, the two problems are distinct. In general merge trees, the values of the minima vary, a feature important for the algorithms presented in the next section. Furthermore, the goal is to compute the tree itself rather than to find the identity of

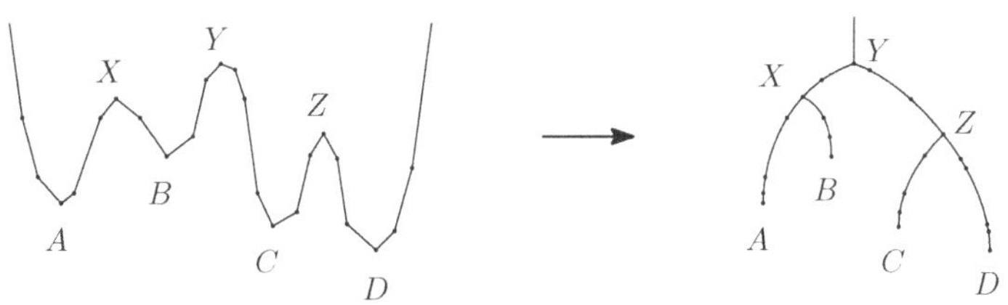

Fig. 1 A function on a graph and its merge tree

the edges. As a consequence, a folklore reduction from sorting[1] shows that the lower bound for merge tree construction is $\Omega(m + n \log n)$, where n, m are the numbers of vertices and edges in the input graph. In contrast, a minimum spanning tree can be built in $O(m \cdot \alpha(m, n))$ time [9], where α is the inverse Ackermann function.

The algorithm of Bremer et al. [6], mentioned in the introduction, works as follows. It initializes the merge tree to be a set of disjoint vertices. It then processes all the edges one by one, in arbitrary order. Given an edge (u, v), it merges, in sorted order, the paths from u and v to the root of the merge tree. The correctness of the algorithm is easy to see: an edge (u, v) certifies that the two vertices belong to the same connected components in all sublevel graphs G_a, with $a \geq \max\{f(u), f(v)\}$; the connected components of the sublevel graphs that contain vertex u are represented by the vertices on the path from u to the root. We use this algorithm for comparison in the next section.

3 Triplet Merge Tree

Merge trees are typically stored in a graph data structure: each node stores a pointer to its parent, or to its children, or both. These pointers allow one to traverse the tree and answer various queries: for example, what is the volume of the component that contains a given vertex; or how many connected components are there at level b that have a vertex with function value below a; etc.

We are interested in an alternative representation that records global information. Instead of storing its immediate parent, each vertex stores the range of values for which it remains its own representative (i.e., the deepest vertex in its connected component), together with the reference to its representative at the end of this range. The necessary information can be thought of as a triplet of vertices (u, s, v), such that u represents itself at levels $a \in [f(u), f(s))$, and v becomes its representative at level $f(s)$. Recursively following the chain of representatives of v, we can find the representative of u at any level, which in turn allows us to answer the queries mentioned before. Figure 2 illustrates the triplet representation of the merge tree in Fig. 1.

A reader familiar with persistent homology will immediately recognize pairs $(f(u), f(s))$ as the birth–death pairs in the 0-dimensional persistence diagram. Similarly, pairs (u, s) are the branches in the decomposition introduced by Pascucci et al. [10] for contour trees. The extra information stored in the triplet—which branch a given branch merges into—is crucial for our main goal: to construct and manipulate the trees directly in the triplet representation. The rest of this

[1]Given a sequence of n values, take a star graph with n leaves and assign the input values to the leaves; assign $-\infty$ to the central vertex. The merge tree of this function is a path, with vertices assigned the input values in sorted order.

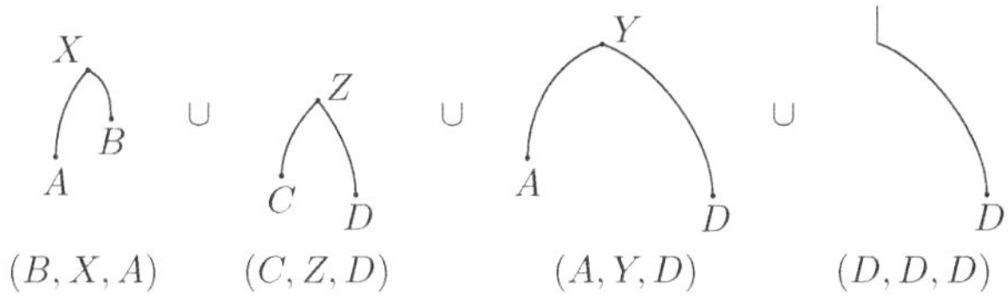

Fig. 2 A subset of the minimal, normalized triplet representation of the merge tree in Fig. 1. The full representation includes triplets (X, X, A), (Z, Z, D), (Y, Y, D), as well as triplets for the unlabeled degree-2 nodes

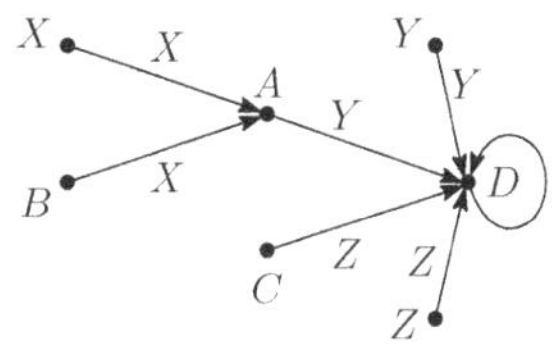

Fig. 3 Directed graph induced by the triplet representation in Fig. 2, with leaves representing the unlabeled degree-2 nodes omitted

section introduces such algorithms and structural formalism necessary to prove their correctness.

Structure Given a graph G with a scalar function $f : \operatorname{Vrt} G \to \mathbb{R}$, we define a *merge triplet* to be a 3-tuple of vertices, (u, s, v), such that u and v are in the same component of $G_{f(s)}$ and $f(v) < f(u) \leq f(s)$, or (u, u, u) if u is the minimum in its component of G. We define the *triplet representation* T to be a set of merge triplets. (A triplet representation is not unique, for a given function. We specify conditions to make it unique below, and we take advantage of this flexibility in our algorithms.) A triplet representation induces a directed graph $D(T)$, with the same vertex set as the graph G; $D(T)$ contains a directed edge (u, v), with label s, if and only if (u, s, v) is a triplet in T. Figure 3 illustrates the directed graph induced by the triplet representation in Fig. 2.

We let $D(T, a)$ be the subgraph of $D(T)$ that consists of all vertices u with $f(u) \leq a$ and all edges (u, v) with label s such that $f(s) \leq a$. The *representative* of vertex u at level a in the triplet representation is vertex v with the lowest function value in the connected component of u in $D(T, a)$.

A representation T is *equivalent* to the graph G if for every vertex u and every level $a \geq f(u)$, the representative of u at a in G is the same as the representative of u at a in T. Similarly, we say that two representations T_1 and T_2 are equivalent, $T_1 \simeq T_2$, if every vertex, at every level, has the same representative in both T_1 and T_2.

A triplet representation is *normalized* if every vertex $u \in \operatorname{Vrt} G$ appears as the first element of exactly one triplet. It follows immediately that the digraph induced by a normalized representation is a forest, if we ignore the self-loops at the roots.

(We note that the internal nodes in such a forest represent extrema of the function, while the internal nodes of the regular merge tree appear as leaves in the digraph; see Figs. 1 and 3.) Furthermore, as the following lemma states, a normalized representation recovers the branches of the merge tree (but not necessarily what they merge into).

Lemma 1 *If representation T is normalized, then $(u, s, v) \in T$ implies that u is its own representative for all levels a, with $f(u) \leq a < f(s)$. Furthermore, u is not its own representative at level $f(s)$. (The only exception are the triplets $(u, u, u) \in T$, corresponding to minima of the connected components. In this case, u represents itself at all levels.)*

Proof If T is normalized, its induced digraph $D(T)$ is a forest (ignoring the self-loops at the roots). The component of the subgraph $D(T, a)$, induced by vertices and edges below $a < f(s)$, that contains vertex u is a subtree rooted at u (unless $u = s$, in which case there is no such component). Since vertex values decrease along the directed edges, u is the minimum of its component, and, therefore, its own representative, by definition. At level $f(s)$, we add the edge (u, v) to the digraph $D(T, f(s))$. Since $f(v) < f(u)$, u stops being its own representative. □

Even for normalized representations, existence of triplet (u, s, v) does not imply that v represents u at level $f(s)$. We say that a representation is *minimal* if along every path in the digraph, the values of the edge labels are increasing. It follows immediately that if T is normalized and minimal, then for every triplet $(u, s, v) \in T$, v represents u at level $f(s)$.

Our main algorithm, Algorithm 3, relies on the following theorem to construct a triplet representation.

Theorem 1 *Suppose a representation T is equivalent to a graph G. Let G' denote graph G with an extra edge (u, v); assume without loss of generality that $f(u) < f(v)$. Then representation $T' = T \cup (v, v, u)$ is equivalent to graph G'.*

Proof If representation T is equivalent to graph G, then at every level $a \in \mathbb{R}$, there is a bijection between the connected components of the sublevel graph G_a and those of the subgraph $D(T, a)$ induced by the representation. If $a < f(v)$, then neither vertex v nor the new edge is present in either, and so there is no change between G_a and G'_a and $D(T, a)$ and $D(T', a)$.

Suppose $a \geq f(v)$. If u and v are in the same connected component of G_a, then they are in the same connected component of the induced subgraph $D(T, a)$. They remain in the same connected component after we add edges (u, v) to G and (v, v, u) to $D(T, a)$. If they are in different connected components, then adding edge (u, v) to G_a merges the two connected components, but edge (v, v, u) merges the corresponding connected components in $D(T, a)$. Thus, there is a bijection between the connected components of G'_a and $D(T', a)$. □

Algorithms It is possible to construct a triplet representation from a graph using Kruskal's algorithm. Algorithm 1 spells out the details. It uses algorithm FindDeepest(T, u), given in Algorithm 2, to find the deepest element in a connected

Algorithm 1: ComputeMergeTree(G)

```
Sort vertices by f value;
foreach vertex u ∈ G do
    T[u] ← (u, u);
    CurDeepest[u] ← u;
    nbrs ← {v | (u, v) ∈ G, f(v) < f(u)};
    if #nbrs > 0 then
        leaves ← {FindDeepest(T, v) : v ∈ nbrs};
        v̂ ← oldest(leaves);
        T[u] ← (u, v̂);
        if #leaves > 1 then
            foreach v ∈ leaves \ {v̂} do
                T[v] ← (u, v̂)
return T
```

Algorithm 2: FindDeepest(T, u)

```
û ← CurDeepest[u]
(_, v) ← T[û]
while û ≠ v do
    û ← CurDeepest[v]
    (_, v) ← T[û]
d ← û
û ← CurDeepest[u]
(_, v) ← T[û]
while û ≠ v do
    û ← CurDeepest[v]
    CurDeepest[v] ← d
    (_, v) ← T[û]
CurDeepest[u] ← d
return d
```

component, using path compression. (Because we use path compression alone, the algorithm takes $O(m \log n)$ steps in the worse case, on a graph with n vertices and m edges.) In the next section, we use this algorithm as a baseline for comparison.

Our main contribution is Algorithm 3 and its auxiliary Algorithms 4, 5, and 6. Together they construct a normalized minimal triplet representation by processing the edges of the input graph, adding them one by one into the triplet representation. Crucially, it does not matter in what order the edges are processed; we harness this flexibility for parallelization in the next section.

Because we maintain a normalized representation, each vertex occurs as the first component of exactly one triplet. In the pseudo-code, we use the associative map notation, $T[u] \leftarrow (s, v)$, to record triplet (u, s, v) and, similarly, $(s, v) \leftarrow T[u]$ to access triplet $(u, s, v) \in T$.

Algorithm 3: ComputeMergeTree2(G)

foreach *vertex* $u \in G$ **do**
    $T[u] \leftarrow (u, u)$;
foreach *edge* $(u, v) \in G$ **do**
    if $f(u) < f(v)$ **then**
        `Merge`(T, v, v, u)
    else
        `Merge`(T, u, u, v)
foreach $u \in T$ **do**
    `Repair`(T, u)
return T

The first loop of Algorithm 3 initializes the vertices of representation T. The result is trivially equivalent to graph G without any edges.

The second loop employs the main workhorse, the operation Merge(T, u, s, v), presented in Algorithm 4. Given a normalized representation T, it finds a normalized representation equivalent to $T \cup (u, s, v)$. Figure 4 illustrates a single transformation performed by the Merge algorithm.

Algorithm 4: Merge(T, u, s, v)

$(u', s_u, u'') \leftarrow$ `Representative`$(T, u, f(s))$;
$(v', s_v, v'') \leftarrow$ `Representative`$(T, v, f(s))$;
if $u' = v'$ **then return**;
if $f(v') < f(u')$ **then**
    `swap`$((u', s_u, u''), (v', s_v, v''))$
$T[v'] \leftarrow (s, u')$;
`Merge`(T, u', s_v, v'') ;

Because Merge(T, u, s, v) does not guarantee to preserve minimality, Algorithm 3 restores this property in the third loop by calling Repair(T, u), presented in Algorithm 6, which finds for each edge the next edge with a higher label.

One could maintain minimality on the fly by keeping back-pointers to the branches that merge into the given branch (making the induced graph undirected), but this would obstruct our goal of lock-free shared-memory parallelism in the next section. So we opt to use the Repair procedure, instead.

Fig. 4 Induced graph of triplet representations before and after a single transformation in the Merge(u, s, v) algorithm

Algorithm 5: Representative(T, u, a)

$(s, v) \leftarrow T[u]$
while $f(s) \leq a$ **and** $s \neq v$ **do**
    $u \leftarrow v$
    $(s, v) \leftarrow T[u]$
return (u, s, v)

Algorithm 6: Repair(T, u)

$(s, \_) \leftarrow T[u]$
$v \leftarrow$ `Representative`$(T, u, f(s))$
if $u \neq v$ **then**
    $T[u] \leftarrow (s, v)$
return $T[u]$

Correctness The core of the algorithm is in the Merge procedure, in Algorithm 4. Assuming that representation T is normalized, we want to show that after Merge(u, s, v) returns, we get a normalized representation equivalent to $T \cup (u, s, v)$.

Lemma 2 *Let T denote the triplet representation at the beginning of Algorithm 4, and T' the representation before the recursive call in step 4. Then, $T \cup (u, s, v) \simeq T' \cup (u', s_v, v'')$.*

Proof Fix a level $a \in \mathbb{R}$. Let $D_1 = D(T \cup (u, s, v), a)$ and $D_2 = D(T' \cup (u', s_v, v''), a)$ denote the directed graphs induced by the representations at the beginning of the algorithm and before the recursive call; see Fig. 4. Let x and y be any two vertices in the same connected component in D_1, and let p be the (undirected) path that connects them. If p does not contain edges (u, s, v) or (v', s_v, v''), then it exists in D_2, and therefore x and y are still connected. If p goes through edge (u, s, v), we can replace that edge by the path $u \ldots u', (v', s, u'), v' \ldots v$ in D_2, since all the edges in the subpaths connecting vertices u and v have values that do not exceed $f(s)$. If p contains edge (v', s_v, v''), we can replace the edge by path $(v', s, u'), (u', s_v, v'')$ since $f(s) < f(s_v)$.

Similarly, in the other direction. If x and y are in the same component of D_2, then either the path connecting them does not contain edges (v', s, u') or (u', s_v, v'') and, therefore, exists in D_1; or we can replace (u', s_v, v'') by $u' \ldots u, (u, s, v), v \ldots v', (v', s_v, v'')$, and (v', s, u') by $v' \ldots v$, $(u, s, v), u \ldots u'$.

In summary, two vertices are connected in D_1 if and only if they are connected in D_2. Therefore, the two triplet representations are equivalent. □

At every point of the Merge algorithm, the representation remains normalized. When the exit condition is detected ($u' = v'$), the extra triplet does not add any new information, i.e., in this case $T \simeq T \cup (u, s, v)$. The recursive calls are made with vertices u' and v''. By definition, $f(v'') < f(v)$. Consequently, the algorithm is always making progress and must eventually terminate. (The progress is most evident in the induced digraph: the algorithm always moves up the directed paths in $D(T)$.) We get the following corollary.

Corollary 1 *Algorithm 4 updates normalized triplet representation T, making it equivalent to representation $T \cup (u, s, v)$. The representation remains normalized.*

It follows that Algorithm 3 computes the correct result, as summarized in the following theorem.

Theorem 2 *Algorithm 3 computes a normalized minimal triplet representation equivalent to the input graph G.*

Proof The first for-loop in Algorithm 3 initializes representation T to be equivalent to a graph on the same vertex set as G, but without any edges. The second for-loop uses Algorithm 4 to update the representation, adding the edges of the graph. The correctness of this loop follows from Theorem 1 and Corollary 1. The third for-loop ensures that the representation is not only normalized, but also minimal by finding the representative of every vertex at the first level where it does not represent itself; its correctness follows from Lemma 1. □

Evaluation Throughout the paper we use three data sets to evaluate the algorithms. The first, Z2, is a 512^3 snapshot of a cosmological simulation. The second, Vertebra, is a 512^3 scan of a head aneurysm, once available online as part of the `volvis.org` collection of data sets. The third, Pumice, is a $640^2 \times 540$ signed distance function to a porous material. We use the vertices and edges of the Freudenthal triangulations of the underlying grids as our input graphs.

All experiments were performed on a compute node with two sockets, each with a 16-core Intel Xeon Processor E5-2698 v3 ("Haswell") at 2.3 GHz.

Figures 5, 6, and 7 illustrate the scaling of Algorithm 1 (labeled "Kruskal's") and Algorithm 3 (labeled "Triplet") as a function of input size; the three inputs were downsampled to the sizes specified on the x-axis. They also show two data points each[2] of the path merging algorithm of Bremer et al. (labeled "Path merging"). The salient point in all cases is that Algorithm 3 is not only competitive with Algorithm 1, but performs significantly better for larger domain sizes, more than five times better for 512^3 Z2 input. The path merging algorithm is slower and scales significantly worse.

[2] We stopped at two data points because by the third, the jobs exhausted the 4-h wallclock request.

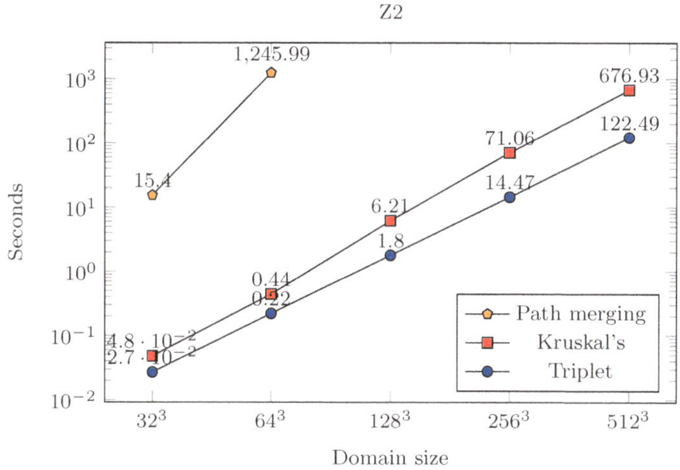

Fig. 5 Running time of the three algorithms as a function of input size, on downsampled Z2 data set

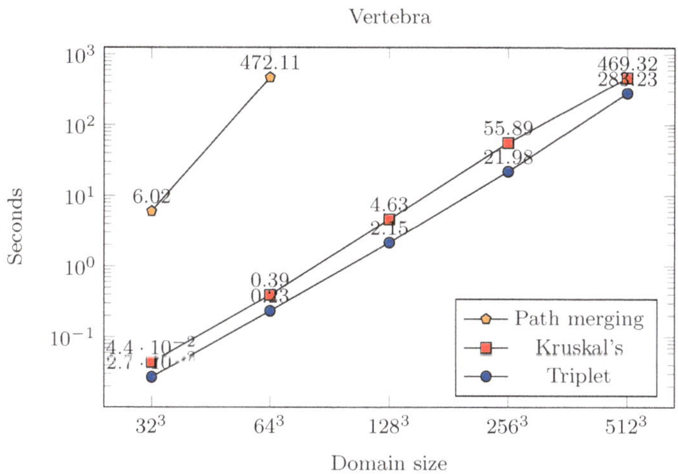

Fig. 6 Running time of the three algorithms as a function of input size, on downsampled Vertebra data set

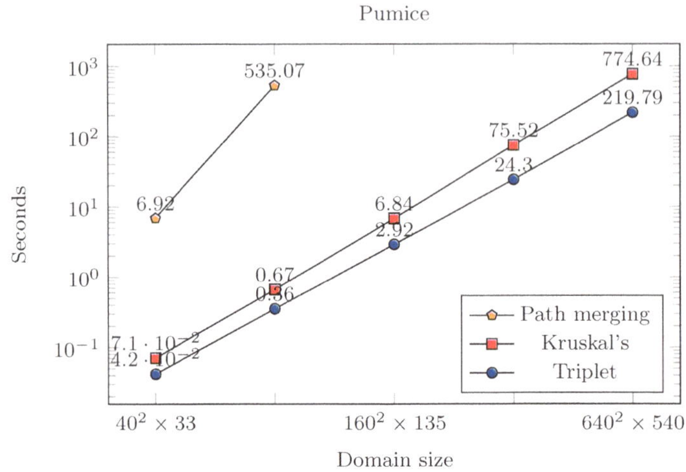

Fig. 7 Running time of the three algorithms as a function of input size, on downsampled Pumice data set

4 Shared Memory

We adapt our new algorithms to allow multiple threads to concurrently modify the triplet representation in shared memory. We use compare-and-swap (CAS) primitive for synchronization between the threads. This primitive atomically checks whether a variable contains a given `expected` value; if it does, CAS replaces it with the given `desired` value. The operation is equivalent to atomically executing Algorithm 7.

Algorithm 7: CAS(v, `expected`, `desired`)

if v = *expected* **then**
  $v \leftarrow$ `desired`;
  return True
else
  return False

Since the variables we need to update atomically store pairs of values, (s, v), we use double-word compare-and-swap (DWCAS) operations. These should not be confused with double compare-and-swap (DCAS) operations, which update two arbitrary words in memory. DWCAS updates two *contiguous* words in memory, and, unlike DCAS, it is supported in modern hardware.[3]

[3]On x86-64 architecture, where we conducted all our experiments, DWCAS is performed by instruction `CMPXCHG16B`. It is automatically emitted by the C++ compilers, when provided with appropriate flags.

Algorithm 8: Parallel ComputeMergeTree2(G)

```
foreach vertex u ∈ G do in parallel
    T[u] ← (u, u);
foreach edge (u, v) ∈ G do in parallel
    if f(v) < f(u) then
        Merge(T, u, u, v)
    else
        Merge(T, v, v, u)
foreach u ∈ T do in parallel
    Repair(T, u)
return T
```

Algorithm 9: Parallel Merge(T, u, s, v)

```
▷ equivalent to (u, s_u, u') ← Representative(T, u, f(s))
(s_u, u') ← T[u];
if f(s_u) < f(s) then
    return Merge(T, u', s, v)
▷ equivalent to (v, s_v, v') ← Representative(T, v, f(s))
(s_v, v') ← T[v];
if f(s_v) < f(s) then
    return Merge(T, u, s, v')
if u = v then return;
if f(v) < f(u) then
    swap((u, s_u, u'), (v, s_v, v'))
if DWCAS(T[v], (s_v, v'), (s, u)) then
    Merge(T, u, s_v, v');
else
    Merge(T, u, s, v);
```

Algorithm 8 adapts Algorithm 3 to the parallel setting by executing its for-loops in parallel. We assume that the triplet representation T is stored in a container that allows concurrent insertion of elements (for example, a concurrent hash map), so that the first for-loop can execute in parallel.

The second for-loop runs in parallel over the edges of the input graph and invokes the Merge algorithm, Algorithm 9, adjusted to use DWCAS for synchronization. The third for-loop performs Repair, as before.

Correctness To understand the correctness of the second for-loop and Algorithm 9, we interpret the state of the data structure as the normalized representation T, together with a complete list of triplets, (u, u, v) or (v, v, u), one for each edge, as prescribed by Theorem 1. Each invocation of Algorithm 9 is assigned one of the outstanding triplets—the algorithm receives the triplet as the arguments u, s, v. In a single instantiation, before or via the recursive call, it transforms the state of the data structure:

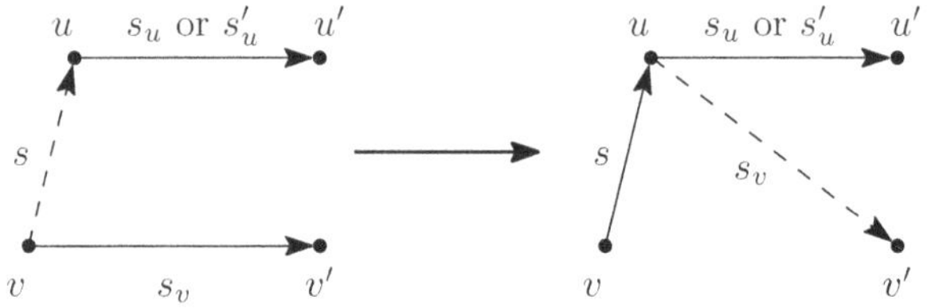

Fig. 8 The effect of the DWCAS transformation in line 10 of Algorithm 9

1. by modifying its assigned triplet, from (u, s, v) to (u', s, v) or (u, s, v'), via the recursive calls in lines 3 or 6;
2. by modifying the triplet representation T in line 10 and replacing the triplet (u, s, v) with (u, s_v, v'), in line 11;
3. by implicitly removing the triplet by returning in line 7.

Each such operation, when applied in isolation, keeps the state before the transformation equivalent to the state after the transformation. This follows from arguments very similar to the proofs of Lemmas 1 and 2.

$$T \cup (u_1, s_1, v_1) \cup \ldots \cup (u, s, v) \cup \ldots \cup (u_m, s_m, v_m)$$

$$\Big\downarrow \simeq$$

$$T' \cup (u_1, s_1, v_1) \cup \ldots \cup (u, s_v, v') \cup \ldots \cup (u_m, s_m, v_m)$$

We claim that if multiple invocations of the algorithm modify the state concurrently, the transformations that they apply are linearizable, in the sense of Herlihy and Wing [11].

To see why this is so, we identify the linearization points, at which the transformations appear to occur atomically. In case of the first two recursive calls, in lines 3 and 6, the preceding reads in lines 9 and 4 act as linearization points. Even though there are no side effects—different invocations of Algorithm 9 do not see each other's triplets, and the triplet representation is not modified—the operation is equivalent to atomically deciding that $T \cup (u, s, v)$ is equivalent to $T \cup (u', s, v)$ or $T \cup (u, s, v')$.

If the condition $u = v$ in line 7 is satisfied, then triplet (u, s, v) is redundant and the return statement in that line is equivalent to its removal, which also appears atomically.

What happens when we reach line 10? In this case, the function values are ordered: $f(u) < f(v) < f(s) < f(s_v)$. When this is true, the representation $T \cup (v, s, u)$ is equivalent to the representation $T' \cup (u, s_v, v')$, where T' has triplet (v, s_v, v') replaced by the triplet (v, s, u); see Fig. 8. (The proof here is similar to the proof of Lemma 2.) If DWCAS in line 10 succeeds, then this transformation appears atomically. If it fails, we simply retry via the recursive call in line 13. We note that in this case we do not guarantee that the triplet (u, s_u, u') is not changed in the representation T, i.e., the edge going out of the vertex u in the induced digraph may change, so that its new label s'_u is such that $f(s'_u) < f(s)$. In this case, the transformation is still valid—the representations remain equivalent—but the transformed representation ceases to be minimal. Since we already do not maintain minimality during the second for-loop of Algorithm 8, but rather restore it via the Repair procedure in the third for-loop, this situation requires no special handling.

Finally, the third for-loop of Algorithm 8 requires no special attention since Repair procedure, Algorithm 6, can already execute in parallel. Since it only changes the target of any edge in the diagraph, i.e., only the third component of any triplet, the result that it obtains is the same even if other threads are simultaneously modifying the triplets it encounters. Suppose Repair changes entry $T[u]$ from (s, v) to (s, v'), and suppose a different Repair procedure queries the entry $T[u]$ as part of its search for a representative of a vertex u' at level a. If $a < f(s)$, then the change is irrelevant, since u is the desired representative. If $a \geq f(s)$, then the search needs to test the entry $T[v']$, since if the original Repair procedure was changing $T[u]$ from (s, v) to (s, v'), then all the edges on the path from v to v' have labels s_i, with $f(s_i) < f(s)$. It makes no difference whether the second Repair procedure goes directly to v' or traverses the original path.

Evaluation We perform experiments on the same datasets, Z2, Vertebra, Pumice, as in the previous section, but now we vary the number of threads used for the computation. We use Intel's Thread Building Blocks library for parallelization, using its `concurrent_unordered_map` to store the triplet representation.

As a baseline, we compare to Algorithm 1 and Algorithm 3. The significance of the latter is that unlike Algorithm 8, it is implemented without atomic variables.[4] Figures 9, 10, and 11 illustrate the results of our experiments. Unfortunately, simply turning on atomics roughly doubles the running time. As the thread count increases, the parallel algorithm outperforms the serial (when using 4 threads in all cases). Although the scaling trails off, we believe the higher thread counts are still worthwhile, especially, in the (common) situation where those cores would remain idle otherwise.

[4]This is important because in C++, once a variable is declared `std::atomic`, all operations on it are protected by atomic primitives and incur the corresponding overheads.

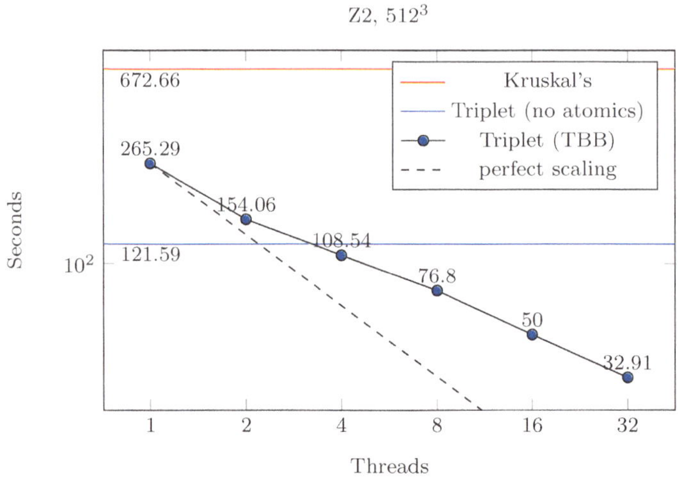

Fig. 9 Scaling of Algorithm 9 as a function of the number of threads, on Z2 dataset

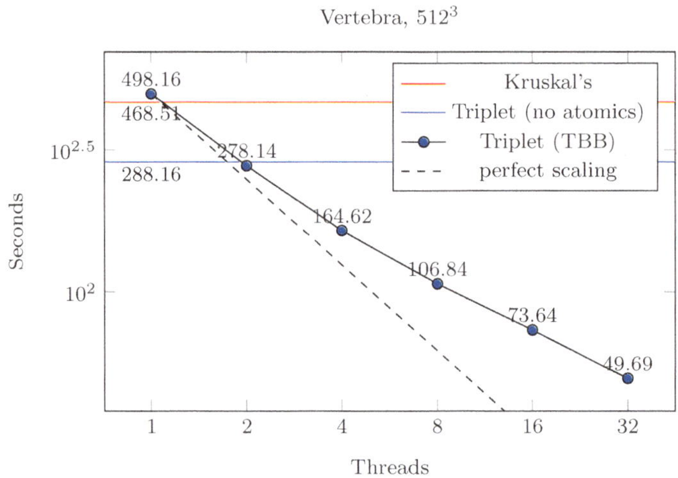

Fig. 10 Scaling of Algorithm 9 as a function of the number of threads, on Vertebra dataset

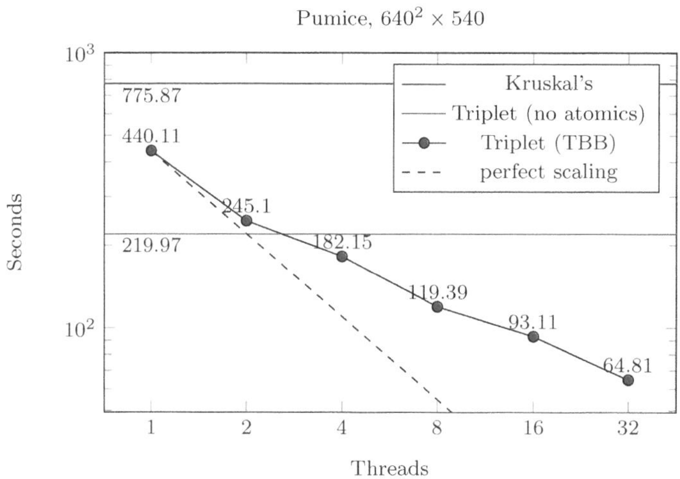

Fig. 11 Scaling of Algorithm 9 as a function of the number of threads, on Pumice dataset

5 Conclusion

We have described a new representation of merge trees. Instead of recording the nesting of sublevel set components, it records the nesting of the branches of merge trees. We have presented algorithms to construct the merge trees in this representations directly, as well as their parallel versions, together with experimental results demonstrating that these algorithms are efficient in practice.

Conspicuously missing from our paper is the complexity analysis of the new algorithms. An $O(mn)$ upper bound for Algorithm 3 is obvious: for each edge, the algorithm touches each vertex at most once (in the amortized sense). It is also not difficult to construct an example on which Algorithm 3 would take quadratic time. However, all such examples that we know of require specifying not only the graph and the function, but also the sequence in which edges must be added. This leaves us hopeful that it is possible to show that the randomized version of Algorithm 3 is also efficient in theory. We view this as the main open question left by our paper.

Acknowledgments This work was supported by the Director, Office of Science, Office of Advanced Scientific Computing Research, of the U.S. Department of Energy under Contract No. DE-AC02-05CH11231, through the grant "Scalable Data-Computing Convergence and Scientic Knowledge Discovery," and by the use of resources of the National Energy Research Scientific Computing Center (NERSC).

References

1. Carr, H., Snoeyink, J., Axen, U.: Computing contour trees in all dimensions. Comput. Geom. Theory Appl. **24**(2), 75–94 (2003)
2. Edelsbrunner, H., Harer, J.: Computational Topology. An Introduction. American Mathematical Society, Providence, Rhode Island (2010)
3. Cormen, T., Leiserson, C.E., Rivest, R.L., Stein, C.: Introduction To Algorithms. MIT Press, Cambridge (2009)
4. Carr, H., Weber, G.H., Sewell, C., Ahrens, J.: Parallel peak pruning for scalable SMP contour tree computation. In: Proceedings of the IEEE Symposium on Large Data Analysis and Visualization (LDAV) (2016)
5. Nešetřil, J., Milková, E., Nešetřilová, H.: Otakar Borůvka on minimum spanning tree problem: translation of both the 1926 papers, comments, history. Discret. Math. **233**, 3–36 (2001)
6. Bremer, P.-T., Weber, G.H., Tierny, J., Pascucci, V., Day, M.S., Bell, J.B.: Interactive exploration and analysis of large scale simulations using topology-based data segmentation. IEEE Trans. Vis. Comput. Graph. **17**(9), 1307–1325 (2011)
7. Pascucci, V., Cole-McLaughlin, K.: Parallel computation of the topology of level sets. Algorithmica **38**(1), 249–268 (2003)
8. Morozov, D., Weber, G.H.: Distributed merge trees. In: Proceedings of the Annual Symposium on Principles and Practice of Parallel Programming (PPOPP), pp. 93–102 (2013)
9. Chazelle, B.: A minimum spanning tree algorithm with inverse-ackermann type complexity. J. Assoc. Comput. Mach. **47**, 1028–1047 (2000)
10. Pascucci, V., Cole-McLaughlin, K., Scorzelli, G.: The Toporrery: Computation and Presentation of Multi-Resolution Topology, pp. 19–40. Springer, Berlin (2009)
11. Herlihy, M.P., Wing, J.M.: Linearizability: a correctness condition for concurrent objects. ACM Trans. Program. Lang. Syst. **12**, 463–492 (1990)

Persistent Intersection Homology for the Analysis of Discrete Data

Bastian Rieck, Markus Banagl, Filip Sadlo, and Heike Leitte

Abstract Topological data analysis is becoming increasingly relevant to support the analysis of unstructured data sets. A common assumption in data analysis is that the data set is a sample—not necessarily a uniform one—of some high-dimensional manifold. In such cases, persistent homology can be successfully employed to extract features, remove noise, and compare data sets. The underlying problems in some application domains, however, turn out to represent *multiple* manifolds with different dimensions. Algebraic topology typically analyzes such problems using intersection homology, an extension of homology that is capable of handling configurations with singularities. In this paper, we describe how the persistent variant of intersection homology can be used to assist data analysis in visualization. We point out potential pitfalls in approximating data sets with singularities and give strategies for resolving them.

1 Introduction

The *manifold hypothesis* is a traditional assumption for the analysis of multivariate data. Briefly put, it assumes that the input data are a sample of some manifold $\mathbb{M}$, whose intrinsic dimension d is much smaller than the ambient dimension D. Typical examples of this assumption are found in dimensionality reduction algorithms [22, 25]. For certain applications, such as image analysis [10] or image recognition [15], we already know this hypothesis to be true—at least with respect to the models that are often used to describe such data. For other applications, there are strategies [13, 19] for testing this hypothesis provided that a sufficient number of samples is available.

B. Rieck (✉) · H. Leitte
TU Kaiserslautern, Kaiserslautern, Germany
e-mail: rieck@cs.uni-kl.de; bastian.rieck@iwr.uni-heidelberg.de; leitte@cs.uni-kl.de

M. Banagl · F. Sadlo
Heidelberg University, Heidelberg, Germany
e-mail: banagl@mathi.uni-heidelberg.de; sadlo@uni-heidelberg.de

H. Carr et al. (eds.), *Topological Methods in Data Analysis and Visualization V*, Mathematics and Visualization, https://doi.org/10.1007/978-3-030-43036-8_3

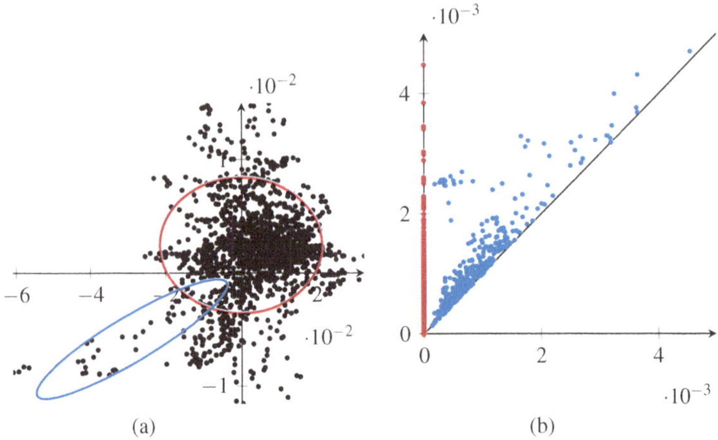

Fig. 1 (**a**) The structure of a central "core" with "flares" emanating from it appears in many data sets (here, 2-year growth rates of Standard & Poor's 500 vs. the U.S. CPI with the core shown in red and one example flare shown in blue). (**b**) The corresponding persistence diagram shows topological features in dimension zero (red) and dimension one (blue)

The *practice* of multivariate data analysis seems to suggest something else, though: Carlsson [4], for example, remarks that many real-world data sets exhibit a central "core" structure, from which different "flares" emanate. Figure 1 illustrates this for a simple 2D data set, generated from 2-year growth rates of Standard & Poor's 500 vs. the U.S. CPI. This structure is irreconcilable with the structure of a single manifold. Novel data analysis algorithms such as MAPPER [24] account for this fact by not making any assumptions about manifold structures and attempting to fit data in a local manner—a strategy that is also employed in low-dimensional manifold learning [23].

In this paper, we argue that some real-world data sets require special tools to assess their structure. Just as persistent homology [11, 12] was originally developed to analyze samples from spaces that are supposed to have the structure of a manifold, we need a special tool to analyze spaces for which this assumption does *not* hold. More precisely, we will tackle the task of analyzing spaces that are composed of different manifolds (with possibly varying dimensions) using *intersection homology* [16] and *persistent intersection homology* [1, 2]. To make it accessible to a wider community of researchers, we devote a large portion of this paper to explaining the theory behind persistent intersection homology. Furthermore, we discuss implementation details and present an open-source framework for its calculation. We also describe pitfalls in "naive" applications of persistent intersection homology and develop strategies to resolve them.

2 Background

We first explain the mathematical tools required to describe spaces that are not composed of a single manifold, but of multiple ones. Next, we introduce (persistent) intersection homology, give a brief algorithm for its computation, and describe how to use it to analyze real-world data sets.

2.1 *Stratifications*

Stratifications are a way of describing spaces that are not a manifold per se, but composed of multiple parts, each of which is a manifold. A common example of such a space is the "pinched torus", which is obtained by collapsing (i.e., pinching) one minor ring of the torus to a single point. Figure 2a depicts an example. The neighborhood of the pinch point is *singular* because it does not satisfy the conditions of a manifold: it does not have a neighborhood that is homeomorphic to a ball. If we remove this singular point, however, the remaining space is just a (deformed) cylinder, i.e., a manifold. Permitting the removal of certain parts of a space may thus be beneficial to describe the manifolds it is composed of. This intuition leads to the concept of stratifications.

Let $X \subseteq \mathbb{R}^n$ be a topological space. A *topological stratification* of X is a filtration of closed subspaces

$$\emptyset \subseteq X_{-1} \subseteq X_0 \subseteq X_1 \subseteq \cdots \subseteq X_{d-1} \subseteq X_d = X, \tag{1}$$

such that for each i and every point $x \in X_i \setminus X_{i-1}$ there is a neighborhood $U \subseteq X$ of x, a compact $(n-1-i)$-dimensional stratified topological space V, and a filtration-preserving homeomorphism $U \simeq \mathbb{R}^i \times CV$, where CV denotes the *open cone* on V, i.e., $CV := V \times [0,1)/(V \times \{0\})$. We refer to $X_i \setminus X_{i-1}$ as the i-dimensional *stratum* of X. Notice that it is always a (smooth) manifold, even though the original

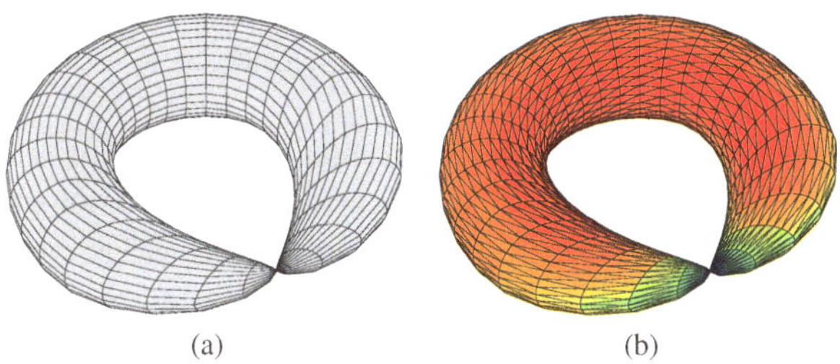

Fig. 2 (**a**) The "pinched torus" is a classical example of an object that is *not* a manifold but composed of parts that are manifolds, provided the singular point that is caused by the "pinch" is ignored. (**b**) The singular point is readily visible when calculating mean curvature estimates

space might not be a manifold. Hence, this rather abstract definition turns out to be a powerful description for a large family of spaces. There are some stratifications with special properties that are particularly suited for analyzing spaces. Goresky and MacPherson [14], the inventors of intersection homology, suggest using a stratification that satisfies $X_{d-1} = X_{d-2}$ so that the $(d-1)$-dimensional stratum is empty, i.e., $X_{d-1} \setminus X_{d-2} = \emptyset$.

2.2 *Homology and Persistent Homology*

Prior to introducing (persistent) intersection homology, we briefly describe simplicial homology and its persistent counterpart. Given a d-dimensional simplicial complex K, the chain groups $\{C_0, \ldots, C_d\}$ contain formal sums (simplicial chains) of simplices of a given dimension. A boundary operator $\partial_p \colon C_p \to C_{p-1}$ satisfying $\partial_{p-1} \circ \partial_p = 0$ (i.e., a closed boundary does not have a boundary itself) then permits us to create a chain complex from the chain groups. This results in two subgroups, namely the cycle group $Z_p := \ker \partial_p$ and the boundary group $B_p := \operatorname{im} \partial_{p+1}$, from which we obtain the pth homology group as

$$H_p := Z_p / B_p, \tag{2}$$

where the /-operator refers to the quotient group. Intuitively, elements in the cycle group Z_p constitute sets of simplicial chains that do not have a boundary, while elements in the boundary group B_p are the boundaries of higher-dimensional simplices. By removing these in the definition of the homology group, we obtain a group that describes high-dimensional "holes" in K.

Homology is a powerful tool to discriminate between different triangulated topological spaces. It is common practice to use the Betti numbers β_p, i.e., the ranks of the homology groups, to obtain a signature of a space. In practice, the Betti numbers turn out to be highly susceptible to noise, which prompted the development of persistent homology [11]. Its basic premise is that the simplicial complex K is associated with a filtration,

$$\emptyset = \mathrm{K}_0 \subseteq \mathrm{K}_1 \subseteq \cdots \subseteq \mathrm{K}_{n-1} \subseteq \mathrm{K}_n = \mathrm{K}, \tag{3}$$

where each K_i is typically assigned a function value, such as a distance. The filtration induces a homomorphism of the corresponding homology groups, i.e., $f_p^{i,j} \colon H_p(\mathrm{K}_i) \to H_p(\mathrm{K}_j)$, leading to the definition of the pth persistent homology group $H_p^{i,j}$ for two indices $i \leq j$ as

$$H_p^{i,j} := Z_p(\mathrm{K}_i) \,/\, \left(B_p\left(\mathrm{K}_j\right) \cap Z_p(\mathrm{K}_i)\right). \tag{4}$$

This group contains all the homology classes of K_i that are still present in K_j. It is possible to keep track of all homology classes within the filtration.

The calculation of persistent homology results in a set of pairs (i, j), which denote a homology class that was created in K_i and destroyed (vanished) in K_j. Letting f_i denote the associated function value of K_i, these pairs are commonly visualized in a persistence diagram [6] as (f_i, f_j). The distance of each pair to the diagonal, measured in the L_∞-norm, is referred to as the *persistence* of a topological feature. It is now common practice in topological data analysis to use persistence to separate noise from salient features in real-world data sets [11, 12]. Figure 1b shows the persistence diagram of an example data set. Since the data set, shown in Fig. 1a, appears to be a "blob", the persistence diagram, as expected, contains few topological features of high persistence in both dimensions.

2.3 Intersection Homology and Persistent Intersection Homology

Despite its prevalence in data analysis, persistent homology exhibits some limitations. In the context of this paper, we are mostly concerned with its lack of duality for non-manifold data sets, and with its inability to detect topological features of data sets consisting of *multiple* manifolds.[1] Recall that for a d-manifold, Poincaré duality means that the Betti numbers satisfy $\beta_k = \beta_{d-k}$. While it is possible to extend persistent homology to obtain something similar for manifolds [7, 9], there is no general duality theorem yet. Additionally, persistent homology cannot detect manifolds of varying dimensionality that are "glued together" in the manner described in Sect. 2.1. For example, we could model the data set from Fig. 1, in which we see a central "core" along with some "flares", as a topological disk to which we added multiple "whiskers". The persistence diagram does not contain evidence of any whiskers, so the data set will have the same persistence diagram as a data set that only contains a topological disk. Carlsson [4] proposes to use filter functions on the data to remedy this situation. While this helps detect the features, it does not detect that the underlying structure does not consist of one single manifold.

Intersection homology faces these challenges by providing a homology theory for such spaces with singularities. We follow the notation of Bendich [1, 2] here, who provided a generic framework for calculating restricted forms of (persistent) homology, of which intersection homology is a special case. In the following, we require a function $\phi\colon K \to \{0, 1\}$ that restricts the usage of simplices. We call

[1] We remark that topologically, this case can often be reduced to the computation of ordinary homology, because a theorem of Goresky and MacPherson [14] ensures that for pseudomanifolds, the intersection homology groups remain the same under normalization, and if they are nonsingular, the intersection homology groups are ordinary homology groups. As it is not clear how to obtain normalizations for real-world data, the calculation of persistent intersection homology is necessary.

a simplex σ *proper* or *allowable* if $\phi(\sigma) = 1$. While $\phi(\mathrm{K})$ is not generally a simplicial complex, we can use it to define a restriction on the chain groups of K by calling a simplicial chain $c \in C_p(\mathrm{K})$ *proper* or *allowable* if both c and $\partial_p c$ can be written as formal sums of proper simplices. We refer to the set of allowable p-chains as $I^\phi C_p(\mathrm{K})$. Since $\partial_{p-1} \circ \partial_p = 0$, the boundary of an allowable p-chain is an allowable chain of dimension $p - 1$, so the boundary homomorphism gives rise to a chain complex on the set of allowable chains. We write $I^\phi H_p(\mathrm{K})$ to denote the pth homology group of this complex, and refer to it as the pth intersection homology group. There is a natural restriction of $\phi(\cdot)$ when K is filtrated, so we can define a set of restricted persistent homology groups $I^\phi H_p^{i,j}$ in analogy to the definition of the persistent homology groups.

2.3.1 Persistent Intersection Homology

To obtain *intersection homology* from this generic framework, we require a few additional definitions: a *perversity*[2] is a sequence of integers

$$\bar{p} = (p_1, p_2, \ldots, p_{d-1}, p_d) \tag{5}$$

such that $-1 \leq p_k \leq k - 1$ for every k. Alternatively, following the original definition of Goresky and MacPherson [14], a perversity is a sequence of integers

$$\bar{p}' = (p'_2, p'_3, \ldots, p'_{d-1}, p'_d) \tag{6}$$

such that $p'_2 = 0$ and either $p'_{k+1} = p'_k$ or $p'_{k+1} = p'_k + 1$. Both definitions permit assessing to what extent a data set deviates from being a manifold. More precisely, the perversity measures how much deviation from full transverse intersections (i.e., intersections of two submanifolds that yield another submanifold) are permitted for a given simplicial complex. Each choice of perversity will yield a different set of restricted (persistent) homology groups. We focus only on low-dimensional perversities in this paper, with $k \leq 3$. Finally, tying all the previous definitions together, we define a function $\phi(\cdot)$ for a given perversity and a given stratification: a simplex σ is considered to be *proper* if

$$\dim(\sigma \cap X_{d-k}) \leq \dim(\sigma) - k + p_k \tag{7}$$

holds for all $k \in \{1, \ldots, d\}$. Intuitively, this inequality bounds the dimensionality of the intersection of a simplex with a given subspace. We set $\dim(\emptyset) := -\infty$ so that simplices without an intersection are considered proper. Larger values for p_k give us more tolerant intersection conditions, whereas smaller values for p_k make the

[2]See the unpublished notes by MacPherson on *Intersection Homology and Perverse Sheaves*, available under http://faculty.tcu.edu/gfriedman/notes/ih.pdf, for the origin of this name.

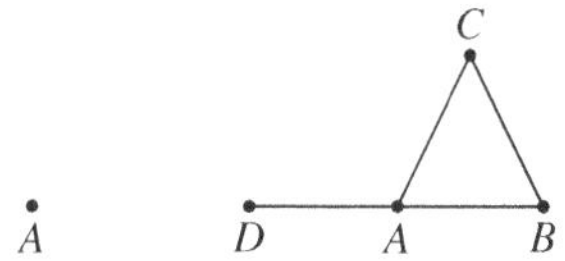

(a) X_0 (b) $X_1 = \mathrm{K}$

Fig. 3 A simple example stratified space (**b**) for which simplicial homology is incapable of detecting the additional "whisker". The *singular stratum* (**a**) only consists of a single vertex, A

intersections more restrictive. This leads to persistent intersection homology groups with a given perversity function.

2.3.2 Simple Example

Figure 3 shows a triangulation of a circle with an additional "whisker". This triangulation is in itself not a manifold: at vertex A, the neighborhood condition that is required for a manifold is violated. However, the space is made up of two manifolds, namely a circle and a line, that are joined at a single point. A natural stratification of such a space thus puts the singular vertex A in X_0 and the full simplicial complex in X_1. With ordinary simplicial homology, we obtain $\beta_0 = 1$, because there is only a single connected component. Intersection homology permits only two different perversities here (we cannot use Goresky–MacPherson perversities because $d = 1$), either $p_1 = -1$ or $p_1 = 0$; as we are only interested in β_0, we do not have to provide a higher-dimensional value for the perversity. For $p_1 = -1$, we obtain $\beta_0 = 2$, because *no* simplex that contains A is proper. This reflects the fact that the simplicial complex is made up of two pieces whose type is different. For $p_1 = 0$, we obtain again $\beta_0 = 1$ because the singular point now leads to a proper connected component: Eq. 7 becomes $\dim(\sigma \cap X_1) \leq \dim(\sigma)$, which is satisfied by every simplex σ.

2.3.3 Implementation

The crucial part of implementing persistent intersection homology lies in an efficient evaluation of Eq. 7: for each simplex σ, the calculating the dimension of the intersection on the left-hand side requires searching through some X_{d-k} and reporting the intersection with the highest dimension. Large speedups can be obtained by (1) restricting the search to l-simplices, where $l := \min(\dim \sigma, d - k)$ is the maximum dimension that can be achieved by the intersection, and (2) enumerating all subsets $\tau \subseteq \sigma$ (in reverse lexicographical order, because we are looking for the largest dimension) and checking whether $\tau \in X_{d-k}$. The second step particularly improves performance when $\dim \sigma$ is small, because we have to enumerate at most

$2^{\dim \sigma}$ simplices and check whether they are part of X_{d-k}. Each check can be done in constant or (at worst) logarithmic time in the size of X_{d-k}. By contrast, calculating all intersections of σ with X_{d-k} takes at least linear time in the size of X_{d-k}. If $2^{\dim \sigma} \cdot \log |X_{d-k}| \ll |X_{d-k}|$, our method will be beneficial for performance. We provide an implementation of persistent intersection homology in `Aleph`,[3] a software library for topological data analysis. We are not aware of any other open-source implementation of persistent intersection homology at this time.

3 Using Persistent Intersection Homology

Prior to using persistent intersection homology in a topological data analysis workflow, we need to discuss one of its pitfalls: the Vietoris–Rips complex is commonly used in topological data analysis to deal with multivariate data sets. For persistent intersection homology, this construction turns out to result in triangulations that yield unexpected results. Figure 4 depicts an example of this issue. Here we see the *one-point union*, i.e., the wedge sum, of two circles, denoted by $S^1 \vee S^1$. Formally, this can be easily modeled as a simplicial complex K (Fig. 4a). The smallest stratification of this space places the singular point x in its own subspace, i.e., $X_0 = \{x\}$, $X_1 = \mathrm{K}$, and uses $\bar{p} = (-1)$. The intersection homology of K results in $\beta_0 = 2$, because of the singular point at which the two circles are connected. Calculating persistent intersection homology of a point cloud that describes this space (Fig. 4b), by contrast, results in $\beta_0 = 1$, regardless of whether we ensure that the triangulation is *flaglike* [17] by performing the first barycentric subdivision (which is guaranteed to make the calculations *independent* of the stratification [14]). The reason for this is that the topological realization of the Vietoris–Rips complex seems to be more closely tied to regular neighborhoods than to the homeomorphism type of $S^1 \vee S^1$. However, the regular neighborhood of a space is always a manifold. It can be thought of as calculating a "thickened" version of the space in which isolated singularities disappear.

As far as we know, Bendich and Harer [2], while discussing other dependencies of persistent intersection homology, did not discuss this aspect. Yet, it is crucial to get persistent intersection homology to "detect" those singularities if we want to understand the manifold structure of a given data set. To circumvent this issue, we propose obtaining additional information about the geometry of a given point cloud in order to determine which points are supposed to be singular. Alternatively, we could try to learn a suitable stratification of the whole space [3] at the cost of reduced performance.

[3]https://github.com/Submanifold/Aleph.

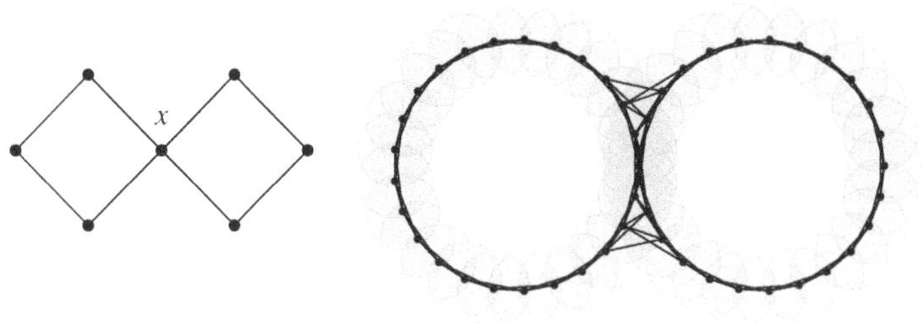

(a) (b)

Fig. 4 Calculating the Vietoris–Rips complex of a point cloud makes it impossible to detect singularities by homological means alone. (**a**) Simplicial complex. (**b**) Vietoris–Rips complex

3.1 Choosing a Stratification

Having seen that the utility and expressiveness of persistent intersection homology hinge upon the choice of a stratification, we now develop several constructions. We restrict ourselves to the detection of isolated singular points, i.e., vertices or 0-simplices, in this paper. A stratification should ideally reflect the existence of singularities in a data set. For the example shown in Fig. 3, a singularity exists at A because the "whisker" will remain a one-dimensional piece regardless of the scale at which we look at the data, while the triangle is a two-dimensional object. This observation leads to a set of stratification strategies, which we first detail before applying them in Sect. 4.

3.1.1 Dimensionality-Based Stratifications

In order to stratify unstructured data according to the local intrinsic dimensionality, we propose the following scheme. We first obtain the k nearest neighbors of every data point and treat them as local patches. For each of these subsets, we perform a principal component analysis (PCA) and obtain the respective set of eigenvalues $\{\lambda_1, \ldots, \lambda_d\}$, where d refers to the maximum number of attributes in the point cloud. We then calculate the largest spectral gap, i.e.,

$$d_i := \underset{j \in \{2,\ldots,d\}}{\arg\max} |\lambda_j - \lambda_{j-1}| - 1, \tag{8}$$

and use it as an estimate of the local intrinsic dimensionality at the ith data point. Points that can be well represented by a single eigenvalue are thus taken to correspond to a locally one-dimensional patch in the data, for example. In practice, as PCA is not robust against outliers, one typically requires some smoothing iterations for the estimates. We use several iterations of smoothing based on nearest neighbors, similar to *mean shift clustering* [5]. The resulting values can then be used to stratify according to local dimensionality.

3.1.2 Density-Based Stratifications

We can also stratify unstructured data according to the behavior of a density estimator, such as a *truncated Gaussian kernel*, i.e.,

$$f(x) := \sum_{y \neq x} \exp\left(-\frac{\text{dist}^2(x, y)}{2h}\right), \tag{9}$$

where h is the bandwidth of the estimator and we define the exponential expression to be 0 if $\text{dist}(x, y) > h$. The density values give rise to a distribution of values so that we can use standard outlier detection methods. Once outliers have been identified, they can be put into the first subset of the filtration. This approach has the advantage of rapidly detecting interesting data points but it cannot be readily extended to higher-dimensional simplices.

3.1.3 Curvature-Based Stratifications

The curvature of a manifold is an important property that can be used to detect differences in local structure. Using a standard algorithm to estimate curvature in meshes [18], we can easily identify a region around the singular point in the "pinched torus" as having an extremely small curvature. Figure 2b depicts this. For higher-dimensional point clouds, we propose obtaining an approximation of curvature by using the curvature of high-dimensional spheres that are fit to local patches of a point cloud. More precisely, we extract the k nearest neighbors of every point in a point cloud and fit a high-dimensional sphere. Such a fit can be accomplished using standard least squares approaches, such as the one introduced by Pratt [20].

4 Results

In the following, we discuss the benefits of persistent intersection homology over ordinary persistent homology by means of several data sets, containing random samples of non-trivial topological pseudomanifolds, as well as experimental data from image processing.

4.1 Wedge of Spheres

We extend the example depicted in Fig. 4 and sample points at random from a wedge of 2-spheres. If no precautions are taken, the resulting data set suffers from

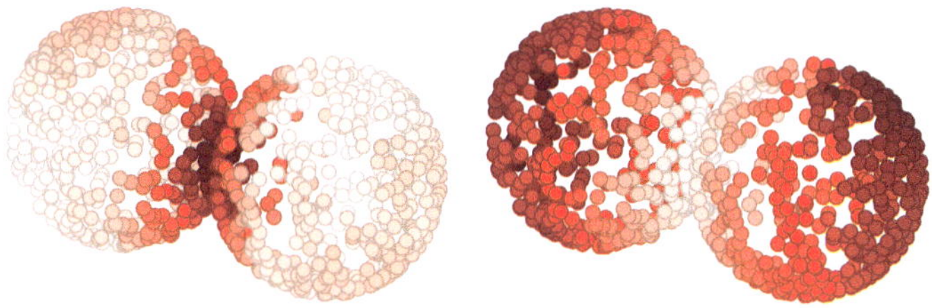

(a) (b)

Fig. 5 A random sample of $S^2 \vee S^2$, color-coded by two stratification strategies. Both descriptors register either extremely high (density) or extremely low (dimensionality) values as we approach the singular part of the data set. The corresponding points are put into X_0. (**a**) Density. (**b**) Local dimension (smoothed)

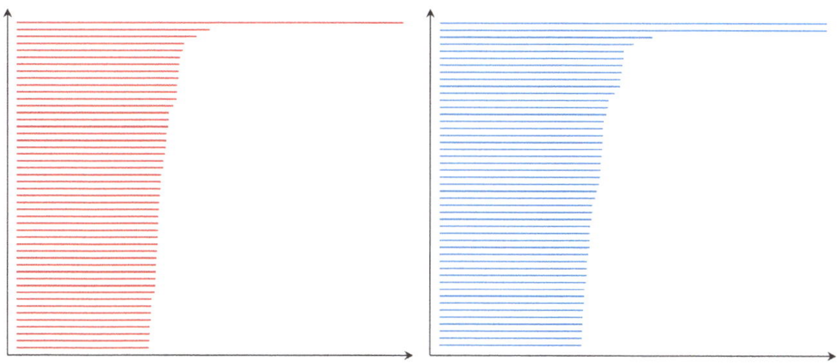

(a) (b)

Fig. 6 Excerpt of the zero-dimensional barcodes for $S^2 \vee S^2$. With persistent homology (**a**), no additional connected component appears, whereas with persistent intersection homology (**b**) with the density-based stratification, the singular point/region results in splitting the data

the problem that we previously outlined. We thus use it to demonstrate the efficacy of our stratification strategies. Figure 5 depicts the data set along with two different descriptors. In both cases, we build a simple stratification in which X_0 contains all singular points, $X_1 = X_0$, and $X_2 = \mathrm{K}$, i.e., the original space. We use the default Goresky–MacPherson perversity $\bar{p}' = (0)$. This suffices to detect that the data set is not a manifold: we obtain $\beta_0 = 2$ for both stratification strategies, whereas persistent homology only shows $\beta_0 = 1$. Figure 6 depicts excerpts of the zero-dimensional barcodes for the data set. The two topological features with infinite persistence are clearly visible in the persistent intersection homology barcode. Since $\beta_2 = 2$, this re-establishes Poincaré duality.

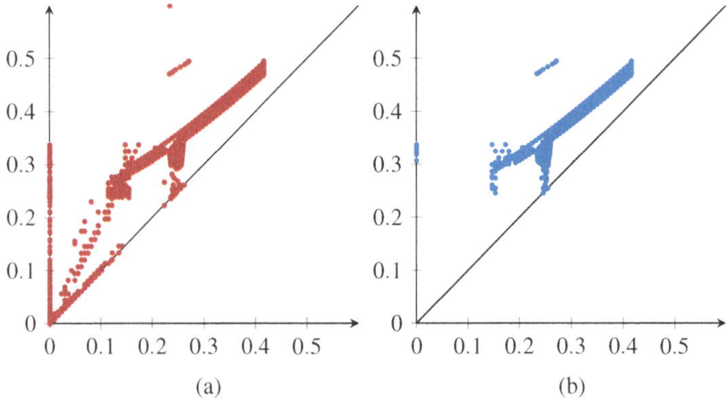

Fig. 7 (**a**) Persistent homology detects more one-dimensional features for the "pinched torus" data set than (**b**) persistent intersection homology

4.2 Pinched Torus

We demonstrate the curvature-based stratification using the "pinched torus" data set. Figure 2b depicts the torus along with curvature estimates. A standard outlier test helps us detect the region around the singular point. We set up the stratification such that X_0 contains all points from the detected region, $X_1 = X_0$, and $X_2 = \mathrm{K}$. Moreover, we use $\bar{p}' = (0)$ because the dimensionality of the input data prevents us from detecting any higher-dimensional features. Persistent homology shows that the point cloud contains a persistent cycle in dimension one. Essentially, the data are considered to be a "thickened circle". Figure 7 depicts the persistence diagrams. We can see that the point with infinite persistence (shown in Fig. 7a at the top border) is *missing* in addition to many other points in the persistent intersection homology diagram (Fig. 7b). The Wasserstein distance [11] between the two diagrams is thus large, indicating the non-manifold structure of the data.

4.3 Synthetic Faces

This data set was originally used to demonstrate the effectiveness of nonlinear dimensionality reduction algorithms [25]. Previous research demonstrated that the data set does not exhibit uniform density [21], which makes the existence of (isolated) singular points possible. It is known that the intrinsic dimension of the data set is three, so we shall only take a look at low-dimensional topological features. More precisely, using the curvature-based stratification, we want to see how persistence diagrams in dimensions 0–2 change when we calculate intersection homology. Note that analyzing three-dimensional features is not expedient, because the stratification cannot detect deviations from "manifoldness" in this dimension.

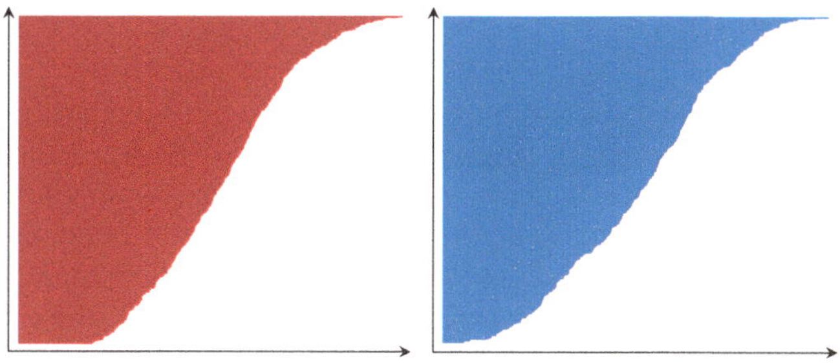
(a) (b)

Fig. 8 Zero-dimensional barcodes for the "Synthetic Faces" data set. Both barcodes are virtually identical, indicating that the singular points do not influence connected components. (**a**) Persistent homology. (**b**) Persistent intersection homology

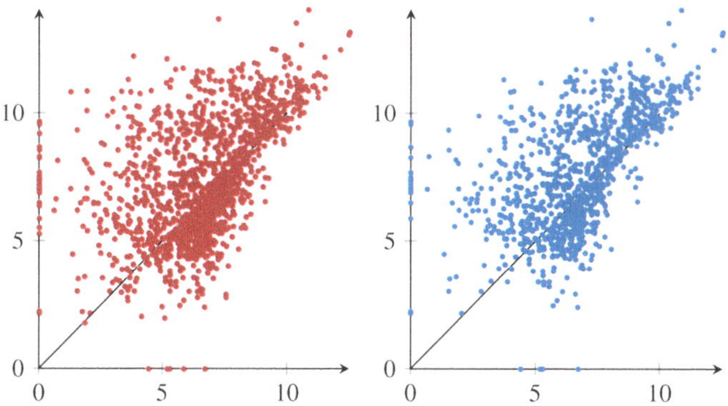

(a) (b)

Fig. 9 Comparison of persistent homology in dimension one (above diagonal) and two (below diagonal) for the "Synthetic Faces" data set. The overall structure is similar, and only few features disappear during the calculation of persistent intersection homology. (**a**) Persistent homology. (**b**) Persistent intersection homology

Figure 8 depicts the zero-dimensional barcodes of the data set. They are virtually identical for both methods (we find that their Wasserstein distance is extremely small), except for some minor shifts in the destruction values, i.e., the endpoints of every interval. This indicates that the singular points only have a very *local* influence on the structure of the data set; they are not resulting in a split, for example. For dimensions one and two, depicted by Fig. 9, we observe a similar behavior. The overall structure of both persistence diagrams is similar, and there is only a slight decrease in total persistence [8] for persistent intersection homology. Likewise, the Wasserstein distance between both diagrams is extremely small.

In summary, we see that we are unable to detect significant differences in zero-dimensional, one-dimensional, and two-dimensional topological features. This lends credibility to the assumption that the data set is a *single* manifold.

5 Conclusion

We showed how to use persistent intersection homology for the analysis of data sets that might not represent a single manifold. Moreover, we described some pitfalls when applying this technique—namely, finding suitable stratifications, and presented several strategies for doing so. We demonstrated the utility of persistent intersection homology on several data sets of low intrinsic dimensionality. Future work could focus on improving the performance of the admissibility condition in Eq. 7 to process data sets with higher intrinsic dimensions. It would also be interesting to extend stratification strategies to higher-dimensional strata, i.e., singular *regions* instead of singular *points*.

References

1. Bendich, P.: Analyzing stratified spaces using persistent versions of intersection and local homology. Ph.D. thesis, Duke University (2009)
2. Bendich, P., Harer, J.: Persistent intersection homology. FoCM **11**(3), 305–336 (2011)
3. Bendich, P., Wang, B., Mukherjee, S.: Local homology transfer and stratification learning. In: Rabani, Y. (ed.) Symposium on Discrete Algorithms, pp. 1355–1370. SIAM, Philadelphia (2012)
4. Carlsson, G.: Topological pattern recognition for point cloud data. Acta Numer. **23**, 289–368 (2014)
5. Cheng, Y.: Mean shift, mode seeking, and clustering. IEEE TPAMI **17**(8), 790–799 (1995)
6. Cohen-Steiner, D., Edelsbrunner, H., Harer, J.: Stability of persistence diagrams. Discret. Comput. Geom. **37**(1), 103–120 (2007)
7. Cohen-Steiner, D., Edelsbrunner, H., Harer, J.: Extending persistence using Poincaré and Lefschetz duality. FoCM **9**(1), 79–103 (2009)
8. Cohen-Steiner, D., Edelsbrunner, H., Harer, J., Mileyko, Y.: Lipschitz functions have L_p-stable persistence. Found. Comput. Math. **10**(2), 127–139 (2010)
9. de Silva, V., Morozov, D., Vejdemo-Johansson, M.: Dualities in persistent (co)homology. Inverse Probl. **27**(12), 124003 (2011)
10. Donoho, D.L., Grimes, C.: Image manifolds which are isometric to Euclidean space. J. Math. Imaging Vision **23**(1), 5–24 (2005)
11. Edelsbrunner, H., Harer, J.: Computational Topology: An Introduction. AMS, Providence (2010)
12. Edelsbrunner, H., Morozov, D.: Persistent homology: theory and practice. In: European Congress of Mathematics. EMS Publishing House, Zürich (2014)
13. Fefferman, C., Mitter, S., Narayanan, H.: Testing the manifold hypothesis. J. Am. Math. Soc. **29**(4), 983–1049 (2016)
14. Goresky, M., MacPherson, R.: Intersection homology theory. Topology **19**(2), 135–162 (1980)
15. Hinton, G.E., Dayan, P., Revow, M.: Modeling the manifolds of images of handwritten digits. IEEE Trans. Neural Netw. **8**(1), 65–74 (1997)
16. Kirwan, F., Woolf, J.: An Introduction to Intersection Homology Theory, 2nd edn. Chapman and Hall/CRC, Boca Raton (2006)
17. MacPherson, R., Vilonen, K.: Elementary construction of perverse sheaves. Invent. Math. **84**(2), 403–435 (1986)

18. Meyer, M., Desbrun, M., Schröder, P., Barr, A.H.: Discrete differential-geometry operators for triangulated 2-manifolds. In: Hege, H.C., Polthier, K. (eds.) Visualization and Mathematics III, pp. 35–57. Springer, Heidelberg (2003)
19. Narayanan, H., Mitter, S.: Sample complexity of testing the manifold hypothesis. In: NIPS 23, pp. 1786–1794. Curran Associates, Inc., Red Hook, NY (2010)
20. Pratt, V.: Direct least-squares fitting of algebraic surfaces. ACM SIGGRAPH Comput. Graph. **21**(4), 145–152 (1987)
21. Rieck, B., Leitte, H.: Persistent homology for the evaluation of dimensionality reduction schemes. Comput. Graph. Forum **34**(3), 431–440 (2015)
22. Roweis, S.T., Saul, L.K.: Nonlinear dimensionality reduction by locally linear embedding. Science **290**(5500), 2323–2326 (2000)
23. Saul, L.K., Roweis, S.T.: Think globally, fit locally: unsupervised learning of low dimensional manifolds. J. Mach. Learn. Res. **4**, 119–155 (2003)
24. Singh, G., Mémoli, F., Carlsson, G.: Topological methods for the analysis of high dimensional data sets and 3D object recognition. In: Eurographics Symposium on Point-Based Graphics. Eurographics Association, Prague (2007)
25. Tenenbaum, J.B., de Silva, V., Langford, J.C.: A global geometric framework for nonlinear dimensionality reduction. Science **290**(5500), 2319–2323 (2000)

Part II
Scalar Topology

Coarse-Graining Large Search Landscapes Using Massive Edge Collapse

Sebastian Volke, Martin Middendorf, and Gerik Scheuermann

Abstract A thorough understanding of discrete optimization problem instances is the foundation for the development of successful solving strategies. For this, the analysis of search spaces is a valuable tool. In particular, networks of solutions—referred to as *search landscapes*—are used in research. Because of the large number of solutions, topological analysis methods are typically restricted to much smaller problem instances than instances that occur in practical applications. In this paper we present a coarse-grained abstraction of search landscapes—the meta landscape—that is accessible for a complete analysis, can be computed easily and preserves relevant properties of the original search landscapes. Thus, detailed topological analysis and visualization become available for problem instances of realistic sizes. We demonstrate the use of our method for search spaces of more than 10^{50} solutions.

Keywords Combinatorial optimization · Fitness landscapes · Topological analysis · Barrier trees

1 Introduction

Combinatorial optimization is a challenging task. One reason is the large search space of typical optimization problems. Therefore, a better understanding of large search spaces is of high interest, but difficult to obtain. It would be helpful if humans could use their visual pattern detection abilities to derive structural properties of search spaces. The knowledge about such properties can then be used to design good (meta-)heuristics for the problem. However, such a visual inspection requires a visualization of the search space. Obviously, any direct visualization of all solutions

S. Volke (✉) · M. Middendorf · G. Scheuermann
Institut für Informatik, Leipzig University, Leipzig, Germany
e-mail: volke@informatik.uni-leipzig.de; middendorf@informatik.uni-leipzig.de; scheuermann@informatik.uni-leipzig.de

H. Carr et al. (eds.), *Topological Methods in Data Analysis and Visualization V*, Mathematics and Visualization, https://doi.org/10.1007/978-3-030-43036-8_4

is usually impossible since even small problems, e.g., the Traveling Salesman Problem with $n \geq 13$ cities, have more solutions than there are pixels on any screen.

Research has found ways of obtaining a structure-preserving, coarse representation of the search space. In this paper we present a novel approach that relies on a massive and implicit collapse of parts of the search space. The method exploits structural properties of the solution set and the chosen search operator, but is independent of the cost function of the optimization problem. The solutions are partitioned into a manageable number of well-describable subsets. On top of them, the coarse-grained *meta landscape* is constructed that approximates the original search space. The meta landscape is small enough such that complete (topological) analyses are feasible. This enables the use of established visualization techniques like dPSO-Vis [18] for the investigation of the search space.

We present a precise mathematical definition of the meta landscape and discuss to what extent topological properties of the meta landscape correspond to topological properties of the original search space. We further discuss how approximate meta landscapes can be constructed for problem instances that are too large for an exhaustive search. Finally, we present and discuss different heuristics for constructing approximate meta landscapes that reveal different aspects of the search space. For the examples of the Traveling Salesman Problem (TSP) and the Quadratic Assignment Problem (QAP), we show how the proposed method can be used in practice to analyze optimization problem instances.

2 Foundations and Related Work

In this section, the mathematical foundations of this work are introduced and related work is reviewed.

2.1 Discrete Optimization Problems

A *discrete optimization problem* (also called *combinatorial optimization problem*) is to find the best solution out of a finite *set of solutions* X where the quality of a solution is determined by a *cost function* $f : X \to \mathbb{R}$, so that $x \in X$ is better than $y \in X$ if $f(x) < f(y)$. Thus, solving a discrete optimization problem is equivalent to finding the global minimum of f. For many discrete optimization problems the set of solutions can be characterized easily, but its size is so large that a complete enumeration is infeasible even for small problem sizes. Many practically relevant discrete optimization problems are NP-hard [5].

Common types of discrete optimization problems are subset problems and permutation problems. Subset problems, like the maximum binary variable saturation problem (MAXSAT), are defined on the power set X over the finite set $\{1, \ldots, n\}$, or equivalently over the set of all binary strings of length n, i.e., $|X| = 2^n$. Permutation problems are defined on the set X of permutations of size n, i.e., $|X| = n!$. Examples are the *Traveling Salesman Problem* (TSP) and related vehicle routing problems, job scheduling problems and the *Quadratic Assignment Problem* (QAP) [5].

The techniques that are proposed in this paper are, in principle, applicable to any type of discrete optimization problem. However, we restrict our presentation to permutation problems and use the TSP and the QAP as particular examples.

The symmetric TSP is: Given a set of n cities and distances $d_{ij} = d_{ji}$ between cities $1 \leq i, j \leq n$, the task is to find the shortest round-trip that visits all cities exactly once. A round-trip can be modeled as a permutation π, where $\pi(i)$ is the city at the i-th position of the round-trip. Then, $f_{TSP}(\pi) = \sum_{i=1}^{n-1} d_{\pi(i),\pi(i+1)} + d_{\pi(n),\pi(1)}$ is the length of a tour. There is no distinguished "first" city, and since d_{ij} is symmetric, the direction of the round-trip is not important. Thus, there are $(n-1)!/2$ equivalence classes of solutions.

The QAP is: given are n facilities with a flow F_{ij} between facilities i and j and n locations with distances d_{ij}, $1 \leq i, j \leq n$. Then, a mapping π of the facilities to the cities is to be found, such that $f_{QAP}(\pi) = \sum_{i=1}^{n} \sum_{j=1}^{n} F_{ij} \cdot d_{\pi(i)\pi(j)}$ is minimized. The QAP does not exhibit any symmetries in general, so that $n!$ equivalence classes of permutations are needed to represent all possible solutions.

2.2 Search Landscapes

The notion of landscapes was first presented in 1932 by Sewall Wright [21] as *fitness landscapes* to model evolutionary processes. The concept has been used many times to study optimization [1, 6, 12, 16, 18, 19]. A (fitness) landscape $L(X, N, f)$ consists of a finite set X, a neighborhood relation $N \subset X \times X$, and a function $f : X \rightarrow \mathbb{R}$ [12]. Depending on the context, f is called fitness function, energy function, or cost function. N induces a graph structure on the set of nodes X. Thus, landscapes can also be defined by a pair $(G_{X,N}, f)$ for a function $f : X \rightarrow \mathbb{R}$ and graph $G_{X,N} = (X, N)$ with vertex set X and edge set N [3]. Landscapes are a valuable tool for the study of local search algorithms [12] where the *neighborhood* of the local search algorithm defines the edge set of the graph. In practice, the neighborhood is often modeled by means of a *search operator* Δ, that is a collection of operator functions $\delta : X \rightarrow X$ [15]. Then, $(x, y) \in N \iff \exists \delta \in \Delta$ such that $\delta(x) = y$.

For a combinatorial optimization problem with set of solutions X, cost function f, and a local search algorithm with neighbourhood N the corresponding *search landscape* is $L(X, N, f) = (G_{X,N}, f)$. In this paper we assume without loss of generality that the graph is connected and consider only symmetric neighbourhoods.

2.3 Landscape Analysis

Landscape analysis is used in different fields of research and various analysis methods have been developed (see [12] for a comprehensive survey). Of particular interest is the investigation of barriers between local minima. Together with the concept of basins of attraction, barriers can be used to structure a landscape topologically. This gives rise to the barrier tree that was described in detail by Flamm et al. [3]. It is a rooted tree that has the local minima and saddle points of the landscape as its nodes and the hierarchical relations between them are represented by the edges. A major drawback of the barrier tree is that it can only be applied if all local minima and all barriers are known. To find them it is often necessary to completely enumerate at least the part of the landscape below a particular barrier. Thus, the barrier tree is often limited to the analysis of smaller problem instances, e.g., TSP instances with up to 13 cities ($2.4 \cdot 10^8$ solutions).

A different approach for the analysis of search landscapes is the computation of the auto-correlation of random walks, as done in [16], and the investigation of distances between local minima, as done in [4, 13]. Both approaches indirectly reveal some structural properties of the search landscape of TSP instances. While not requiring a complete enumeration of all solutions, these approaches have the drawback that no comprehensive topological structure of the search landscape is determined. Only indications about the shape of the search landscape can be obtained. The methods have been applied to problems instances up to approximately 400 cities [4, 16].

Local Optima Networks (LON) [8, 10, 11] have been introduced to investigate search methods. The idea is to reduce the search space to a (sampled) selection of local minima. The connectivity is defined through means of a so called escape perturbation operator that mutates a local optimum in an attempt to escape from its basin of attraction. Being defined on minima, LONs do not provide context for arbitrary solutions in the search space. Although topological analyses are possible on LONs [7], the results only describe the behavior of the escape operator on the set of local optima in the search landscape with respect to the search operator.

In this work, we propose a method that allows to approximate the topological structure for search landscapes that cannot be enumerated completely. It applies to single-operator landscapes and incorporates all solutions of the search space.

2.4 Landscape Visualization

There exist direct visualization approaches that place individual landscape nodes in the Euclidean plane by using multidimensional scaling techniques. This can be based, e.g., on the probability of transitions between solutions [9], or on distances within the landscape [20]. Potential disadvantages of these approaches comprise the error inherent to projection techniques as well as the reliance on samples of the landscape.

There exist also topological visualization approaches. A graph layout of the barrier tree is used, e.g., by Hallam and Prügel-Bennet [6]. In this work, we build upon the visualization tool dPSO-Vis by Volke et al. [18, 19]. In the latter approach, a 1D height function is computed that has the same barrier tree as the original landscape. The focus of that work was the study of folding landscapes of RNA molecules, but the idea has also been applied to study search operators for the TSP [1].

3 Coarse-Grained Search Landscapes: The *Meta Landscape*

A main problem with search landscapes of combinatorial optimization is the exponential number of landscape nodes. This makes a topological analysis infeasible even for problem instances of small size. We tackle this problem by creating a coarse-grained abstraction of the search landscape that is amenable to complete computational analysis. In the following, we define the *meta landscape* that is the foundation for the coarse-graining. Then, some properties of meta landscapes and how their topology relates to the topology of the original search landscape are discussed.

3.1 Definition

Given a search landscape $L(X, N, f)$, we introduce the *meta landscape* $L(\tilde{\mathbf{X}}, \tilde{N}, \tilde{f})$. Basically, the approach can be regarded as a simplification of the search landscape by means of collapsing edges. Formally, a *meta solution* $\tilde{X} \subset X$ is defined as a set of solutions where the subgraph $G[\tilde{X}]$ of the neighborhood graph with vertex set $\tilde{X}$ is connected. A set of meta solutions $\tilde{\mathbf{X}} \subset 2^X$ *represents* X if $\tilde{\mathbf{X}}$ is a partition of X, i.e., the meta solutions are pairwise disjoint and together cover X: $\bigcup_{\tilde{X} \in \tilde{\mathbf{X}}} \tilde{X} = X$.

We define a neighborhood relation $\tilde{N}$ between meta solutions by $(\tilde{X}, \tilde{Y}) \in \tilde{N} \iff \exists x \in \tilde{X}, y \in \tilde{Y} : (x, y) \in N$. The cost function f is transferred to the meta solutions by constructing the function $\tilde{f} : \tilde{\mathbf{X}} \to \mathbb{R}, \tilde{X} \mapsto \min\{f(x) \mid x \in \tilde{X}\}$. Thus, every meta solution is represented by the local optimum that it contains. Now, $L(\tilde{X}, \tilde{N}, \tilde{f}) = (G_{\tilde{\mathbf{X}}, \tilde{N}}, \tilde{f})$ is called *meta landscape*. Observe, that the meta landscape is itself a search landscape, so that topological analysis approaches for search landscapes can be applied to meta landscapes as well.

3.2 Properties

Meta landscapes have some useful properties which indicate that meta landscapes can be considered a valid topological compression method for search landscapes. First, we note that for a maximal partition of X, i.e., $\forall \tilde{X} : |\tilde{X}| = 1$, the meta landscape is equivalent to the original search landscape and thus possesses the same barrier tree. Hence, the approximation through the meta landscape converges towards exactness with the degree of its granularity. If the partition is coarse-grained, the meta landscape conceals topological features of the search landscape. The following theorem shows a relation between the local minima of a meta landscape and the local minima of its underlying landscape.

Theorem 1 *Every local minimum of the meta landscape contains at least one local minimum of the original search landscape and this local minimum is the minimal solution contained in the meta solution.*

Proof Consider a local minimum $\tilde{M}$ from the meta landscape and let $m \in \tilde{M}$ be the solution from $\tilde{M}$ that has the smallest value of f. Hence, $\tilde{f}(\tilde{M}) = f(m)$. Suppose that m is not a local minimum in $L(X, N, f)$. Then, a solution $s \in X$ exists with $f(s) < f(m)$ and $(m, s) \in N$. By the construction $s \notin \tilde{M}$. Hence, there exists a meta solution $\tilde{S} \neq \tilde{M}$ with $s \in \tilde{S}$. From $(m, s) \in N$ follows $(\tilde{M}, \tilde{S}) \in \tilde{N}$, and from $f(s) < f(m)$ we have $\tilde{f}(\tilde{S}) \leq f(s) < f(m) = \tilde{f}(\tilde{M})$. Thus, $\tilde{M}$ is not a local minimum in the meta landscape, which is a contradiction and the result follows.

Note that there is no guarantee that all local minima of the original search landscape are contained in different meta solutions. Thus, a meta landscape can contain significantly fewer local minima than the original search landscape.

The barrier tree is defined with the concept of saddle height [3]. The saddle height between two minima m_1 and m_2 is the minimal cost $h(m_1, m_2)$, such that m_1 and m_2 are still in the same connected component of the subgraph of the search landscape consisting only of the solutions $\{x \in X \mid f(x) \leq h(m_1, m_2)\}$. The following theorem shows how saddle height in the meta landscape is related to saddle height in the original landscape:

Theorem 2 *The saddle height in the meta landscape is a lower bound for the saddle height in the original search landscape.*

Proof Consider two local minima m_1 and m_2 in the search landscape. If there exists a single meta solution $\tilde{M}$ that contains both m_1 and m_2 then the saddle height $\tilde{h}(\tilde{M}, \tilde{M})$ is bounded by $\min\{f(m_1), f(m_2)\}$ and thus, $\tilde{h}(\tilde{M}, \tilde{M}) \leq h(m_1, m_2)$. Now consider the case that there exist two meta solutions $\tilde{M}_1$ and $\tilde{M}_2$ with $m_1 \in \tilde{M}_1$ and $m_2 \in \tilde{M}_2$. Consider a path p between m_1 and m_2 in the original search landscape with $\max_{s \in p}(f(s)) = h(m_1, m_2)$. Path p induces a path $\tilde{p}'$ in the meta landscape by replacing every solution in p with the meta solution that contains it (and by contracting subpaths where all nodes are equal to a same single node). By construction of the meta landscape, the maximal value of f along path $\tilde{p}'$ is less than or equal to $h(m_1, m_2)$. Then, $\tilde{h}(\tilde{M}_1, \tilde{M}_2) \leq \max_{\tilde{s} \in \tilde{p}'}(\tilde{f}(\tilde{s})) \leq \max_{s \in p}(f(s)) = h(m_1, m_2)$.

For investigators of search landscapes, this is important as it allows to draw conclusions about the non-existence of connections below a fixed cost value. The meta landscape might suffer from false positives, i.e., it may show connections when there are none. But there are no false negatives, so that statements about separation in the meta landscape are also valid in the original search landscape.

3.3 *Relaxation*

In practice, the cost function $\tilde{f}$ cannot be determined exactly without solving the optimization problem. But then an approximation of $\tilde{f}$ might help. We formalize the approximation by allowing for a maximal deviation from the exact values. The relaxed cost function $r\tilde{f} : \tilde{X} \in \tilde{\mathbf{X}} \rightarrow \mathbb{R}$ is required to satisfy $\forall \tilde{X} \in \tilde{\mathbf{X}}$: $\left|\tilde{f}(\tilde{X}) - r\tilde{f}(\tilde{X})\right| < \epsilon/2$ for some $\epsilon > 0$.

Relaxed cost functions correspond to so-called ϵ-approximate solutions [5], which are available for many optimization problems. The relaxation makes it much harder to draw reliable conclusions from the meta landscape. For the discussion, we introduce the notion of persistence of a minimum as the strength of a perturbation of the cost function that is needed to eliminate the local minimum. This roughly corresponds to the cost difference between the local minimum and a saddle that connects a local minimum with lower costs.

Every minimum of the relaxed meta landscape with a persistence of at least ϵ is still guaranteed to contain a minimum of the original search landscape. Also, the saddle heights can only be interpreted with an additional uncertainty of ϵ. Thus, two solutions m_1 and m_2 are guaranteed to have no connection below a cost value c in the search landscape, if they have no connection below a cost value of $c + \epsilon$ in the meta landscape.

4 Meta Landscapes for Permutation Problems

The use of the meta landscape approach to analyze a search space requires addressing two problem-specific tasks. First, the meta solutions have to be defined in a way that allows to compute the neighborhood relation and the optimal cost value within each meta solution efficiently. Clearly, this depends on the specific optimization problem and the used search operators. Second, the approximation quality of the meta landscape depends on the particular partition of the solutions into meta solutions. In a sense, the quality improves with the granularity of the partition. Thus, the number of meta solutions should be chosen as large as possible while still retaining computational feasibility for both the generation of the meta landscape and its analysis. Further, the partition has to preserve important topological features of the search landscape. Consequently, structures with particularly important features should be placed into different meta solutions. This is a very important property for analysis applications, although it is difficult to guarantee in the individual case and requires profound domain knowledge.

For solving the first task we introduce in the following the permutation tree that allows to easily generate meta solutions for permutation problems. To address the second task, we demonstrate how to control the generation of meta solutions in a way that can be tailored towards specific analysis goals by domain experts, e.g., by applying a heuristic.

4.1 Permutation Trees

For the convenient generation of partitions of the set of permutations of size n, we organize the permutations into a tree structure $PT(n)$ with a height of n (cf. Fig. 1). The tree is called *permutation tree* for further reference. It is a decision tree where at each level one position in the permutation is set. Thus, each leaf of the tree represents one permutation of size n. In practice, this tree is pruned so that in case of symmetries in the cost function of the permutation problem every equivalence class is only present once in the tree.

Every subtree of the permutation tree represents the set of permutations that match in the positions that have been fixed further up in the permutation tree, and differ in the remaining positions. In the following, we characterize each such set of permutations by its fixed, or equivalently by its variable positions. Every set of mutually disjoint subtrees that together include all leaves of the permutation tree, represents a partition of the set of permutations. Thus, we can identify meta solutions with subtrees of the permutation tree.

If we remove a set of k edges from $PT(n)$, the tree is decomposed into $k + 1$ connected components, each of which possibly contains some leaf nodes of $PT(n)$. Components without leaf nodes are discarded. Thus, each such edge set defines a partition of permutations. Likewise, every partition of permutations can be identified with a set of edges that would decompose $PT(n)$ into the corresponding subtrees.

To obtain sets of permutations that are well-suited for use in meta landscapes, we require that for every level i of the permutation tree the variable positions of the

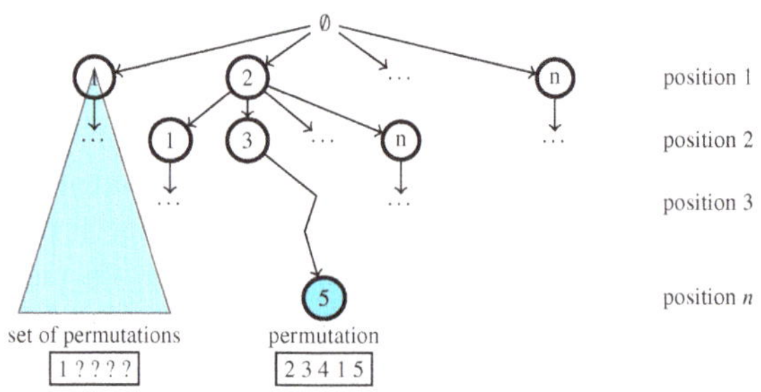

Fig. 1 Permutation tree $PT(n)$. At every level of the tree, one position in the permutations is fixed to the specified element. Every leaf of the tree represents one permutation. A subtree corresponds to a set of permutation that agree at the fixed positions and differ at the remaining ones

permutations form a consecutive range of indices $p_{i+1}, \ldots, p_n$. Given a solution x that belongs to the meta solutions $\tilde{X}$, we consider its set of neighbors $N_x = \{y \in X \mid (x, y) \in N\}$ and investigate the cardinality of $N_x^{in} = N_x \cap \tilde{X}$ and $N_x^{out} = N_x \setminus \tilde{X}$. We want N_x^{in} to be maximal, so that connectivity within a meta solution is as strong as possible. At the same time, we want N_x^{out} to be minimal, so that meta solutions are as separated from each other as possible.

Consider a meta solution $\tilde{X}$ with i fixed positions and $k = (n - i)$ consecutive variable positions. For three common operators on permutations—consecutive swaps, interchange of two positions, 2opt (for details see Schiavinotto et al. [15])—we obtain the following quantities:

Operator	$\lvert N_x^{in} \rvert$	$\lvert N_x^{out} \rvert$
Consecutive swap	$k - 1$	$n - k$
Interchange, 2opt	$(k \cdot (k - 1))/2$	$(n \cdot (n - 1) - k \cdot (k - 1))/2$

These numbers are optimal for sets of permutations with i fixed positions. When changing the partitioning scheme and allowing non-consecutive variable positions in the permutations, we end up with much worse ratios in particular for the consecutive swap and the 2opt operators.

4.2 Heuristic Generation of Partitions by Branching

The generation of a partition by cutting a number of edges in the permutation tree facilitates a branching approach to define and refine such partitions. Initially, we select the out edges of the root node for cutting (cf. Fig. 2). Then, we iteratively select one of these edges and branch it, i.e., we replace it by the out edges of its target node. This is repeated until a partition of sufficient resolution is generated, such that the number of meta solutions stays manageable and at the same time the size of the meta solutions is small enough to provide a meaningful approximation

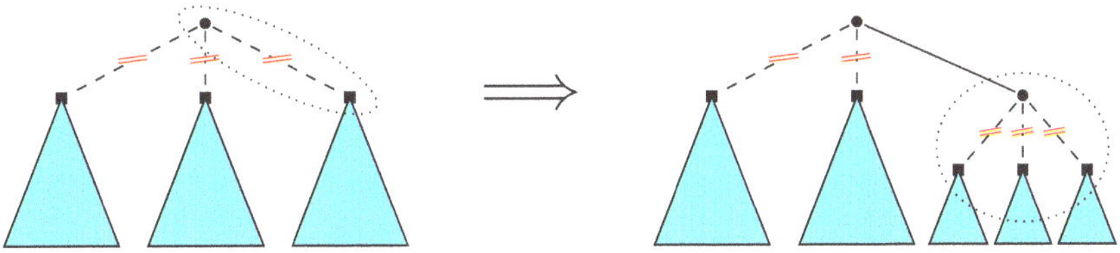

Fig. 2 Branching step in a permutation tree. Initially (left side), some edges (dashed) are selected for cutting. Among these, a branching candidate is determined (circled edge). In the branching step (right side), this candidate is replaced by the outgoing edges of its target

of the original landscape. The process of selecting edges for further refinement can be steered heuristically with respect to the needs of the expert.

For a general analysis of the search landscape, we propose the following branching criteria:

- Branch evenly: Branching is performed so that the resulting meta solutions have nearly equal size.
- Branch by cost: For every meta solution a lower bound for the optimal solution in the meta solution is computed. The meta solution with the lowest bound is branched.
- Branch by cost range: For every meta solution we compute a lower bound and an upper bound. The meta solution with the largest difference between upper and lower bound is branched.

Branching criteria can also be combined, e.g., a smaller size of a meta solution can be weighted against lower costs of a meta solution. However, it seems reasonable to restrict the branching to meta solutions that have a certain minimum size, so that too small meta solutions are avoided.

Furthermore, for an analysis of a landscape it might be interesting to incorporate the global optimum into the branching. In that case the optimum needs to be known or determined beforehand by using an exact solving method. Then, the global optimum can be pre-branched by iteratively branching the meta solution that contains the global optimum until the meta solution falls under a certain size.

5 Results

In this section, we investigate of the proposed meta landscape experimentally and show how it can be applied to TSP and QAP instances. We present the results by using a topological visualization tool for meta landscapes that is a variant of the tool dPSO-Vis [18] with the visual modifications of Bin et al. [1]. dPSO-Vis computes a one-dimensional landscape outline that has the same barrier tree as the meta landscape. We compensate for the differing sizes of the meta solutions by introducing a width for every solution into the visualization (dPSO-Vis considers only individual solutions that naturally are of equal size). As every solution is laid out in an individual horizontal interval, this can be easily achieved by scaling these intervals so their size is proportional to the size of the meta solution. This makes the layout of the topological landscape more robust against changes of the applied branching. It also allows the visual comparison of the visual sizes of different landscape parts—particularly of topological substructures—even if the sizes of the individual meta solutions differ.

5.1 *Validation of the Approach*

Though we have some theoretical results pertaining to the correctness of the topological approximation with the meta landscape, the quality and amount of detail can only be verified experimentally. For that purpose, we considered small TSP instances with at most 12 cities, so that the complete search landscape was available as a ground truth. The results are demonstrated with the example of a 10-city problem that was generated by randomly placing cities in the plane (cf. Fig. 3). A general drop in the overall landscape height can be observed when using large meta solutions. This is an effect of using the local optimum within a meta solution as the reference cost value for the whole meta solution. However, even when using a meta landscape with only few meta solutions the most persistent topological features are preserved.

5.2 *Branching Strategies*

We show the results of different branching strategies (cf. Sect. 4.2) for the example of the 52 city TSP instance berlin52 from the TSPLIB [14]. We expect a complex topology, because of the huge magnitude of the search space for this instance. Extrapolating experiences with smaller problem instances, we also expect that no branches in the barrier tree occur outside a small percentage of the best solutions. To estimate the minimal costs within each meta solution, we used the branch-and-bound scheme of Volgenant and Jonker [17].

Figure 4 shows the meta landscapes for the different branching strategies. The meta landscapes differ in the number of branches and the resolution within the interesting parts of the landscape. From the visual size of the meta solutions and

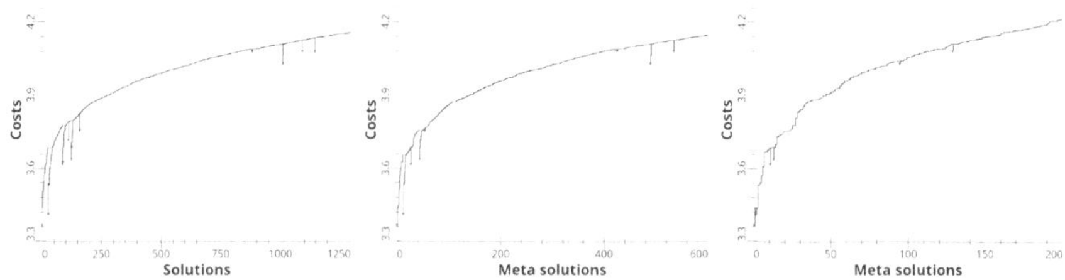

(a) (b) (c)

Fig. 3 Interchange operator landscapes of a randomly generated TSP instance with 10 cities (181 440 solutions). Only the lower part of the search landscape is shown as higher parts are topologically not so interesting. The meta landscape were constructed by cutting the permutation tree at the 7-th and 5-th level, respectively, resulting in 30,240 and 1512 meta solutions. (**a**) Ground truth from complete enumeration. (**b**) 7 levels of the permutation tree. (**c**) 5 levels of the permutation tree

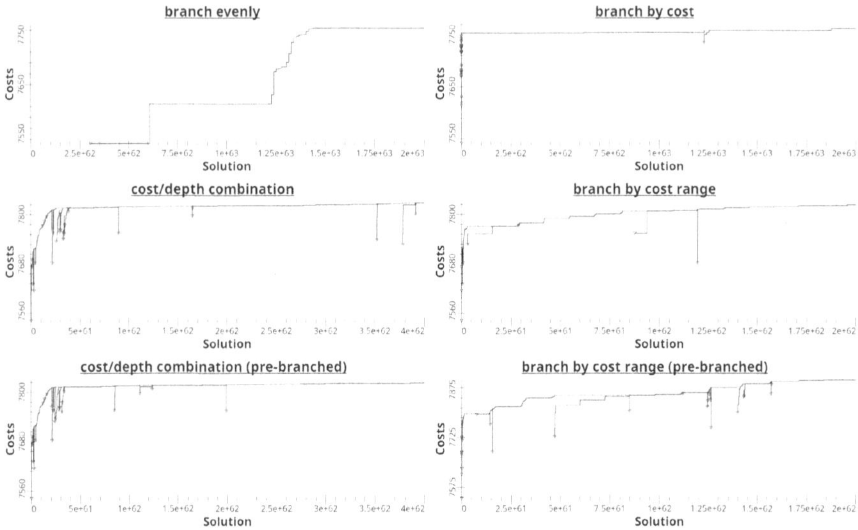

Fig. 4 Swap operator landscapes of the berlin52 instance from the TSPLIB [14] that have been generated with different branching strategies (cf. Sect. 4.2). The complete landscape contains $7.76 \cdot 10^{65}$ solutions. All meta landscapes consist of 50,000 meta solutions with at least 3 and at most 17 fixed permutation positions. Only the topologically relevant part is shown here

their cost values, it can be seen that some branching strategies generate too many meta solutions in parts of the landscape that can be considered as less interesting, i.e., parts with solutions of high costs. This is particularly visible for the strategies branching by cost values and branching evenly. Branching by cost, however, is able to detect a high number of local minima, so that it appears reasonable to include some cost-related metric into the branching. A combined branching strategy that allows level differences between different parts of the permutation tree in relation to the corresponding cost differences (depth/cost combination in Fig. 4) results in the more balanced images and also tends to reveal most branches around the global optimum when compared to other branching strategies. A minimization of the cost intervals for each meta solution appears reasonable from a theoretical point of view (we gain a well-defined persistence per meta solution). However, it does not perform as well as the depth/cost combination. We tested pre-branching the global optimum for the cost/depth combination and the cost range strategy. In both cases, a slightly better focus on the near-optimal part of the landscape can be noticed, but the effect is not very strong.

While the figure shows results for only one example instance, it should be noted that similar differences between the branching strategies have also been found for other TSP instances. Therefore, we suggest to use branching strategies that combine different criteria, e.g., the depth/cost combination, or a branching by cost ranges, and to omit pre-branching.

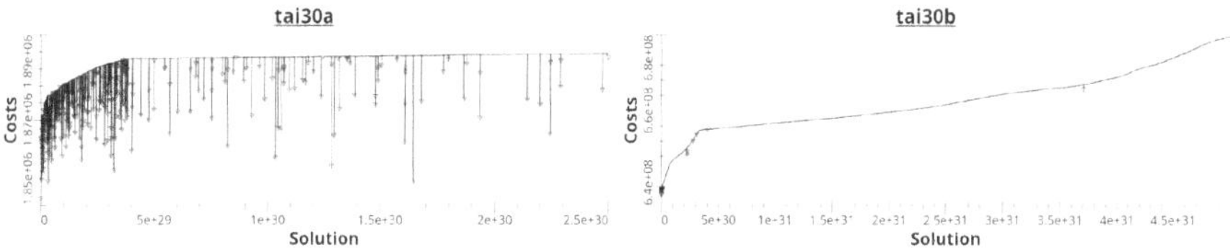

Fig. 5 Landscapes of the problem instances tai30a and tai30b from the QAPLIB [2]. The landscapes have been created with the same parameter set, using branching by a combination of cost and depth with a size of 50,000 meta solutions

5.3 Comparison of QAP Instances

Ongoing research tries to categorize QAP instances in order to identify structural properties of them [7], as there can be huge differences between instances. There exist small, but still not optimally solved problem instances, e.g., instance tai30a of Taillard (included in the QAPLIB [2]). However, for the similar instances tai30b which is the sibling instance in the QAPLIB, the global optimum is known. We computed and compared meta landscapes for these two instances. As no well-performing branch-and-bound algorithm for QAP is available, we used probing with multiple local searches to estimate the cost minimum within each meta solution.

Figure 5 contrasts the meta landscapes of both instances. Instance tai30a possesses a much more complicated barrier tree than tai30b. There are many local minima with nearly equal persistence. There also appears a plateau-like structure in the landscape which is indicated by the almost horizontal part of the landscape in the center and the right of the landscape visualization. All this indicates that a bad performance of local search is to be expected. Differently, tai30b contains only few local minima, which also branch away at well-separated cost values. Further, the shape of the fitness landscape might be an indicator for individual structural properties that could be exploited for a well-performing search algorithm.

6 Conclusion

In this paper the concept of meta landscapes was proposed as a well-defined coarse-grained representation of search landscapes. Some properties were shown that allow an interpretation of meta landscapes and to transfer the results back to the original search landscape. Some advice for the computation and usage of meta landscapes was given. In particular, we demonstrated how a heuristically controlled branching scheme could be used to define an application-specific meta landscape.

Since meta landscapes are completely enumerable, topological methods like the barrier tree can be applied to it. Then, topological visualization methods, like dPSO-Vis [18, 19] that build upon the barrier tree, can be used to visually analyse the search landscape. The approach makes permutation problem instances with sizes over 50 accessible for a (visual) analysis of the topological structure of their search landscapes. Thus, the sizes of analyzable search landscapes are increased by several orders of magnitude in comparison to completely enumerable landscapes. This was demonstrated for the example of the TSP and the QAP.

In future work, we want to focus on improved branching strategies for partitioning the search landscape since the branching strategy has a strong influence on the quality of the meta landscape. One possibility would be to pre-sample local minima and to incorporate their distribution within the search landscape into the branching. Also, theoretical results about the quality of the branching are missing. In this paper we focused on permutation problems. However, for subset (or bitstring) problems like MAXSAT, a decision tree can be defined that is similar in structure to the permutation tree. The branching strategies are not dependent on the solutions being permutations, so that the approach should be applicable to subset problems with minimal adaptation. Here, we expect to be able to analyze problems of a size up to 225 variables as these possess similar landscape sizes.

Acknowledgments This work was partly supported by the German Federal Ministry of Education and Research (BMBF) within the project Competence Center of Scalable Data Services and Solutions (ScaDS) Dresden/Leipzig (BMBF grant 01IS14014B).

References

1. Bin, S., Volke, S., Scheuermann, G., Middendorf, M.: Comparing the optimization behaviour of heuristics with topology based visualization. In: Theory and Practice of Natural Computing. Lecture Notes in Computer Science, vol. 8890, pp. 47–58. Springer, Heidelberg (2014)
2. Burkard, R., Karisch, S., Rendl, F.: QAPLIB – a quadratic assignment problem library. J. Glob. Optim. **10**(4), 391–403 (1997)
3. Flamm, C., Hofacker, I.L., Stadler, P.F., Wolfinger, M.T.: Barrier trees of degenerate landscapes. Z. Phys. Chem. **216**, 1–19 (2002)
4. Fonlupt, C., Robilliard, D., Preux, P., Talbi, E.G.: Fitness landscapes and performance of meta-heuristics. In: Meta-Heuristics: Advances and Trends in Local Search Paradigms for Optimization, pp. 257–268. Kluwer Academic Publishers, Berlin (1999)
5. Garey, M.R., Johnson, D.S.: Computers and Intractability: A Guide to the Theory of NP-Completeness. W. H. Freeman, New York (1979)
6. Hallam, J., Prügel-Bennett, A.: Large barrier trees for studying search. IEEE Trans. Evol. Comput. **9**(4), 385–397 (2005)
7. Herrmann, S., Ochoa, G., Rothlauf, F.: Coarse-grained barrier trees of fitness landscapes. In: International Conference on Parallel Problem Solving from Nature, pp. 901–910. Springer, Heidelberg (2016)
8. Iclanzan, D., Daolio, F., Tomassini, M.: Data-driven local optima network characterization of QAPLIB instances. In: Proceedings of the 2014 Annual Conference on Genetic and Evolutionary Computation, pp. 453–460. ACM, New York (2014)
9. McCandlish, D.M.: Visualizing fitness landscapes. Evolution **65**(6), 1544–1558 (2011)
10. Ochoa, G., Tomassini, M., Vérel, S., Darabos, C.: A study of NK landscapes' basins and local optima networks. In: Proceedings of the 10th Annual Conference on Genetic and Evolutionary Computation, pp. 555–562. ACM, New York (2008)

11. Ochoa, G., Verel, S., Daolio, F., Tomassini, M.: Local optima networks: a new model of combinatorial fitness landscapes. In: Recent Advances in the Theory and Application of Fitness Landscapes, pp. 233–262. Springer, Berlin (2014)
12. Pitzer, E., Affenzeller, M.: A comprehensive survey on fitness landscape analysis. In: Recent Advances in Intelligent Engineering Systems, pp. 161–191. Springer, Heidelberg (2012)
13. Preux, P., Robilliard, D., Fonlupt, C., Karp, R., Steele, J.: Fitness Landscapes of Combinatorial Problems and the Performance of Search Algorithms (1997)
14. Reinelt, G.: TSPLIB—a traveling salesman problem library. ORSA J. Comput. **3**(4), 376–384 (1991)
15. Schiavinotto, T., Stützle, T.: A review of metrics on permutations for search landscape analysis. Comput. Oper. Res. **34**(10), 3143–3153 (2007)
16. Stadler, P.F., Schnabl, W.: The landscape of the traveling salesman problem. Phys. Lett. A **161**(4), 337–344 (1992)
17. Volgenant, T., Jonker, R.: A branch and bound algorithm for the symmetric traveling salesman problem based on the 1-tree relaxation. Eur. J. Oper. Res. **9**(1), 83–89 (1982)
18. Volke, S., Middendorf, M., Hlawitschka, M., Kasten, J., Zeckzer, D., Scheuermann, G.: dPSO-Vis: topology-based visualization of discrete particle swarm optimization. Comput. Graph. Forum **32**(3), 351–360 (2013)
19. Volke, S., Bin, S., Zeckzer, D., Middendorf, M., Scheuermann, G.: Visual analysis of discrete particle swarm optimization using fitness landscapes. In: Recent Advances in the Theory and Application of Fitness Landscapes. Emergence, Complexity and Computation, vol. 6, pp. 487–507. Springer, Berlin (2014)
20. Volke, S., Zeckzer, D., Scheuermann, G., Middendorf, M.: A visual method for analysis and comparison of search landscapes. In: Proceedings of the 2015 Annual Conference on Genetic and Evolutionary Computation, pp. 497–504. ACM, New York (2015)
21. Wright, S.: The roles of mutation, inbreeding, crossbreeding and selection in evolution. In: Proceedings of the Sixth International Congress of Genetics, pp. 356–366 (1932)

Adjusting Control Parameters of Topology-Accentuated Transfer Functions for Volume Raycasting

Yuriko Takeshima, Shigeo Takahashi, and Issei Fujishiro

Abstract The design of automatic transfer functions for volume rendering is a perennial problem in volume visualization. Over the last three decades, a variety of design methodologies have been proposed. However, sensitive adjustment of related control parameters remains entrusted to users, because rendering conditions, such as the thickness of emphasized subvolumes in the ray direction and the size of a target dataset, differ on a case-by-case basis. Our group previously proposed one-dimensional transfer functions to accentuate topological changes in the scalar field of the target dataset. However, the method forces us to determine the actual control parameter values for the transfer functions in an empirical manner. In this paper, we propose a supplementary mechanism with which to judiciously define an appropriate profile of the opacity values. More specifically, the height and width of the hat opacity transfer functions that accentuate feature isosurfaces are determined according to the number of voxels belonging to the relevant topologically equivalent scalar field interval. The feasibility of the proposed method is evaluated by its application to five kinds of volume datasets.

1 Introduction

Volume rendering is one of the traditional visualization techniques for scalar fields. Over the last three decades, various methodologies have been proposed for designing transfer functions [1]. Data-centric approaches, which perform a

Y. Takeshima (✉)
Tokyo University of Technology, Tokyo, Japan
e-mail: takeshimayrk@stf.teu.ac.jp

S. Takahashi
University of Aizu, Fukushima, Japan
e-mail: takahashis@acm.org

I. Fujishiro
Keio University, Yokohama, Japan
e-mail: fuji@fj.ics.keio.ac.jp

H. Carr et al. (eds.), *Topological Methods in Data Analysis and Visualization V*, Mathematics and Visualization, https://doi.org/10.1007/978-3-030-43036-8_5

mathematical analysis of the dataset before pertinent rendering, have become well established. Kindlmann and Durkin [2] defined a histogram volume consisting of first and second partial derivatives of the volume dataset, which is used to design transfer functions in order to emphasize the boundaries between different materials contained in the target volume dataset. Kniss et al. [3] generalized the histogram volume-based method in order to design three-dimensional transfer functions. Hladůvka et al. [4] and Kindlmann et al. [5] independently proposed multi-dimensional transfer functions based on isosurface curvatures in the target volume dataset. Weber et al. [6] proposed a topology-based method that defines the opacity at a sample point based on the topological characterization of a target dataset. Zhou et al. [7] proposed an automatic transfer function design using a residue flow model controlled by a contour tree [8]. Shape-based transfer functions, proposed by Praßni et al. [9], are based on the shape of the surfaces to be visualized. Xiang et al. [10] proposed a graph cut segmentation method and localized transfer function, which is suitable for datasets with a comparatively clear boundary rather than for simulation datasets whose values are distributed smoothly. In addition, it is conceivable to design the transfer functions based on other structures of the target dataset, such as representative isosurfaces [11].

Our group previously proposed yet another topology-based method [12], which designs one-dimensional transfer functions to accentuate topological changes in the scalar field of the target dataset. However, the method forces us to determine the actual control parameter values for the transfer functions in an empirical manner; the users have to adjust the control parameter values according to a target dataset. In this paper, therefore, we propose a supplementary mechanism with which to judiciously define an appropriate profile of opacity values. More specifically, the height and width of the hat opacity transfer functions that accentuate the feature isosurfaces are determined according to the number of voxels belonging to the relevant topologically equivalent scalar field interval. We focus on a traditional one-dimensional transfer function that assigns an opacity to a field value and a multi-dimensional transfer function that assigns an opacity to a field value as well as other attributes, such us inclusion level [13]. To evaluate the proposed method, we use our transfer functions to visualize five different kinds of volume datasets. Furthermore, we introduce an evaluation mechanism for measuring the quality of the volume-rendered images with the designed transfer functions and also discuss the processing speed of our method.

2 Volume Skeletonization

First, we extract the topological structure from a target dataset using the topological volume skeletonization algorithm [12, 13].

2.1 Volume Skeleton Tree

The level-set graph, known as the volume skeleton tree (VST), allows us to evaluate the topological attributes of each voxel by illuminating both the global and local features of the volume dataset. A node of the VST represents a critical point that displays a change either in the number of connected isosurface components or in the genus of each of the isosurface components. Critical points are classified into four groups: maxima (C_3), saddles (C_2, C_1), and minima (C_0), which represent isosurface appearance, merging, splitting, and disappearance, respectively, as the scalar field value decreases. A link of the VST represents a topology-preserving connected component of interval volume (transition of isosurface) [14]. A link is defined as *solid* if its isosurface component expands as the scalar field value decreases; it is defined as *hollow* if it shrinks.

Both the isosurfaces merging at C_2 and splitting at C_1 have four topological transition paths with different isosurface spatial configurations, as shown in Fig. 1. In what follows, the VST uses the notation for the critical points with its own connectivity, as indicated in Fig. 1, where the solid incident link represents a solid isosurface and the broken link a hollow isosurface. The saddle points of $C_i\,(i = 1, 2)$ are classified into 3-C_i and 2-C_i, according to their degree (valence). For later convenience, all the boundary voxels are assumed to be connected to the virtual minimum having $-\infty$ as its scalar field value [12]. Note that the link to the C_0 node is solid when the node is the virtual minimum, as shown in Fig. 1.

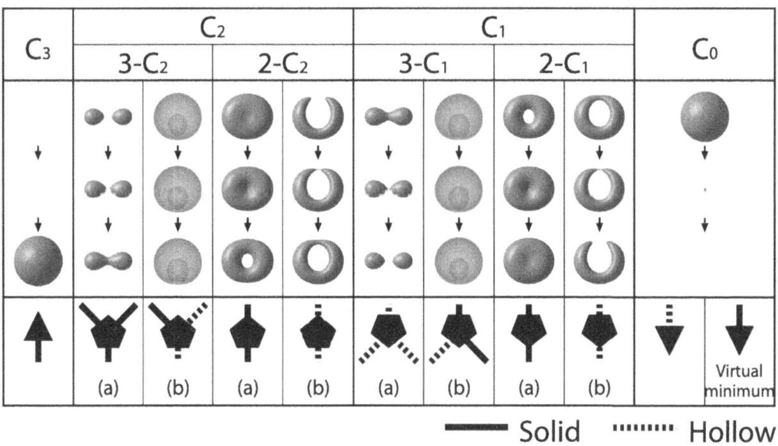

Fig. 1 The connectivity of critical points in the VST

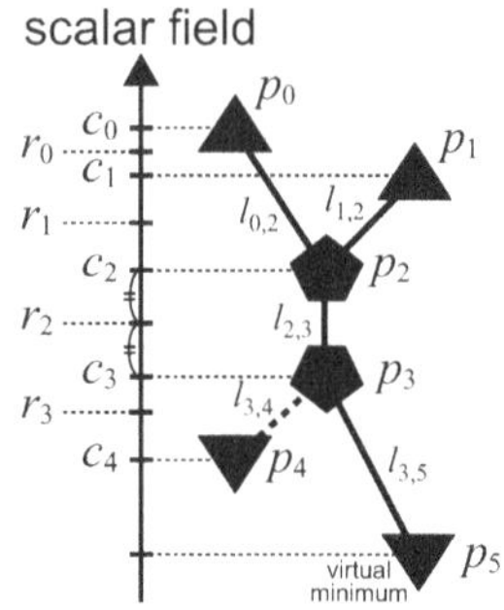

Fig. 2 An example of a VST. The critical field values and the representative field values are denoted by c_i $(i = 0, \cdots, 4)$ and r_i $(i = 0, \cdots, 3)$, respectively. The inclusion level of the link $l_{3,4}$ is 1, whereas those of the outer links are 0

2.2 Feature Values in the VST

In a VST, a node p_i has coordinates $\boldsymbol{x}_i$ and a scalar field value c_i $(i = 0, \cdots, m-1)$, which is referred to as *critical field value*, denoted in the monotonically decreasing order of the scalar field. *Representative field value* is also defined as the mid-value $r_i = (c_{i+1} + c_i)/2$ of each interval $[c_{i+1}, c_i](i = 0, \ldots, m-2)$ bounded by consecutive critical field values, as shown in Fig. 2.

A link $l_{i,j}$ has the genus of the corresponding isosurface component, the index of adjacent nodes p_i, and p_j, and its subvolume as the sweep of a connected component of a topologically equivalent isosurface. We denote a volume ratio belonging to the subvolume corresponding to link $l_{i,j}$ as $v_{i,j}$, calculated as $n_{i,j}/N$, where N denotes the total number of voxels in the target volume and $n_{i,j}$ denotes the number of voxels in the subvolume representing link $l_{i,j}$.

Because the VST is sensitive to small changes in field values, it may contain a large number of minor critical points if the volume dataset involves high-frequency noise. In order to capture the global features of the entire volume dataset, it is necessary to eliminate some minor links from the VST. To control the level of detail (LoD) of the VST, we define a persistency value of $l_{i,j}$ as $v_{i,j}/|c_i - c_j|$. In our algorithm, the VST is simplified by removing links in ascending order until the value reaches a specified threshold [12].

Furthermore, we can extract an inclusion level that represents the depth of its associated isosurface in the nested structure at the corresponding scalar field value and serves as an additional variable for the multi-dimensional transfer functions that emphasize these nested structures. Figure 1 clearly illustrates that isosurface nested structures originate only from the transition paths in 3-C_2(b) and 3-C_1(b). This motivates us to locate such isosurface inclusions directly from the VST if we can identify all the nodes that correspond to the previously mentioned transition paths. Indeed, the inclusion level can be systematically extracted by tracing the VST from the virtual minimum, because its incident link is known to be solid [13]. In Fig. 2, it has a nested structure in the field interval $[c_4, c_3]$ where the isosurfaces belonging to $l_{3,5}$ exist outside the isosurfaces belonging to $l_{3,4}$.

3 Transfer Function Design

Next, we design a transfer function reflecting the topological structures in the VST (Sect. 2). The basic principle is to accentuate the topological change of an isosurface within the volumetric domain [12].

3.1 *Color Transfer Function*

In our previously proposed method [13], wherein the color transfer function is defined piecewise using the range of the HSV hex-cone $[0, 2/3\pi]$, so that the function value decreases linearly over an evenly divided hue interval for each of the field intervals $[c_{i+1}, c_i]$. Our transfer functions can be designed based on blue-white-red, blue-white-yellow, heat object, as well as rainbow colormap. In every case, we define the color transfer function based on the above principle of dividing a given hue range. Figure 3a shows the design of our color transfer function, which allows us to assign a steep color gradation to the regions where the consecutive critical field values are in closer proximity to one another.

3.2 *One-Dimensional Opacity Transfer Function*

For opacity transfer functions, we propose two methods, accentuating critical or representative field values. The first method assigns local hat functions centered at the critical field values c_i, whereas the second is centered at representative field values r_i. When the critical field values are emphasized, topological changes, such as splitting and merging isosurfaces, can be visualized more clearly. On the other hand, emphasizing the representative field values reveals the representative topological structures. In our framework, because the outermost isosurface does not shrink as the field value decreases, we minimize the occlusion artifacts induced by the isosurface nested structure by decreasing the base elevation of the hat functions

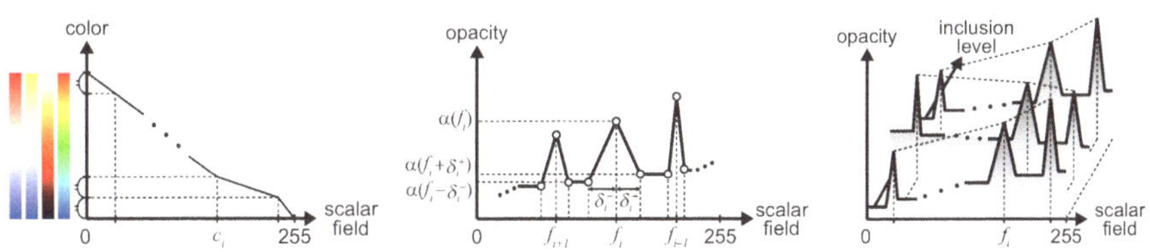

(a) (b) (c)

Fig. 3 Basic principle of designing transfer functions (TFs). (**a**) Color TF. (**b**) 1D opacity TF. (**c**) 2D opacity TF

for c_{m-1} through c_0, or r_{m-2} through r_0, in a stepwise fashion, as shown in Fig. 3b. Note that in this figure, c_i and r_i are expressed commonly as f_i.

In our previously proposed method [12], we determined the actual opacity values empirically, because the opacity of the object projected on the screen was affected by the rendering conditions, such as the thickness of the emphasized subvolumes in the ray direction and the size of a target dataset.

In order to emphasize the internal structure of the volume dataset more clearly, it is necessary to set the opacity value higher for the area closer to the center of the dataset. In our previously proposed method [12], we assumed that the isosurfaces whose field value is higher, exist inside the dataset; thus, a higher opacity value is assigned to them, without any consideration of the thickness of the emphasized subvolumes in the ray direction. If the sampling distance is larger than the thickness, it will not be visualized regardless of the height of the corresponding opacity value. However, if the sampling distance is small, many sampling points are generated in unnecessary regions, which may cause a computational burden. To address this problem, we calculate the thickness of the emphasized region in the ray direction in a pseudo manner, and we introduce the following variable G_i into the opacity calculation formula:

$$G_i = \frac{1.0 - V_{0,i}^{1/3}}{d}, \tag{1}$$

where $V_{i,i+1}$ denotes the volume ratio of the subvolume belonging to the corresponding interval volume at $[c_{i+1}, c_i]$, and d the average of the number of grid points in the x-, y-, and z-axis directions of the target dataset. Note that $V_{i,i+1}$ and $v_{i,i+1}$ are different when multiple links of the VST are included in the field interval $[c_{i+1}, c_i]$. For example, as Fig. 2 shows, $V_{1,2}$ is calculated as the sum of the subvolume $v_{1,2}$ corresponding to link $l_{1,2}$ and a part of the subvolume $v_{0,2}$ whose scalar field is included in the interval $[c_2, c_1]$. Assuming that the homeomorphic regions are spherically distributed, $V_{0,i}$ is proportional to the radius of the sphere. Moreover, because $\Sigma_{i=0}^{m-2} V_{i,i+1}$ equals 1, $1 - V_{0,i}^{1/3}$ can be thought of as a pseudo distance from the data boundary to the corresponding region. This is further divided by d, and it is used as the thickness per grid interval.

Herein, we define the opacity value α at three control points, such as the top and two bottoms of the hat function, using the following equations:

$$\alpha(f_i) = \frac{m-1-i}{m-1}\mu G_i, \quad \alpha(f_i + \delta_i^+) = \frac{m-1-i}{m-1} G_{i-1}, \quad \alpha(f_i - \delta_i^-) = \frac{m-1-i}{m-1} G_i, \tag{2}$$

where μ denotes a coefficient for the degree of emphasis. From this setting, the opacity values decrease as the number of voxels corresponding to the subvolume increases.

The opacity value of a scalar field value s other than the control points is calculated using Eqs. (3):

$$\alpha(s) = \begin{cases} \dfrac{m-1-i}{m-1} G_i & (f_{i+1} + \delta_{i+1}^{+} \leq s < f_i - \delta_i^{-}) \\ \dfrac{(s - f_i + \delta_i^{-})\alpha(f_i) - (s - f_i)\alpha(f_i - \delta_i^{-})}{\delta_i^{-}} & (f_i - \delta_i^{-} \leq s < f_i) \\ \dfrac{(s - f_i)\alpha(f_i + \delta_i^{+}) - (s - f_i - \delta_i^{+})\alpha(f_i)}{\delta_i^{+}} & (f_i \leq s < f_i + \delta_i^{+}) \end{cases} . \quad (3)$$

If the field interval $[-\delta_i^{-}, \delta_i^{+}]$ to be emphasized is too narrow, the sampling point of volume rendering may not appear in the region whose scalar values belong to the interval. Therefore, to secure a certain number of voxels there, our transfer function controls the width of the hat functions, as follows:

- For the critical field values ($f_i = c_i$),

$$\delta_i^{-} = \eta \frac{c_i - c_{i+1}}{V_{i,i+1}}, \qquad \delta_i^{+} = \eta \frac{c_{i-1} - c_i}{V_{i-1,i}}. \quad (4)$$

- For the representative field values ($f_i = r_i$),

$$\delta_i^{+} = \delta_i^{-} = \eta \frac{r_i - r_{i+1}}{V_{i,i+1}}, \quad (5)$$

where η denotes another coefficient representing the degree of emphasis. To avoid an overlap between different hat functions, for the critical field values, if $\delta_i^{-} > |c_i - r_i|$ then let $\delta_i^{-} = |c_i - r_i|$, and if $\delta_i^{+} > |c_i - r_{i-1}|$ then let $\delta_i^{+} = |c_i - r_{i-1}|$. For the representative field values, if $\delta_i^{-} > |r_i - c_{i+1}|$ then let $\delta_i^{-} = |r_i - c_{i+1}|$, and if $\delta_i^{+} > |r_i - c_i|$ then let $\delta_i^{+} = |r_i - c_i|$.

3.3 *Multi-Dimensional Opacity Transfer Function*

Generally, a one-dimensional transfer function is used to assign an opacity value to a field value. However, if isosurface components with the same field value have different meanings, it is impossible to obtain a visualization result that emphasizes them separately. To address this problem, we proposed a design method for multi-dimensional transfer functions based on topological attributes [13].

In this paper, we extend a two-dimensional opacity transfer function that depends on the scalar field value and the inclusion level [13]. The inclusion level is assigned for the isosurface component that has the same field value, with larger values for a more inner component. The inclusion level of the outermost isosurface component is set to 0.

Figure 3c shows an overview of our design of a two-dimensional transfer function. When the inclusion level u is even, because the isosurface component becomes outer as the scalar field value decreases, the opacity values $\alpha_m(s, u)$ are defined as $(u + 1)\alpha(s)/u_{max}$ using a one-dimensional opacity transfer function, where u_{max} denotes the maximum inclusion level. When the inclusion level u is odd, because the isosurface component becomes inner as the scalar field value decreases, opacity values $\alpha_m(s, u)$ are defined as follows:

$$\alpha_m(s, u) = \begin{cases} \dfrac{u+1}{u_{max}} \dfrac{i}{m-1} G_i & (f_{i+1} + \delta_{i+1}^{+} < s \leq f_i - \delta_i^{-}) \\ \dfrac{(s - f_i + \delta_i^{-})\alpha_m(f_i, u) - (s - f_i)\alpha_m(f_i - \delta_i^{-}, u)}{\delta_i^{-}} & (f_i - \delta_i^{-} < s < f_i) \\ \dfrac{u+1}{u_{max}} \dfrac{i}{m-1} \mu G_i & (s = f_i) \\ \dfrac{(s - f_i)\alpha_m(f_i + \delta_i^{+}, u) - (s - f_i - \delta_i^{+})\alpha_m(f_i, u)}{\delta_i^{+}} & (f_i < s < f_i + \delta_i^{+}) \\ \dfrac{u+1}{u_{max}} \dfrac{i}{m-1} G_{i-1} & (s = f_i + \delta_i^{+}) \end{cases} . \quad (6)$$

4 Empirical Evaluation

From the aspects of the size of the dataset, the parameter values of a transfer function, and the LoD of the VST, we assessed their impact on the visualization results. All datasets were found to be affinely mapped to the range [0, 255]. The platform used for our experiments was a standard PC (OS: Redhat ES 7.2; CPU: Intel Xeon; Clock: 2.50 Hz; RAM: 64 GB; GPU: NVIDIA Quadro K6000).

4.1 Sensitivity to μ and η

To obtain appropriate visualization results, we can control μ and η in our transfer function design. From the definition shown in Eqs. (2), (4), and (5), μ and η affect the opacity of the accentuated isosurfaces and the width of the accentuated field interval. First, we investigated the effect of μ and η, which are the parameter values of our transfer function.

Figure 4 shows the visualization results when our method was applied to a tooth dataset ($161 \times 161 \times 161$) [15]. We used a blue-white-yellow colormap. As seen in these images, opacity increases as a μ and η increase. However, it should also be noted that the accentuated isosurfaces are not significantly emphasized when only the μ value is changed. Because a region of the accentuated field interval is narrow, it is difficult to emphasize the target isosurfaces simply by increasing the height

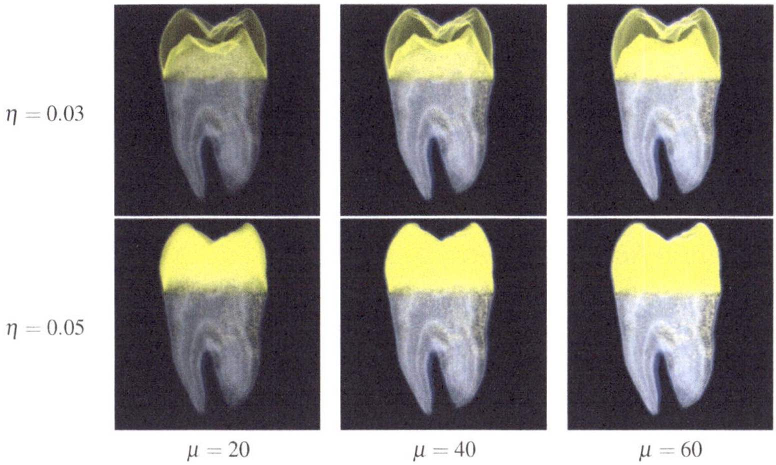

Fig. 4 Visualizing the tooth dataset with different values of μ and η, which affect the degree of emphasis of the accentuated isosurfaces and the width of the accentuated field intervals, respectively. These images show that the accentuated isosurfaces are not significantly emphasized when only the μ value is changed

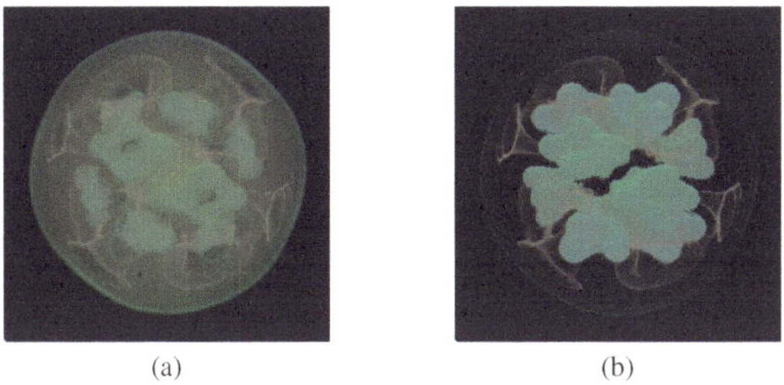

Fig. 5 Visualizing the implosion dataset with our one-dimensional transfer function and two-dimensional transfer function. These images shows our two-dimensional transfer function makes it possible to clearly visualize the inner structures by removing unnecessary surface. (**a**) One-dimensional TF. (**b**) Two-dimensional TF

of the hat function of the opacity transfer function. Therefore, an η value becomes important for this type of dataset.

Next, in order to evaluate our two-dimensional transfer function, we applied it to a laser fusion implosion dataset ($225 \times 225 \times 225$) [16], where small bubble-spike structures evolve around a contact surface between a fuel ball and pusher during the stagnation phase. The contact surface is occluded by the other outer component residing in the pusher domain, which is nothing but a phantom surface created by the action-reaction effect.

Figure 5 shows the visualization results ($\mu = 20$, $\eta = 0.05$) of the laser fusion implosion dataset, which emphasizes the representative field values with a rainbow colormap. Figure 5a does not clearly show the inner isosurface components of interest, because these are indeed occluded by the outer phantom surface. On the other hand, Fig. 5b clearly illustrates that our two-dimensional transfer function makes it possible to clearly visualize the inner structures by removing unnecessary phantom surface.

4.2 Dataset Size Sensitivity

Next, we evaluated whether similar visualization results can be obtained from the datasets with different sizes. In order to demonstrate the effectiveness of our method, we applied it to two types of volume datasets.

The first is an analytical dataset [12], which is formulated as follows:

$$f(x,y,z) = 4c^2\left((x-R)^2+(z-R)^2\right) - \left((x-R)^2+y^2+(z-R)^2+c^2-d^2\right)^2$$
$$+4c^2\left((x+R)^2+(z+R)^2\right) - \left((x+R)^2+y^2+(z+R)^2+c^2-d^2\right)^2, \quad (7)$$

where $0 < d < c$ and $c^2 + d^2 \geq 6R^2$. This dataset has two maxima, three saddles, and a virtual minimum. Figure 6 shows the VST of the analytical function and the visualization results applied to three analytical datasets with different sizes: $33 \times 33 \times 33$, $65 \times 65 \times 65$, and $129 \times 129 \times 129$. As Fig. 6 shows, our transfer functions to accentuate the critical field values with a blue-white-red colormap ($\mu = 30$, $\eta = 0.035$). The volume ratio corresponding to each link differs according to the dataset size, as shown in Table 1. It is clear that the volume ratios are nearly equal regardless of the size of dataset. The results presented in Figs. 6b–d show that the proposed transfer functions provide similar visualization results regardless of the size of the dataset.

The second is a three-dimensional head dataset, which was provided courtesy of Siemens Medical Systems, Inc. (Iselin, NJ). Figure 7 shows the visualization results applied to two head datasets with different sizes: $65 \times 65 \times 65$ and $129 \times 129 \times 129$. These images use our transfer functions to accentuate the critical field values with a blue-white-yellow colormap ($\mu = 40$, $\eta = 0.05$). The figure illustrates that the visualization results obtained for a real-world dataset are similar to those obtained for the analytical dataset, regardless of the size of the dataset.

These results demonstrate that the effectiveness of our method does not depend on the size of datasets, even for different types of datasets.

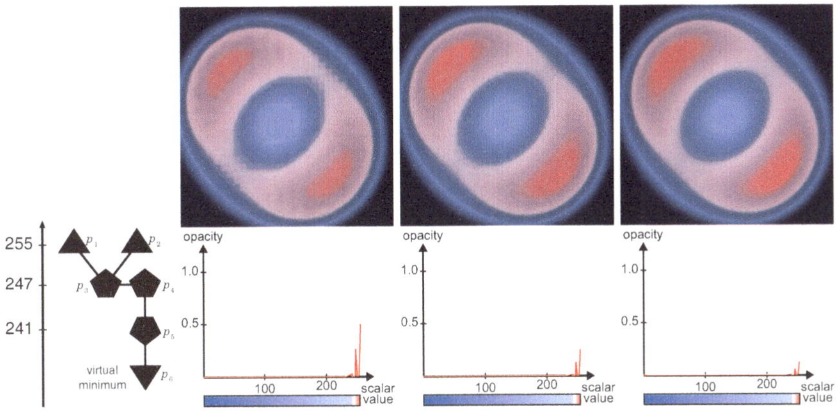

Fig. 6 VST and visualization results for the analytical datasets with our opacity transfer functions ($\mu = 30, \eta = 0.035$). These images show that our transfer functions provide similar visualization results regardless of the size of the dataset. (**a**) VST. (**b**) 33 × 33 × 33. (**c**) 65 × 65 × 65. (**d**) 129 × 129 × 129

Table 1 Volume ratios corresponding to each link of the analytical datasets shown in Fig. 6

Link	33 × 33 × 33	65 × 65 × 65	129 × 129 × 129
l_{1-3}	4.67e−2	4.86e−2	4.96e−2
l_{2-3}	4.67e−2	4.86e−2	4.96e−2
l_{3-4}	0.0	0.0	0.0
l_{4-5}	1.73e−1	1.82e−1	1.86e−1
l_{5-6}	7.33e−1	7.21e−1	7.15e−1

4.3 *Response to the LoD of the VST*

Finally, in order to evaluate whether our method is effective even if the number of the accentuated surfaces increases, we examined how our transfer function responds to differences in the LoD of the VST. When the VST extracts the topological structure from the target dataset, it may contain a large number of minor critical points. Because they may hide the important global structure of the dataset, we need to simplify the extracted VST by removing them.

We applied our method to a nucleon dataset (41 × 41 × 41) [17] to evaluate the effect of the LoD of the VST. Figure 8 shows the visualization results with a heat object colormap ($\mu = 20, \eta = 0.04$); the results emphasize the representative field values for different LoDs of the VST. Figure 8a shows a visualization result based on the original VST, which has 1006 critical points and 21 critical field values. Note that we cannot show the original VST because it is too complicated to draw. Figures 8b, c show the results based on simplified VSTs that have ten and six critical

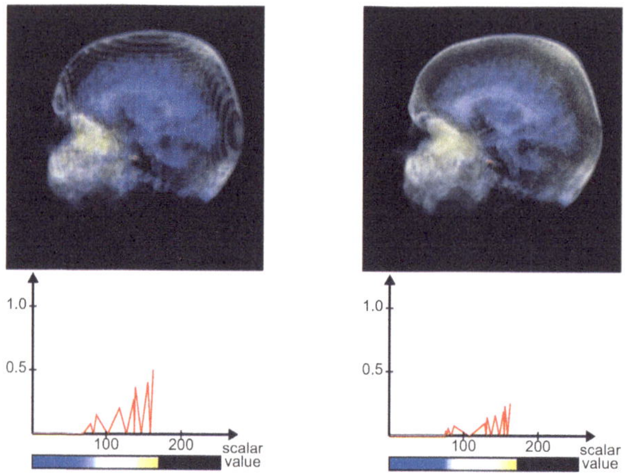

Fig. 7 Visualization results of the head datasets with our opacity transfer functions ($\mu = 40$, $\eta = 0.05$). These images show that our transfer functions provide similar visualization results regardless of the size of the dataset. (**a**) 65 × 65 × 65. (**b**) 129 × 129 × 129

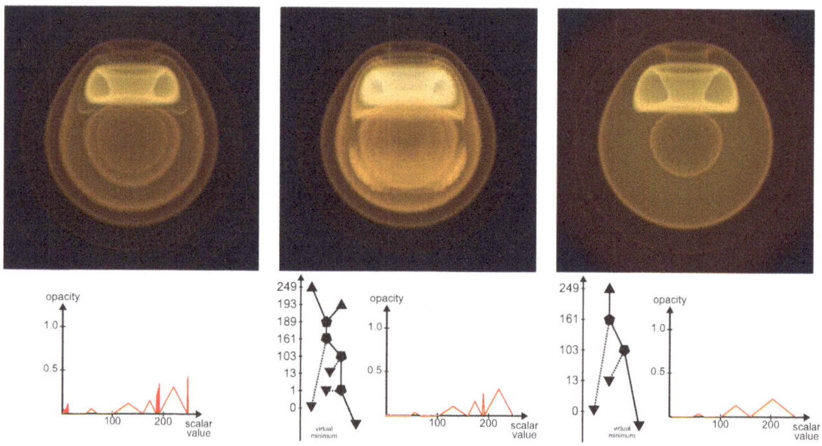

Fig. 8 Visualizing the nucleon dataset applied for different LoDs of the VST. These images show that our transfer function can effectively visualize the topological structure even if the number of critical points decreases. (**a**) Original VST (1006 critical points). (**b**) Simplified VST (10 critical points). (**c**) Simplified VST (6 critical points)

points and six and five critical field values, respectively. These images illustrate that our transfer function can effectively visualize the topological structure, regardless of the number of critical points.

5 Further Controllability

In this section, we evaluate our method further from the perspective of image quality and processing speed.

5.1 *Image Quality*

In order to evaluate the quality of a visualization result, we defined an evaluation function based on a normalized Shannon information entropy as follows:

$$Entropy = \frac{-\sum_{i=0}^{M} p(A_i)\log_2 p(A_i)}{\log_2 M}, \tag{8}$$

where M and $p(A_i)$ represent the number of levels and the probability of pixels of level i, respectively. When the value of entropy is high, there are many pixels at various levels, which means that the information entropy of the visualized image is large. Conversely, when there are numerous pixels with the same levels, the information entropy is small. Therefore, we assume that the higher the entropy, the better the quality of the image. Note that the evaluation function can also potentially lead to misleading results in the presence of noise in the dataset. We evaluate the quality of images in terms of their hue, saturation, and luminance values. When the entropies of hue, saturation, and luminance are represented by E_h, E_s, and E_l, respectively, our evaluation function is defined as follows:

$$\begin{cases} E = k_H E_H + k_S E_S + k_L E_L \\ k_H + k_S + k_L = 1 \qquad (0 \leq k_H, k_S, k_L \leq 1) \end{cases}, \tag{9}$$

where k_H, k_S, and k_L are the weight coefficients for hue, saturation, and luminance, respectively. Users can control these coefficients based on their preferences.

Figure 9a shows the heatmap that represents our evaluation function, where $k_H = k_S = k_L = 1/3$ of the analytical dataset ($129 \times 129 \times 129$). The value for μ was increased by 10 increments ranging from 10 to 50. The value for η was increased by 0.005 increments, and its range was set so that the field interval that emphasizes the critical field value does not exceed the adjacent representative field values. The cells surrounded by the black box and the white box represent the minimum value and the maximum value of the corresponding entropy, respectively. Figures 9b, c show the best and worst visualization results. The values of μ and η in the best case are 30 and 0.035, respectively; in the worst case, they are 10 and 0.005. Moreover, in the best case, the value of our evaluation function is 0.88; in the worst case, it is 0.79. In this example, the images in the worst cases are darker than the images representing the best cases. By introducing our evaluation mechanism, we can automatically obtain an appropriate visualization result, which is well balanced among hue, saturation,

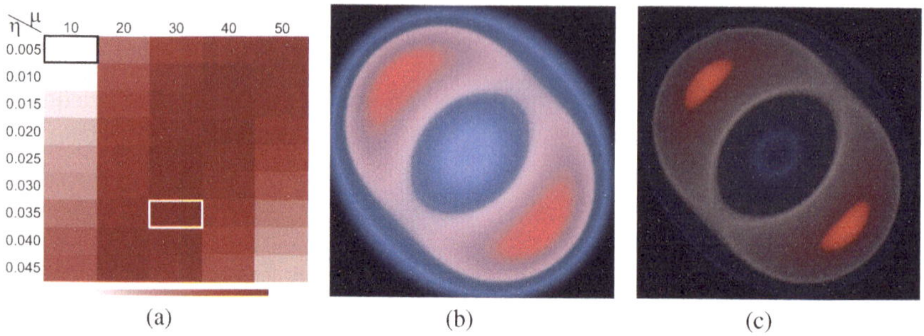

Fig. 9 Entropy heatmap of E and the best and worst visualization results of the analytical dataset. We can automatically obtain an appropriate visualization result referring to the entropy heatmap. (**a**) Entropy heatmap of E ($k_H = k_S = k_L = 1/3$). (**b**) Best ($\mu = 30, \eta = 0.035$). (**c**) Worst ($\mu = 10, \eta = 0.005$)

and luminance, from the entropy heatmap, as shown in Fig. 9a. Note that μ and η in the best case were used to obtain Fig. 6.

5.2 *Processing Speed*

In our method, the time required for the transfer function design, excluding the VST extraction, was 0.01 CPU seconds for any dataset throughout this paper. However, the computation time required for VST extraction also increases as the dataset size increases. Our extraction algorithm [12, 18] can adjust the accuracy of the VST extraction with interactive operations, although it may take several minutes to several tens of minutes of processing time in some cases. Though a discussion of the extraction algorithm of the VST is beyond the scope of this paper, we had to accelerate our algorithm to extract the topological structures from large-scale datasets.

In order to shorten the processing time required for the VST extraction, it is conceivable to use downsized datasets to obtain the VST and to reduce the processing time. To visualize a large-scale dataset, we design the transfer function based on the VST extracted from the downsized dataset and then visualize the original large dataset using that function. The experiments presented in Sect. 4.2 that similar visualization results can be obtained even if the VST extracted from the downsized dataset is used.

Figure 10 shows the VST and the visualization result of the stag beetle dataset ($832 \times 832 \times 494$) [19], which emphasizes the critical isosurfaces. The VST was extracted from the small stag beetle dataset ($208 \times 208 \times 123$) and simplified until the number of the critical points became 14. As the image illustrates we can obtain an effective visualization result even if the transfer function is designed based on the VST extracted from the downsized dataset.

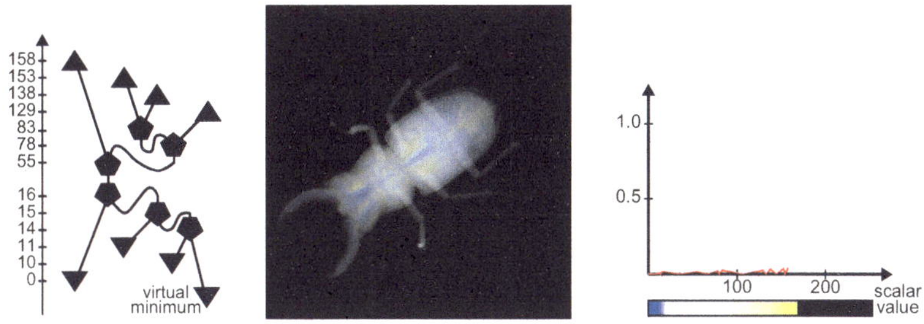

Fig. 10 VST, visualization result, and transfer function of the stag beetle dataset. We can obtain effective visualization results even if the transfer function is designed based on the VST extracted from the downsized dataset. (**a**) VST. (**b**) Visualization result. (**c**) Transfer function

6 Conclusion

In this paper, we proposed a method to define the opacity values of topology-accentuated transfer functions. Controlling the height and width of the hat functions made it possible to emphasize the feature isosurfaces regardless of the kind of dataset. According to our empirical evaluations so far, μ and η should range from 10 to 50 and from 0.01 to 0.05, respectively, because the internal structure becomes invisible outside of the range in many cases. Our multi-dimensional transfer function is also able to more clearly emphasize the inner structures in a target dataset. In addition, our empirical evaluations suggest that the proposed transfer function design can automatically provide guaranteed results regardless of the size of the target dataset and the LoD of the VST. Furthermore, by introducing the evaluation function with different combinations of color components, we can anticipate appropriate visualization results based on users' preferences.

The present transfer function design can incorporate sampling distance as a variable of its definition in order to determine the region to be emphasized around accentuated isosurfaces. In order to evaluate the effects of the designed transfer functions in more detail, we will apply them to more complicated and larger datasets in future studies. Moreover, we should take up the challenge of considering parameter settings for a multi-dimensional transfer function based on multi-variate topological structures [20].

Acknowledgments This work was supported, in part, by MEXT KAKENHI under the Grant-in-Aid for Scientific Research on Innovative Areas 25120014 and JSPS KAKENHI under the Grant-in-Aid for Science Research (A) No. 26240015 and JP17H00737 and Scientific Research (C) No. 26330142 and JP17K00173. The authors would like to thank the anonymous reviewers for their valuable suggestions for revising our paper.

References

1. Ljung, P., Krüger, J., Gröller, E., Hadwiger, M., Hansen, C.D., Ynnerman, A.: State of the art in transfer functions for direct volume rendering. CGF **35**(3), 669–691 (2016)
2. Kindlmann, G., Durkin, J.W.: Semi-automatic generation of transfer functions for direct volume rendering. In: Proceedings of the IEEE Symposium on Volume Visualization, pp. 79–86 (1998)
3. Kniss, J., Kindlmann, G., Hansen, C.: Multidimensional transfer functions for interactive volume rendering. IEEE TVCG **8**(3), 270–285 (2002)
4. Hladůvka, J., König, A., Gröller, E.: Curvature-based transfer functions for direct volume rendering. In: Proceedings of the Spring Conference on Computer Graphics 2000, pp. 58–65 (2000)
5. Kindlmann, G., Whitaker, R., Tasdizen, T., Möller, T.: Curvature-based transfer function for direct volume rendering: methods and applications. In: Proceedings of the IEEE Vis 2003, pp. 513–520 (2003)
6. Weber, G., Dillard, S., Carr, H., Pascucci, V., Hamann, B.: Topology-controlled volume rendering. IEEE TVCG **13**(2), 330–341 (2007)
7. Zhou, J., Takatsuka, M.: Automatic transfer function generation using contour tree controlled residue flow model and color harmonics. IEEE TVCG **15**, 1481–1488 (2009)
8. Carr, H., Snoeyink, J., Panne, M.V.D.: Flexible isosurfaces: simplifying and displaying scalar topology using the contour tree. Comput. Geom. Theory Appl. **43**(1), 42–58 (2010)
9. Praßni, J.-S., Ropinski, T., Mensmann, J., Hinrichs, K.H.: Shape-based transfer functions for volume visualization. In: Proceedings of the IEEE PacificVis 2010, pp. 9–16 (2010)
10. Xiang, D., Tian, J., Yang, F., Yang, Q., Zhang, X., Li, Q., Liu, X.: Skeleton cuts—an efficient segmentation method for volume rendering. IEEE TVCG **17**(9), 1295–1306 (2011)
11. Fernandes, O., Frey, S., Ertl, T.: Interpolation-based extraction of representative isosurfaces. In: Advances in Visual Computing, pp. 403–413. Springer International Publishing, Cham (2016)
12. Takahashi, S., Takeshima, Y., Fujishiro, I.: Topological volume skeletonization and its application to transfer function design. Graph. Models **66**(1), 24–49 (2004)
13. Takeshima, Y., Takahashi, S., Fujishiro, I., Nielson, G.M.: Introducing topological attributes for objective-based visualization of simulated datasets. In: Proceedings of the International Workshop on Volume Graphics, pp. 137–236 (2005)
14. Fujishiro, I., Maeda, Y., Sato, H., Takeshima, Y.: Volumetric data exploration using interval volume. IEEE TVCG **2**(2), 144–155 (1996)
15. Pfister, H., Lorensen, B., Bajaj, C., Kindlmann, G., Schroeder, W., Avila, L.S., Martin, K., Machiraju, R., Lee, J.: The transfer function bake-off. IEEE Comput. Graph. Appl. **21**(3), 16–22 (2001)
16. Sakagami, H., Murai, H., Seo, Y., Yokokawa, M.: 14.9 TFLOPS three-dimensional fluid simulation for fusion science with HPF on the Earth Simulator. In: Proceedings of the Supercomputing 2002 (2002)
17. Sonderforschungsbereiche 382 of the German Research Council. Nucleon.
18. Takahashi, S., Nielson, G.M., Takeshima, Y., Fujishiro, I.: Topological volume skeletonization using adaptive tetrahedralization. In: Proceedings of the GMP 2004, pp. 227–236 (2004)
19. Gröller, M.E., Glaeser, G., Kastner, J.: Stag beetle (2005). https://www.cg.tuwien.ac.at/research/publications/2005/dataset-stagbeetle/.
20. Carr, H., Duke, D.: Joint contour nets. IEEE TVCG **20**(8), 1077–2626 (2014)

Topological Machine Learning with Persistence Indicator Functions

Bastian Rieck, Filip Sadlo, and Heike Leitte

Abstract Techniques from computational topology, in particular persistent homology, are becoming increasingly relevant for data analysis. Their stable metrics permit the use of many distance-based data analysis methods, such as multidimensional scaling, while providing a firm theoretical ground. Many modern machine learning algorithms, however, are based on kernels. This paper presents *persistence indicator functions* (PIFs), which summarize persistence diagrams, i.e., feature descriptors in topological data analysis. PIFs can be calculated and compared in linear time and have many beneficial properties, such as the availability of a kernel-based similarity measure. We demonstrate their usage in common data analysis scenarios, such as confidence set estimation and classification of complex structured data.

1 Introduction

Persistent homology [9–11], now over a decade old, has proven highly relevant in data analysis. The last years showed that the usage of topological features can lead to an increase in, e.g., classification performance of machine learning algorithms [20]. The central element for data analysis based on persistent homology is the *persistence diagram*, a data structure that essentially stores related critical points (such as minima or maxima) of a function, while providing two stable metrics, namely the *bottleneck distance* and the pth *Wasserstein distance*. Certain stability theorems [7, 8] guarantee that the calculations are robust against perturbations and the inevitable occurrence of noise in real-world data.

B. Rieck (✉) · H. Leitte
TU Kaiserslautern, Kaiserslautern, Germany
e-mail: rieck@cs.uni-kl.de; bastian.rieck@iwr.uni-heidelberg.de; leitte@cs.uni-kl.de

F. Sadlo
Heidelberg University, Heidelberg, Germany
e-mail: sadlo@uni-heidelberg.de

H. Carr et al. (eds.), *Topological Methods in Data Analysis and Visualization V*, Mathematics and Visualization, https://doi.org/10.1007/978-3-030-43036-8_6

This stability comes at the price of a very high runtime for distance calculations between persistence diagrams: both metrics have a complexity of at least $\mathcal{O}\left(n^{2.5}\right)$, or, if naively implemented, $\mathcal{O}\left(n^3\right)$ [9, p. 196]. Using randomized algorithms, it is possible to achieve a complexity of $\mathcal{O}\left(n^{\omega}\right)$, where $\omega < 2.38$ denotes the best matrix multiplication time [18]. Further reductions in runtime complexity are possible if approximations to the correct value of the metric are permitted [14]. Nevertheless, these algorithms are hard to implement and their performance is worse than $\mathcal{O}\left(n^2\right)$, meaning that they are not necessarily suitable for comparing larger sets of persistence diagrams.

In this paper, we describe a summarizing function for persistence diagrams, the persistence indicator function (PIF). PIFs were informally introduced in a previous publication [22]. Here, we give a more formal introduction, demonstrate that PIFs can be easily and rapidly calculated, derive several properties that are advantageous for topological data analysis as well as machine learning, and describe example usage scenarios, such as hypothesis testing and classification. We make our implementation, experiments, and data publicly available.[1]

2 Related Work

The *persistence curve* is a precursor to PIFs that is widely used in the analysis of Morse–Smale complexes [4, 13, 17]. It counts the number of certain critical points, such as minima or maxima, that either occur at a given persistence threshold or at given point in time. The curve is then used to determine a relevant threshold, or cut-off parameter for the simplification of the critical points of a function. To the best of our knowledge, no standardized variant of these curves appears to exist.

Recognizing that persistence diagrams can be analyzed at multiple scales as well in order to facilitate hierarchical comparisons, there are some approaches that provide approximations to persistence diagrams based on, e.g., a smoothing parameter. Among these, the stable kernel of Reininghaus et al. [20] is particularly suited for topological machine learning. Another approach by Adams et al. [1] transforms a persistence diagram into a finite-dimensional vector by means of a probability distribution. Both methods require choosing a set of parameters (for kernel computations), while PIFs are fully parameter-free. Moreover, PIFs also permit other applications, such as mean calculations and statistical hypothesis testing, which pure kernel methods cannot provide.

Recently, Bubenik [5] introduced *persistence landscapes*, a functional summary of persistence diagrams. Within his framework, PIFs can be considered to represent a summary (or projection) of the *rank function*. Our definition of PIFs is more straightforward and easier to implement, however. Since PIFs share several properties of persistence landscapes—most importantly the existence of simple

[1] https://github.com/Submanifold/topological-machine-learning.

function-space distance measures—this paper uses similar experimental setups as Bubenik [5] and Chazal et al. [6].

3 Persistence Indicator Functions (PIFs)

Given a persistence diagram $\mathscr{D}$, i.e., a descriptor of the topological activity of a data set [9], we can summarize topological features by calculating an associated persistence indicator function of $\mathscr{D}$ as

$$\begin{aligned} \mathbb{1}_{\mathscr{D}} \colon \mathbb{R} &\longrightarrow \mathbb{N} \\ \epsilon &\longmapsto \left|\left\{(c,d) \in \mathscr{D} \mid \epsilon \in [c,d]\right\}\right|, \end{aligned} \tag{1}$$

i.e., the number of points in the persistence diagram that, when being treated as a closed interval, contain the given parameter ϵ. Equivalently, a PIF can be considered to describe the rank of the pth homology group of a filtration of a simplicial complex. A PIF thus measures the amount of topological activity as a function of the threshold parameter ϵ. This parameter is commonly treated as the "range" of a function defined *on* the given data set, e.g., a distance function [11] or an elevation function [2]. Figure 1 demonstrates how to calculate the persistence indicator function $\mathbb{1}_{\mathscr{D}}(\cdot)$ from a persistence diagram or, equivalently, from a persistence barcode. For the latter, the calculation becomes particularly easy. In the barcode, one only has to check the number of intervals that are intersected at any given time by a vertical line for some value of ϵ.

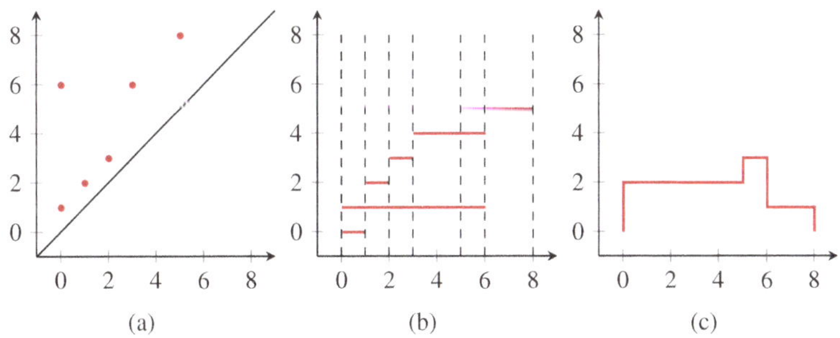

Fig. 1 A persistence diagram (**a**), its persistence barcode (**b**), and its corresponding persistence indicator function (**c**). Please note that the interpretation of the axes changes for each plot

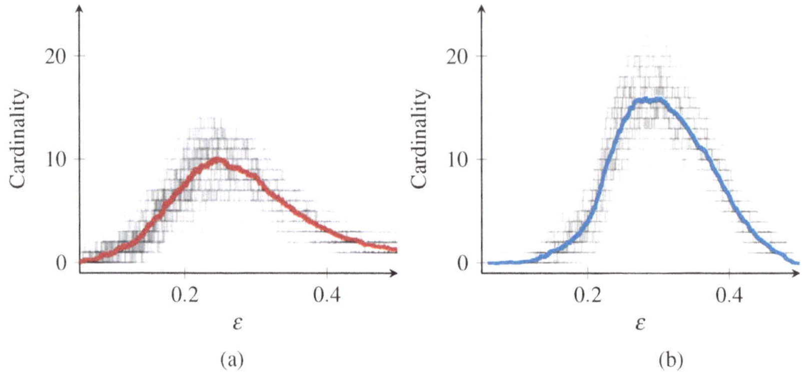

Fig. 2 Mean persistence indicator function of the one-dimensional persistence diagrams of a sphere and of a torus. Both data sets have been sampled at random and are scaled such that their volume is the same. (**a**) Sphere ($r \approx 0.63$). (**b**) Torus ($R = 0.025$, $r = 0.05$)

3.1 Properties

We first observe that the PIF only changes at finitely many points. These are given by the creation and destruction times, i.e., the x- and y-coordinates, of points in the persistence diagram. The PIF may thus be written as a sum of appropriately scaled *indicator functions* (hence the name) of the form $\mathbb{1}_{\mathscr{I}}(\cdot)$ for some interval $\mathscr{I}$. Within the interval $\mathscr{I}$, the value of $\mathbb{1}_{\mathscr{D}}(\cdot)$ does *not* change. Hence, the PIF is a *step function*. Since step functions are compatible with addition and scalar multiplication, PIFs form a vector space. The addition of two PIFs corresponds to calculating the union of their corresponding persistence diagrams, while taking multiplicities of points into account. As a consequence, we can calculate the *mean* of a set of PIFs $\{\mathbb{1}^1_{\mathscr{D}}, \ldots, \mathbb{1}^n_{\mathscr{D}}\}$ as

$$\overline{\mathbb{1}_{\mathscr{D}}}(\cdot) := \frac{1}{n}\sum_{i=1}^{n} \mathbb{1}^i_{\mathscr{D}}(\cdot), \tag{2}$$

i.e., the standard pointwise mean of set of elements. In contrast to persistence diagrams, for which a mean is not uniquely defined and hard to calculate [26], this calculation only involves addition (union) and scalar multiplications of sets of intervals. Figure 2 depicts mean persistence indicator functions for two randomly-sampled data sets. We see that the resulting mean persistence indicator functions already introduce a visual separation between the two data sets.

As a second derived property, we note that the absolute value of a step function (and that of a PIF) always exists; one just calculates the absolute value for every interval in which the step function does *not* change. The absolute value of a step function is again a step function, so the Riemann integral of a PIF is well-defined, giving rise to their 1-norm as

$$\|\mathbb{1}_{\mathscr{D}}\|_1 := \int_{\mathbb{R}} |\mathbb{1}_{\mathscr{D}}(x)| \mathrm{d}x, \tag{3}$$

which is just the standard norm of an L^1-space. The preceding equation requires the use of an absolute value because linear operations on PIFs may result in negative values. The integral of a PIF (or its absolute value) decomposes into a sum of integrals of individual step functions, defined over some interval $[a, b]$. Letting the value of the step function over this interval be l, the integral of the step function is given as $l \cdot |b - a|$, i.e., the volume of the interval scaled by the value the step function assumes on it. We can also extend this norm to that of an L-space, where $p \in \mathbb{R}$. To do so, we observe that the pth power of any step function is well-defined—we just raise the value it takes to the pth power. Consequently, the pth power of a PIF is also well-defined and we define

$$\|\mathbb{1}_{\mathscr{D}}\|_p := \left(\int_{\mathbb{R}} |\mathbb{1}_{\mathscr{D}}(x)|^p \mathrm{d}x \right)^{\frac{1}{p}}, \tag{4}$$

which is the standard norm of an L-space. Calculating this integral again involves decomposing the range of the PIF into individual step functions and calculating their integral. We have $\|\mathbb{1}_{\mathscr{D}}\|_p < \infty$ for every $p \in \mathbb{R}$ because there are only finitely many points in a persistence diagram, so the integrals of the individual step functions involved in the norm calculation are bounded.

Hypothesis Testing Treating the norm of a PIF as a random variable, we can perform topology-based hypothesis testing similar to persistence landscapes [5]. Given two different samples of persistence diagrams, $\{\mathscr{D}_1^1, \ldots, \mathscr{D}_n^1\}$ and $\{\mathscr{D}_1^2, \ldots, \mathscr{D}_n^2\}$, we calculate the 1-norm of their respective mean PIFs as Y_1 and Y_2, and the variances

$$\sigma_i^2 := \frac{1}{n-1} \sum_{j=1}^{n} \left(|\mathbb{1}_{\mathscr{D}_j^i}| - Y_i \right)^2, \tag{5}$$

for $i \in \{1, 2\}$. We may then perform a standard two-sample z-test to check whether the two means are likely to be the same. To this end, we calculate the z-score as

$$z := \frac{Y_1 - Y_2}{\sqrt{s_1^2/n - s_2^2/n}} \tag{6}$$

and determine the critical values at the desired α-level, i.e., the significance level, from the quantiles of a normal distribution. If z is outside the interval spanned by the critical values, we *reject* the null hypothesis, i.e., we consider the two means to be different. For the example shown in Fig. 2, we obtain $Y_1 \approx 2.135$, $s_1^2 \approx 0.074$,

$Y_2 \approx 2.79$, and $s_2^2 \approx 0.093$. Using $\alpha = 0.01$, the critical values are given by $z_1 \approx -2.58$ and $z_2 \approx 2.58$. Since $z \approx -11.09$, we *reject* the null hypothesis with $p \approx 1.44 \times 10^{-28} \ll 0.01$. Hence, PIFs can be used to confidently discern random samples of a sphere from those of a torus with the same volume.

Stability The 1-norm of a PIF is connected to the *total persistence* [8], i.e., the sum of all persistence values in a persistence diagram. We have

$$\|\mathbb{1}_{\mathscr{D}}\|_1 = \int_{\mathbb{R}} |\mathbb{1}_{\mathscr{D}}(x)| \mathrm{d}x = \sum_{I \in \mathscr{I}} c_I \operatorname{vol}(I), \tag{7}$$

where $\mathscr{I}$ denotes a set of intervals for which the number of active persistence pairs does not change, and c_I denotes their count. We may calculate this partition from a persistence barcode, as shown in Fig. 1b, by going over all the dotted slices, i.e., the intervals between pairs of start and endpoints of each interval. The sum over all these volumes is equal to the total persistence of the set of intervals, because we can split up the volume calculation of a single interval over many sub-interval and, in total, the volume of every interval is only accumulated once. Hence,

$$\|\mathbb{1}_{\mathscr{D}}\|_1 \int_{\mathbb{R}} |\mathbb{1}_{\mathscr{D}}(x)| \mathrm{d}x = \sum_{(c,d) \in \mathscr{D}} |d - c| = \sum_{(c,d) \in \mathscr{D}} \operatorname{pers}(c, d) = \operatorname{pers}(\mathscr{D}), \tag{8}$$

where $\operatorname{pers}(\mathscr{D})$ denotes the total persistence of the persistence diagram. According to a stability theorem by Cohen-Steiner et al. [8], the 1-norm of a PIF is thus stable with respect to small perturbations. We leave the derivation of a similar equation for the general L-norm of a PIF for future work.

3.2 The Bootstrap for Persistence Indicator Functions

Developed by Efron and Tibshirani [12], the bootstrap is a general statistical method for—among other applications—computing confidence intervals. We give a quick and cursory introduction before showing how this method applies to persistence indicator functions; please refer to Chazal et al. [6] for more details.

Assume that we have a set of independent and identically distributed variables $X_1, \ldots, X_n$, and we want to estimate a real-valued parameter θ that corresponds to their distribution. Typically, we may estimate θ using a statistic $\widehat{\theta} := s(X_1, \ldots, X_n)$, i.e., some function of the data. A common example is to use θ as the population mean, while $\widehat{\theta}$ is the sample mean. If we want to calculate *confidence intervals* for our estimate $\widehat{\theta}$, we require the distribution of the difference $\theta - \widehat{\theta}$. This distribution, however, depends on the unknown distribution of the variables, so we have to approximate it using an empirical distribution. Let $X_1^*, \ldots, X_n^*$ be a sample of the original variables, drawn with replacement. We can calculate $\widehat{\theta}^* := s(X_1^*, \ldots, X_n^*)$

and approximate the unknown distribution by the empirical distribution of $\widehat{\theta} - \widehat{\theta}^*$, which, even though it is not computable analytically, can be approximated by repeating the sampling procedure a sufficient number of times. The quantiles of the approximated distribution may then be used to construct confidence intervals, leading to the following method:

1. Calculate an estimate of θ from the input data using $\widehat{\theta} := s(X_1, \dots, X_n)$.
2. Obtain $X_1^*, \dots, X_n^*$ (sample *with* replacement) and calculate $\widehat{\theta}^* := s(X_1^*, \dots, X_n^*)$.
3. Repeat the previous step B times to obtain $\widehat{\theta}_1^*, \dots, \widehat{\theta}_B^*$.
4. Given α, compute an approximate $(1 - 2\alpha)$ quantile interval as

$$[\widehat{\theta}_1, \widehat{\theta}_2] \approx [\widehat{\theta}_B^{*(\alpha)}, \widehat{\theta}_B^{*(1-\alpha)}], \tag{9}$$

   where $\widehat{\theta}_B^{*(\alpha)}$ refers to the αth empirical quantile of the bootstrap replicates from the previous step.

This procedure yields both a lower bound and an upper bound for $\widehat{\theta}$. It is also referred to as the percentile interval for bootstraps [12, pp. 168–176]. More complicated versions—yielding "tighter" confidence intervals—of this procedure exist, but the basic method of sampling with replacement and computing the statistic $s(\cdot)$ on the bootstrap samples remains the same.

In order to apply the bootstrap to *functions*, we require empirical processes [16]. The goal is to find a *confidence band* for a function $f(x)$, i.e., a pair of functions $l(x)$ and $u(x)$ such that the probability that $f(x) \in [l(x), u(x)]$ for $x \in \mathbb{R}$ is at least $1 - \alpha$. Given a function f, let $Pf := \int f \mathrm{d}P$ and $P_n f := n^{-1} \sum_{i=1}^{n} f(X_i)$. We obtain a *bootstrap empirical process* as

$$\{\mathbb{P}f\}_{f \in \mathcal{F}} := \{\sqrt{n}\left(P_n^* f - P_n f\right)\}, \tag{10}$$

where $P_n^* := n^{-1} \sum_{i=1}^{n} f(X_i^*)$ is defined on the bootstrap samples (as introduced above). Given the convergence of this empirical process, we may calculate

$$\widehat{\theta} := \sup_{f \in \mathcal{F}} |\mathbb{P}f|, \tag{11}$$

which yields a statistic to use for the bootstrap as defined above. From the corresponding quantile, we ultimately obtain $[\widehat{\theta}_1, \widehat{\theta}_2]$ and calculate a confidence band

$$C_n(f) := \left[P_n f - \frac{\widehat{\theta}_1}{n}, P_n f + \frac{\widehat{\theta}_2}{n}\right] \tag{12}$$

for the empirical mean of a set of PIFs. Figure 3 depicts an example of confidence bands for the mean PIF of the sphere and torus samples. We can see that the confidence band for the torus is extremely tight for $\epsilon \in [0.2, 0.3]$, indicating that

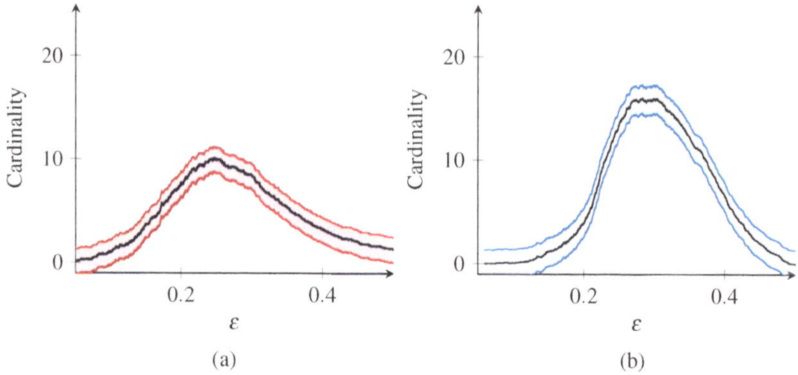

Fig. 3 Confidence bands at the $\alpha = 0.05$ level for the mean persistence indicator functions of a sphere and a torus. The confidence band is somewhat tighter for $\epsilon \geq 0.2$. (**a**) Sphere. (**b**) Torus

the limit behavior of samples from a torus is different at this scale from the limit behavior of samples from a sphere.

3.3 Distances and Kernels

Given two persistence diagrams $\mathscr{D}_i$ and $\mathscr{D}_j$, we are often interested in their dissimilarity (or distance). Having seen that linear combinations and norms of PIFs are well-defined, we can define a family of distances as

$$\operatorname{dist}_p(\mathbb{1}_{\mathscr{D}_i}, \mathbb{1}_{\mathscr{D}_j}) := \left(\int_{\mathbb{R}} |\mathbb{1}_{\mathscr{D}_i}(x) - \mathbb{1}_{\mathscr{D}_j}(x)|^p \mathrm{d}x \right)^{\frac{1}{p}}, \tag{13}$$

with $p \in \mathbb{R}$. Since the norm of a PIF is well-defined, this expression is a metric in the mathematical sense. Note that its calculation requires essentially only evaluating all individual step functions of the difference of the two PIFs once. Hence, its complexity is *linear* in the number of sub-intervals.

Example Figure 4 depicts pairwise distance matrices for random samples of a sphere and of a torus. The first matrix of each group is obtained via dist_p for PIFs, while the second matrix in each group is obtained by the pth Wasserstein distance. We observe two groups of data sets in all matrices, meaning that both classes of distances are suitable for detecting differences.

We can also employ the distance defined above to obtain a *kernel* [23]. To this end, we define

$$k_p(\mathscr{D}_i, \mathscr{D}_j) := -\operatorname{dist}_p(\mathbb{1}_{\mathscr{D}_i}, \mathbb{1}_{\mathscr{D}_j}), \tag{14}$$

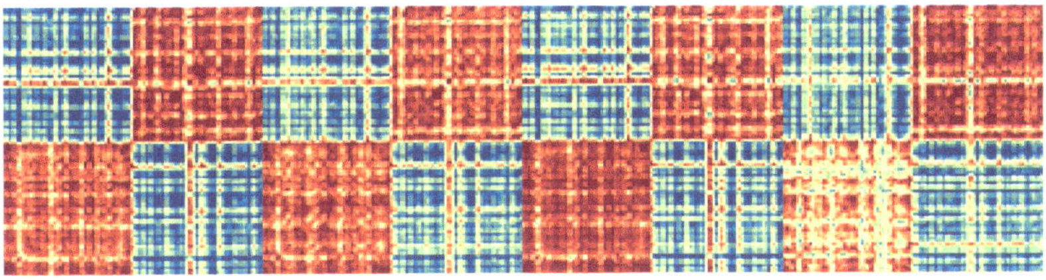

(a) (b)

Fig. 4 A comparison of distance matrices obtained using the distance measure dist_p for PIFs, and the corresponding p^{th} Wasserstein distance W_p. The left matrix of each group shows dist_p, while the right matrix depicts W_p. Red indicates close objects (small distances), while blue indicates far objects (large distances). (**a**) $p = 1$. (**b**) $p = 2$

where $p \in \{1, 2\}$ because we need to make sure that the kernel is *conditionally positive definite* [23]. This kernel permits using PIFs with many modern machine learning algorithms. As an illustrative example, we will use *kernel support vector machines* [23] to separate random samples of a sphere and a torus.

Example Again, we use 100 random samples (50 per class) from a sphere and a torus. We only use one-dimensional persistence diagrams, from which we calculate PIFs, from which we then obtain pairwise kernel matrices using both k_1 and k_2. Finally, we train a support vector machine using nested stratified 5-fold cross-validation. With k_1, we obtain an average accuracy of 0.98 ± 0.049, whereas with k_2, we obtain an average accuracy of 0.95 ± 0.063. The decrease in accuracy is caused by the additional smoothing introduced in this kernel.

4 Applications

In the following, we briefly discuss some potential application scenarios for PIFs. We use only data sets that are openly available in order to make our results comparable. For all machine learning methods, we use SCIKIT-LEARN [19].

4.1 Analysis of Random Complexes

It is often useful to know to what extent a data set exhibits random fluctuations. To this end, we sampled 100 points from a unit cube in $\mathbb{R}^3$, which has the topology of a point, i.e., no essential topological features in dimensions >0. We calculated the Vietoris–Rips complex at a scale such that no essential topological features remain in dimensions ≥ 1, and obtained PIFs, which are depicted in Fig. 5 along with their corresponding mean. All functions use a common axis in order to simplify their

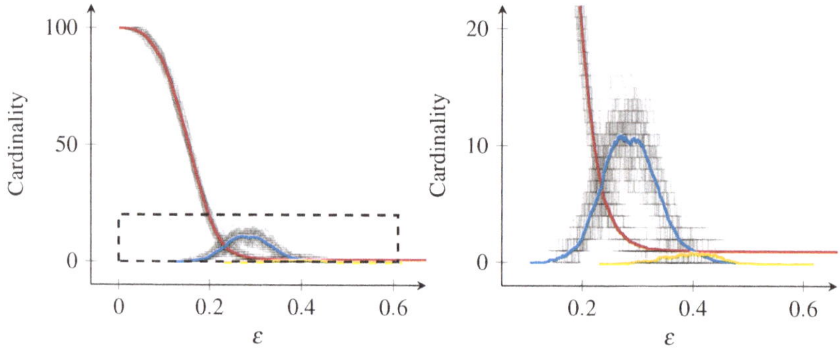

Fig. 5 PIFs for random complexes sampled over a unit cube in $\mathbb{R}^3$. Dimensions zero (red), one (blue), and two (yellow) are shown. To show the peak in dimension two better, the right-hand side shows a "zoomed" version of the first chart (dashed region)

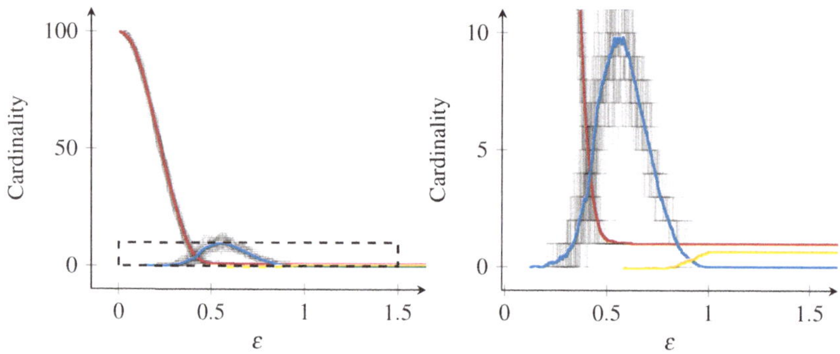

Fig. 6 PIFs for random samples of a sphere with $r = 1.0$. Again, dimensions zero (red), one (blue), and two (yellow) are shown, along with a "zoomed" version of the first chart (dashed region)

comparison. We first comment on the dynamics of these data. Topological activity is "shifted", meaning that topological features with a different dimensionality are not active at the same scale. The maximum of topological activity in dimension one (blue curve) is only reached when there are few zero-dimensional features. The maximum in dimension two (yellow curve) also does not coincide with the maximum in dimension one. These results are consistent with a limit theorem of Bobrowski and Kahle [3], who showed that (persistent) Betti numbers follow a Poisson distribution.

By contrast, for a data set with a well-defined topological structure, such as a 2-sphere, the PIFs exhibit a different behavior. Figure 6 depicts all PIFs of random samples from a sphere. We did not calculate the Vietoris–Rips complex for all possible values of ϵ but rather selected an ϵ that is sufficiently large to capture the correct Betti numbers of the sphere. Here, we observe that the stabilization of topological activity in dimension zero roughly coincides with the maximum of topological activity in dimension one. Topological activity in dimension two only starts to increase for larger scales, staying stable for a long interval. This activity corresponds to the two-dimensional void of the 2-sphere that we detect using persistent homology.

PIFs can thus be used to perform a test for "topological randomness" in real-world data. This is useful for deciding whether a topological approximation is suitable or needs to be changed (e.g., by calculating a different triangulation, using α-shapes, etc.). Moreover, we can use a PIF to detect the presence or absence of a shared "scale" in data sets. For the random complexes, there is *no* value for ϵ in which stable topological activity occurs in more than one dimension, whereas for the sphere, we observe a stabilization starting from $\epsilon \approx 0.75$.

4.2 Shakespearean Co-occurrence Networks

In a previous work, co-occurrence networks from a machine-readable corpus of Shakespeare's plays [21] have been extracted. Their topological structure under various aspects has been analyzed, using, for example, their clique communities [22] to calculate two-dimensional embeddings of the individual networks. The authors observed that comedies form clusters in these embeddings, which indicates that they are more similar to each other than to plays of another category. Here, we want to show that it is possible to differentiate between comedies and non-comedies by using the kernel induced by PIFs. Among the 37 available plays, 17 are comedies, giving us a baseline probability of 0.46 if we merely "guess" the class label of a play. Since the number of plays is not sufficiently large to warrant a split into test and training data, we use various cross-validation techniques, such as *leave-one-out*. The reader is referred to Kohavi [15] for more details. Table 1 reports all results; we observe that k_1 outperforms k_2. Since k_2 emphasizes small-scale differences, the number of topological features in two networks that are to be compared should be roughly equal. This is *not* the case for most of the comedies, though. We stress that these results are only a demonstration of the capabilities of PIFs; the comparatively low accuracy is partially due to the fact that networks were extracted automatically. It is interesting to note *which* plays tend to be mislabeled. For k_1, ALL'S WELL THAT ENDS WELL, CYMBELINE, and THE WINTER'S TALE are mislabeled more than all other plays. This is consistent with research by Shakespeare scholars who suggest different categorization schemes for these (and other) *problem plays*.

Table 1 Classifier performance for Shakespearean co-occurrence networks

Kernel	5-fold	LOO	LPO ($p = 2$)	LPO ($p = 3$)	LPO ($p = 4$)	Split
k_1	0.84	0.83	0.83	0.80	0.79	0.80
k_2	0.64	0.00	0.67	0.68	0.68	0.68

Classification based on k_1 outperforms the second kernel k_2

4.3 Social Networks

Yanardag and Vishwanathan [27] crawled the popular social news aggregation website reddit.com in order to obtain a set of graphs from online discussions. In each graph, the nodes correspond to users and an edge signifies that a certain user responded to a another user's comment. The graphs are partitioned into two classes, one of them representing discussion-based forums (in which users typically communicate freely among each other), the other representing communities based on a question–answer format (in which users typically only respond to the creator of a topic). The data set is fully-balanced.

Here, we want to find out whether it is possible to classify the graphs using nothing but topological information. We use the *degree*, i.e., the number of neighbors, of a node in the graph to obtain a filtration, assigning every edge the maximum of the degrees of its endpoints. We then calculate one-dimensional persistent homology and our kernel for $p = 1$ and $p = 2$. Figure 7 shows the results of applying *kernel principal component analysis* (k-PCA) [24], which each point in the embedding corresponding to a single graph. A small set of outliers appears to "skew" the embedding (Fig. 7a), but an inspection of the data shows that these graphs are extremely small (and sparse) in contrast to the remaining graphs. After removing them, the separation between both classes is visibly better (Fig. 7b). Progress from "left" to "right" in the embedding, graphs tend to become more dense.

As a second application on these data, we use a kernel support vector machine to classify all graphs, without performing outlier removal. We split the data into training (90%) and test (10%) data, and use 4-fold cross validation to find the best hyperparameters. The average accuracy for k_1 is 0.88, while the average accuracy for k_2 is 0.81. PIFs thus manage to surpass previous results by Yanardag and Vishwanathan [27], which employed a computationally more expensive strategy,

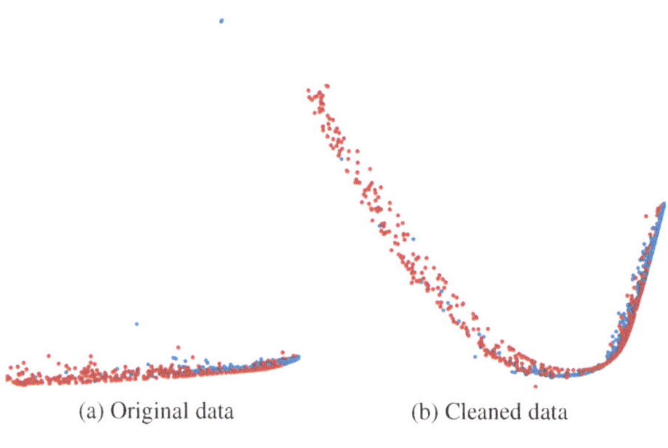

(a) Original data (b) Cleaned data

Fig. 7 Embeddings based on k-PCA for the k_1 kernel. (**a**) Every node represents a certain graph. The color indicates a graph from a discussion-based forum (red) or a Q/A forum (blue). (**b**) We removed some outliers to obtain a cleaner output. It is readily visible that the two classes suffer from overlaps, which influence classification performance negatively

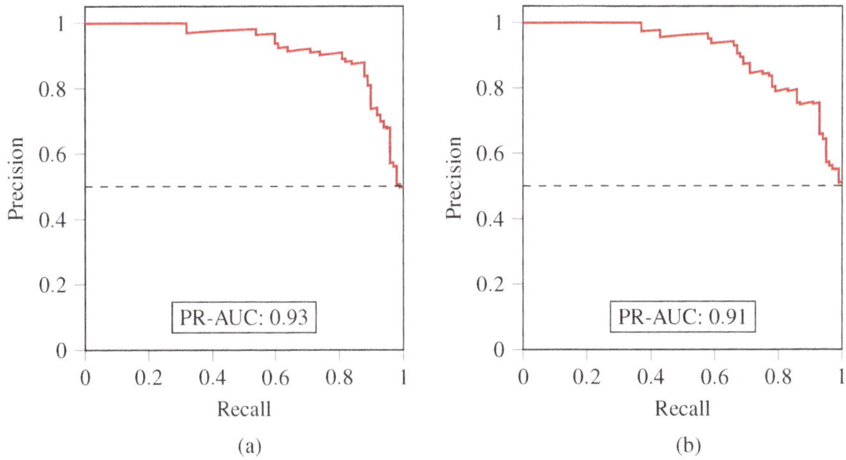

Fig. 8 Precision–recall curves for both kernels on the social networks data set. Each curve also includes the area-under-the-curve (AUC) value. (**a**) k_1. (**b**) k_2

i.e., graph kernels [25] based on learned latent sub-structures, and obtained an average accuracy of 0.78 for these data. Figure 8 depicts precision–recall curves for the two kernels. The kernel k_1 manages to retain higher precision at higher values of recall than k_2, which is again due to its lack of smoothing.

5 Conclusion

This paper introduced persistence indicator functions (PIFs), a novel class of summarizing functions for persistence diagrams. While being approximative by nature, we demonstrated that they exhibit beneficial properties for data analysis, such as the possibility to perform bootstrap experiments, calculate distances, and use kernel-based machine learning methods. We tested the performance on various data sets and illustrated the potential of PIFs for topological data analysis and topological machine learning. In the future, we want to perform a more in-depth analysis of the mathematical structure of PIFs, including detailed stability theorems, approximation guarantees, and a description of their statistical properties.

References

1. Adams, H., Emerson, T., Kirby, M., Neville, R., Peterson, C., Shipman, P., Chepushtanova, S., Hanson, E., Motta, F., Ziegelmeier, L.: Persistence images: a stable vector representation of persistent homology. J. Mach. Learn. Res. **18**(8), 1–35 (2017)
2. Agarwal, P.K., Edelsbrunner, H., Harer, J., Wang, Y.: Extreme elevation on a 2-manifold. Discr. Comput. Geom. **36**(4), 553–572 (2006)
3. Bobrowski, O., Kahle, M.: Topology of random geometric complexes: a survey (2014). https://arxiv.org/abs/1409.4734
4. Bremer, P.T., Edelsbrunner, H., Hamann, B., Pascucci, V.: A topological hierarchy for functions on triangulated surfaces. IEEE Trans. Vis. Comput. Graph. **10**(4), 385–396 (2004). https://doi.org/10.1109/TVCG.2004.3
5. Bubenik, P.: Statistical topological data analysis using persistence landscapes. J. Mach. Learn. Res. **16**, 77–102 (2015)
6. Chazal, F., Fasy, B.T., Lecci, F., Rinaldo, A., Singh, A., Wasserman, L.: On the bootstrap for persistence diagrams and landscapes. Model. Anal. Inf. Syst. **20**(6), 111–120 (2013)
7. Cohen-Steiner, D., Edelsbrunner, H., Harer, J.: Stability of persistence diagrams. Discr. Comput. Geom. **37**(1), 103–120 (2007)
8. Cohen-Steiner, D., Edelsbrunner, H., Harer, J., Mileyko, Y.: Lipschitz functions have L_p-stable persistence. Found. Comput. Math. **10**(2), 127–139 (2010)
9. Edelsbrunner, H., Harer, J.: Computational Topology: An Introduction. AMS, New York (2010)
10. Edelsbrunner, H., Morozov, D.: Persistent homology: theory and practice. In: European Congress of Mathematics. EMS Publishing House, Zürich (2014)
11. Edelsbrunner, H., Letscher, D., Zomorodian, A.J.: Topological persistence and simplification. Discr. Comput. Geom. **28**(4), 511–533 (2002)
12. Efron, B., Tibshirani, R.J.: An Introduction to the Bootstrap. Monographs on Statistics and Applied Probability, vol. 57 . Chapman & Hall/CRC, Boca Raton, FL (1993)
13. Günther, D., Boto, R.A., Contreras-Garcia, J., Piquemal, J.P., Tierny, J.: Characterizing molecular interactions in chemical systems. IEEE Trans. Vis. Comp. Graph. **20**(12), 2476–2485 (2014)
14. Kerber, M., Morozov, D., Nigmetov, A.: Geometry helps to compare persistence diagrams. In: Goodrich, M., Mitzenmacher, M. (eds.) Proceedings of the 18th Workshop on Algorithm Engineering and Experiments (ALENEX), pp. 103–112. SIAM, Philadelphia, PA (2016)
15. Kohavi, R.: A study of cross-validation and bootstrap for accuracy estimation and model selection. In: Proceedings of the IJCAI, vol. 2, pp. 1137–1143 (1995)
16. Kosorok, M.R.: Introduction to Empirical Processes and Semiparametric Inference. Springer, New York, NY (2008)
17. Laney, D., Bremer, P.T., Mascarenhas, A., Miller, P., Pascucci, V.: Understanding the structure of the turbulent mixing layer in hydrodynamic instabilities. IEEE Trans. Vis. Comput. Graph. **12**(5), 1053–1060 (2006)
18. Mucha, M., Sankowski, P.: Maximum matchings via Gaussian elimination. In: 45th Annual IEEE Symposium on Foundations of Computer Science, pp. 248–255 (2004)
19. Pedregosa, F., Varoquaux, G., Gramfort, A., Michel, V., Thirion, B., Grisel, O., Blondel, M., Prettenhofer, P., Weiss, R., Dubourg, V., Vanderplas, J., Passos, A., Cournapeau, D., Brucher, M., Perrot, M., Duchesnay, É.: SCIKIT-LEARN: machine learning in Python. J. Mach. Learn. Res. **12**, 2825–2830 (2011)
20. Reininghaus, J., Huber, S., Bauer, U., Kwitt, R.: A stable multi-scale kernel for topological machine learning. In: IEEE Conference on Computer Vision and Pattern Recognition (CVPR), pp. 4741–4748. Curran Associates, Inc., Red Hook, NY (2015)
21. Rieck, B., Leitte, H.: Shall I compare thee to a network?—Visualizing the topological structure of Shakespeare's plays. In: Workshop on Visualization for the Digital Humanities at IEEE VIS. Baltimore, MD (2016)

22. Rieck, B., Fugacci, U., Lukasczyk, J., Leitte, H.: Clique community persistence: a topological visual analysis approach for complex networks. IEEE Trans. Vis. Comput. Graph. **22**(1), 822-831 (2018). https://doi.org/10.1109/TVCG.2017.2744321
23. Schölkopf, B., Smola, A.J.: Learning with Kernels. The MIT Press, Cambridge, MA (2002)
24. Schölkopf, B., Smola, A.J., Müller, K.R.: Nonlinear component analysis as a kernel eigenvalue problem. Neural Comput. **10**(5), 1299–1319 (1998)
25. Sugiyama, M., Ghisu, M.E., Llinares-López, F., Borgwardt, K.: `graphkernels`: R and Python packages for graph comparison. Bioinformatics **34**(3), 530–532 (2017)
26. Turner, K., Mileyko, Y., Mukherjee, S., Harer, J.: Fréchet means for distributions of persistence diagrams. Discr. Comput. Geom. **52**(1), 44–70 (2014)
27. Yanardag, P., Vishwanathan, S.V.N.: Deep graph kernels. In: Proceedings of the 21th ACM SIGKDD International Conference on Knowledge Discovery and Data Mining, pp. 1365–1374. ACM, New York, NY (2015)

Pathological and Test Cases for Reeb Analysis

Hamish Carr, Julien Tierny, and Gunther H. Weber

Abstract After two decades of computational topology, it is clearly a computationally challenging area. Not only do we have the usual algorithmic and programming difficulties with establishing correctness, we also have a class of problems that are mathematically complex and notationally fragile. Effective development and deployment therefore requires an additional step—construction or selection of suitable test cases. Since we cannot test all possible inputs, our selection of test cases expresses our understanding of the task and of the problems involved. Moreover, the scale of the data sets we work with is such that, no matter how unlikely the behavior mathematically, it is nearly guaranteed to occur at scale in every run. The test cases we choose are therefore tightly coupled with mathematically pathological cases, and need to be developed using the skills expressed most obviously in constructing mathematical counter-examples. This paper is therefore a first attempt at reporting, classifying and analyzing test cases previously used for algorithmic work in Reeb analysis (contour trees and Reeb graphs), and the expression of a philosophy of how to test topological code.

H. Carr (✉)
School of Computing, University of Leeds, Leeds, UK
e-mail: h.carr@leeds.ac.uk

J. Tierny
Sorbonne Universities, UPMC Univ Paris 06, CNRS, LIP6 UMR 7606, Paris, France
e-mail: julien.tierny@lib6.fr

G. H. Weber (✉)
Lawrence Berkeley National Laboratory, Berkeley, CA, USA
University of California at Davis, Davis, CA, USA
e-mail: ghweber@lbl.gov; ghweber@ucdavis.edu

H. Carr et al. (eds.), *Topological Methods in Data Analysis and Visualization V*, Mathematics and Visualization, https://doi.org/10.1007/978-3-030-43036-8_7

1 Introduction

Computational topology began in the 1980s for scalar fields [13] in geographic information systems, and for vector fields [14] for scientific visualization, with scalar field analysis then developing for the analysis of 3D (volumetric) data.

Over time, topological analysis in scientific visualization has included techniques based on Reeb Analysis, Morse-Smale Analysis, Persistent Homology, Vector Field Analysis, and Tensor Field Analysis. While articles commonly describe new algorithms, they rarely describe testing strategies, although tools exist that could be applied, particularly for vector fields [20, 25]. We will, however, restrict our attention to Reeb Analysis, where we have prior experience. Similarly, while the strategies for debugging visualization described by Laramee [17] can be applied, we focus primarily on the test cases we use for topological algorithms.

While these techniques are powerful for understanding data, they are conceptually complex. They are also particularly difficult to implement, as they are susceptible to a wide range of errors during program construction. Thus, in addition to the normal struggle to frame an algorithm accurately, robustly and efficiently, we have to contend with problems due to the difficulty of the underlying mathematics.

Between us, we have accumulated over 40 years of experience in working with topological code. As a result, we have developed and employed a variety of strategies for constructing, testing and debugging programs. These strategies, however, rarely form part of publications, since there is usually barely room for all the technical details. Since these strategies are of value to other researchers or programmers attempting to grapple with complex algorithms, we therefore aim to start the discussion of test cases and testing strategies.

We do not have space for the details of all the algorithmic work, so we start with a quick overview instead in Sect. 2. We then sketch a number of conceptual approaches to test cases in Sect. 3 and introduce two types of pathological cases, flat regions in Sect. 4 and the *W-structure* in Sect. 5. Section 6 then gives some concrete examples of test sets that we have used, and Sect. 7 discusses some of the techniques that we use for visualizing intermediate results during debugging. Then Sect. 8 summarizes our experience and presents some conclusions.

2 Reeb Analysis

Reeb Analysis studies the relationships between isocontours to extract knowledge from a mathematical function or data set. Consider a *scalar field*, i.e. a function of the form $f : \mathscr{R}^d \rightarrow \mathscr{R}$. Since the data we wish to analyze is normally spatial in nature, we shall assume $d \in \{2, 3, 4\}$. A level set or inverse image of f is defined by choosing an *isovalue* $h \in \mathscr{R}$, then extracting all points in the domain of the function with function value h, i.e. $f^{-1}(h) = \{x \in \mathscr{R}^d : f(x) = h \in \mathscr{R}\}$. These level sets are often referred to as *isocontours* (*isolines* where $d = 2$, *isosurfaces* where $d = 3$).

Any given isocontour may have multiple connected components, which are ambiguously referred to as isolines and isosurfaces. We therefore use *isocontour components* to refer to the individual surfaces, in line with the literature.

If we contract each isocontour component to a single point, we construct the *contour tree* [2]. For more general functions, where the domain is a general manifold $\mathcal{M}$, the same construction gives the *Reeb graph* [18]. Although a special case of the Reeb graph, the contour tree is easier to compute [4], and the fastest Reeb graph algorithm reduces the input to a simple domain, computes the contour tree over that domain, then reconnects the domain (and the tree) to build the Reeb graph [22].

More recently, Reeb Analysis has been extended to functions of the form $f : \mathcal{M} \rightarrow \mathcal{R}^r$, where $r > 1$. These cases are covered by the mathematics of fiber topology, and we replace isocontours with *fibers* representing inverse images of the form $f^{-1}(h) = \{x \in \mathcal{R}^d : f(x) = h \in \mathcal{R}^r\}$. Continuous contraction of these fibers then results in the Reeb space [11]. This can be constructed approximately [3] for the general case, or precisely [21] for the case $f : \mathcal{R}^3 \rightarrow \mathcal{R}^2$.

Rather than recapitulate all of these algorithms, we refer the reader to the original papers, and assume some degree of familiarity with the details, as we are presently interested in describing debugging practice and test cases for them.

3 Approaches

Generally speaking, debugging complex code depends on testing representative types of data, since exhaustive testing of all possible inputs is combinatorially impossible. Within this, test sets may be analytic, stochastic, empirical, or synthetic, but the choice normally depends on the specific problem domain.

Analytic Frequently, computation replicates an existing mathematical method, and as a result, test cases can be constructed from known mathematical examples, which were generally developed during mathematical debugging of an idea. Since these are likely to display interesting or challenging behavior, they are commonly used as test functions for which the ground truth result is already well understood.

For our work in computational topology, this ideal approach has been less useful than it might seem. This occurs because mathematical development generally considers smooth infinitely differentiable functions. Since most code assumes simplicial or cubic meshes with linear, trilinear or ad hoc interpolation, the sampled data rarely captures the original mathematical function exactly unless sampled at high resolution. This is of particular concern when debugging, as manual validation of intermediate stages for anything over 10^3 is time-consuming and wearisome.

Moreover, mathematical reasoning is reductive, and attempts to deal with a small number of simple cases, in order to stay within the reasoning abilities of a human being. As a result, analytic examples tend to have simple topology—i.e. relatively small numbers of topological events. However, the sampling necessary to capture this causes them to be medium scale in terms of data, which makes them unattractive for early stage testing. Later on, simple topology has typically already been tested, and medium scale examples are used to test combinations of simple topology. At this point, analytic functions are rarely complex enough to provide good medium scale tests. We therefore tend to avoid analytic functions except at the conceptual stage.

Stochastic A second approach is to generate data sets stochastically—i.e. to choose randomly from all possible data sets. While this has the merit that it does not prefer any particular data sets, it fails to guarantee that challenging topology will be tested early, or indeed ever. As a result, we tend to avoid stochastic approaches.

Curiously, however, as the data scales up, stochastic effects mean that *every* pathological mathematical case will occur multiple times, leading to the problem that we refer to as **too much topology**. In practice, 1 GB of data means that there may be tens of millions of topological events. If the isovalues at which they occur are independent, then even for double precision floating point, it is highly likely that multiple topological events will happen at the same isovalue, which means that robust handling of complex topology is always required. This has also driven much of the work on topological simplification, so paradoxically, while avoiding stochastic approaches in the abstract, we rely heavily on them in practice.

Empirical Since the goal of computation is to process data, the third approach therefore looks to existing data, either from prior experience or from a current data problem. While it is generally simple to obtain data from a variety of sources, there are at least four problems with empirical data:

1. Scale: as with mathematical test data, empirical data is often at too large a scale to be useful in the early stages of development, although we commonly use empirical data for testing at the medium to large scale.
2. Noise: many acquired data sets, particularly medical data, are noisy due to the original acquisition process, and this tends to result in large numbers of topological events, which are undesirable for small scale testing. Noisy data types therefore tend to be of more use at the medium to large scale.
3. Blandness: clean empirical data can suffer the opposite problem: that the number of topological events is much smaller than the data set, again hampering manual validation. Clean simulation data is particularly prone to this.
4. Clumping: some types of empirical data, such as medical, tend to have heavily clumped values, as for example where isovalues correspond to different tissue types. This tends to result in many topological events over a narrow range of values, again hampering manual validation.

Having said that, empirical data becomes particularly useful at medium to large scales, since many data types naturally result in large numbers of topological events, providing a useful test of the scalability of the underlying approach. Moreover, we have found that terrain data, which is self-similar at different scales, is often useful for debugging, as discussed below.

Synthetic Since neither analytic nor empirical approaches give good test sets for early stages, we find that early stage testing relies heavily on synthetic examples for algorithm development and debugging. We aim to keep input sizes small, and to exhibit a rich topological behavior. We depend in particular on pathological cases and counter-example construction. Scalable examples are then built algorithmically for data construction, by copying a known pathology, by working backwards from the desired output structure, or by replicating copies of smaller-scale features.

Although mathematical functions would seem to be the best strategy, these are most commonly C^{∞}, and are a poor fit to the demands of algorithmic development. We therefore tend to start with small synthetic examples, then scale by judicious selection of empirical data, usually starting either with small terrain examples or clean simulation data, and progressing to large complex data sets.

As we have noted, we tend to rely on experience with previous problems to choose data sets that have previously caused algorithmic, theoretical or interpretational difficulties. While not exclusive, there are two major types of data which we know routinely present these difficulties even at small to medium scales: flat regions and W-structures.

4 Flat Regions

Morse Theory assumes that critical points occur at unique values, and that there are no flat regions—i.e. regions with gradient of 0 but dimensionality >0. While this considerably simplifies the mathematics, it tends to have the reverse effect in practical data. And, although perturbation through simulation of simplicity [10] allows us to reduce the problem to the mathematically tractable, it imposes both algorithmic and interpretational costs.

To make matters worse, flat regions are frequently observed in quantized data sets. Many types of data have a narrow range of interesting values, with multiple topological events clustering tightly together. Even a small amount of quantization tends to result in flat regions. Moreover, for algorithmic purposes, flat regions are often broken up by symbolic perturbation [10], which adds a different mathematical ϵ to each value to guarantee unique values throughout the data. This induces additional ϵ-persistent edges, which must then be suppressed in user interfaces and/or accounted for through simplification.

The converse of this is that these data sets are often valuable test cases of whether symbolic perturbation is implemented correctly and consistently. A good example of this occurs in the "nucleon" data set from VolVis. It contains small to moderately sized regions of constant value that can be resolved using symbolic perturbation.

Equally, the hydrogen data set [19] has extremely large constant regions that stress test symbolic perturbation implementations. For instance, most of the region

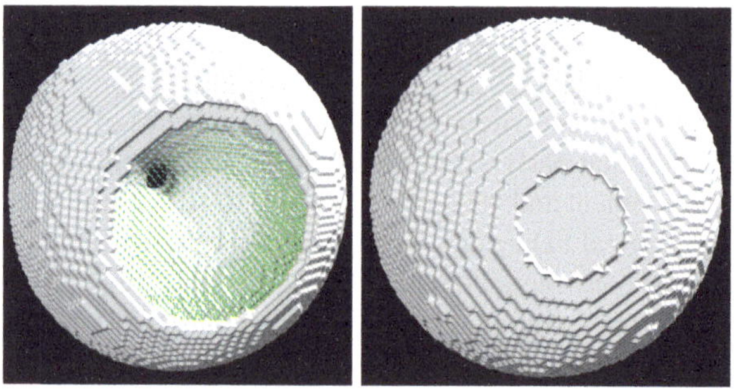

Fig. 1 Flat region in the hydrogen data set that is likely a quantization artifact and should be removed by simulation of simplicity. In the continuous function the hole would likely be filled smoothly until it disappears at a critical point

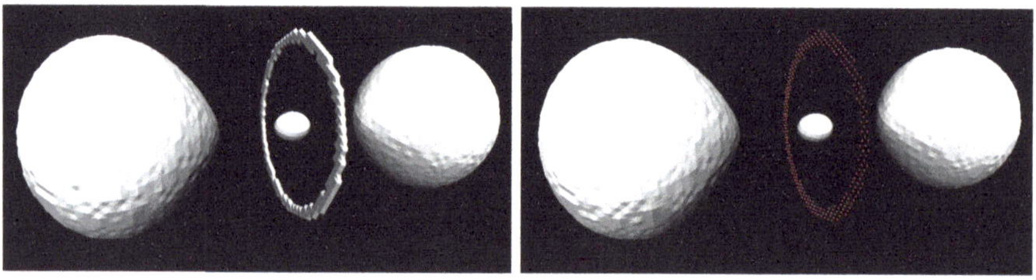

Fig. 2 Flat region in the hydrogen data set that probably corresponds to a feature of interest and should be characterized in its entirety. The ring-structure likely has a correspondence in the smooth, real-valued function (but sampling on a grid without quantization would break it apart)

around the hydrogen atom has a constant value, as shown in Fig. 1. Here, the flat region is likely spurious and should be removed by symbolic perturbation. In the quantized data, a whole "cap" like structure forms around a flat region and subsequently closes an isosurface component. In this case, symbolic perturbation leads to the "correct" behavior: the component closes off smoothly.

However, not all flat regions are unimportant or spurious: they can be the regions of most interest in the data. Again considering the hydrogen data set, Fig. 2 shows a flat region of particular interest where two protons interact. Here, even a continuous function would likely contain a region of constant value, i.e. the circle around which the ring forms. Since this behavior is intrinsic to the underlying phenomenon, suppressing the region to ensure mathematically clean behavior may actually mislead interpretation of results.

Ideally, we would be able to distinguish between "spurious" and real features, and to define stability for them in a mathematically sound framework. This may involve consideration of the difference between persistent simplification and geometric simplification [6], but is broadly speaking beyond the scope of the current discussion.

Phenomena like these have also led to approaches that aim at avoiding symbolic perturbation and identify critical regions directly [1, 9, 16, 24]. For example, in the hydrogen atom, the "ring structure" shown in Fig. 2 appears around a region of constant function value. Symbolic perturbation breaks the ring up into at least a maximum (around which the ring starts to form in the perturbed version) and a saddle (where the ring closes in the perturbed version). One may argue that in this instance it is desirable to detect the entire ring structure as a critical entity, i.e. a circle around which the ring forms.

5 W Structures

While flat regions test our ability to reconcile quantized data with mathematical formalisms, our other pathological case, the *W structure* can be constructed either mathematically or by observation in data.

This structure was first shown as an illustration of a potential case by Carr et al. [5] (as Figure 2). However, the implications of this structure were not fleshed out, and the illustration was omitted from the later journal paper [6] for reasons of space. Since then, it has caused difficulties both in proofs (L. Arge, Personal communication to H. Carr), and in algorithmic analysis of recent parallel approaches [8].

We refer to this as a *W structure* since it consists of a horizontal zigzag of edges (or paths) in the contour tree. In 2D, these can be constructed as a sequence of nested volcanic caldera. Similarly, in 3D, nested shells alternating between minima and maxima will also display this behavior. However, once boundary conditions are taken into account, an alternating sequence of ridges and valleys stretching across the data set will also result in a W structure.

Once we realize the impact of boundary conditions, it is easy to construct W structures with any desired complexity, as illustrated in Fig. 3. Here, we start with alternating ridges and valleys. Next, we insert saddle points in each ridge and valley to divide them into multiple extrema each. Finally, we assign values to each location, making sure we stay consistent with the assignment of saddle points. As a result, we see a horizontal zigzag emerge in the contour tree, and is clear that we can use this construction to make arbitrarily complex W structures.

We note that there are many variations possible. For example, we have chosen to make all downwards saddles lower than all upwards saddles. While this is not necessary, it is easy to enforce. Similarly, the exact ordering along the ridges and valleys can be altered: in larger examples, we can have multiple extrema for each. We have also chosen to place the saddles in the middle—there are boundary effects when they are at the edge of the data. Moreover, in practice we tend to use diagonal ridges and valleys, in order to pack more features into a small space. But the basic strategy is clear: alternate ridges and valleys generate W-structures, and this gives us useful test cases at any desired scale.

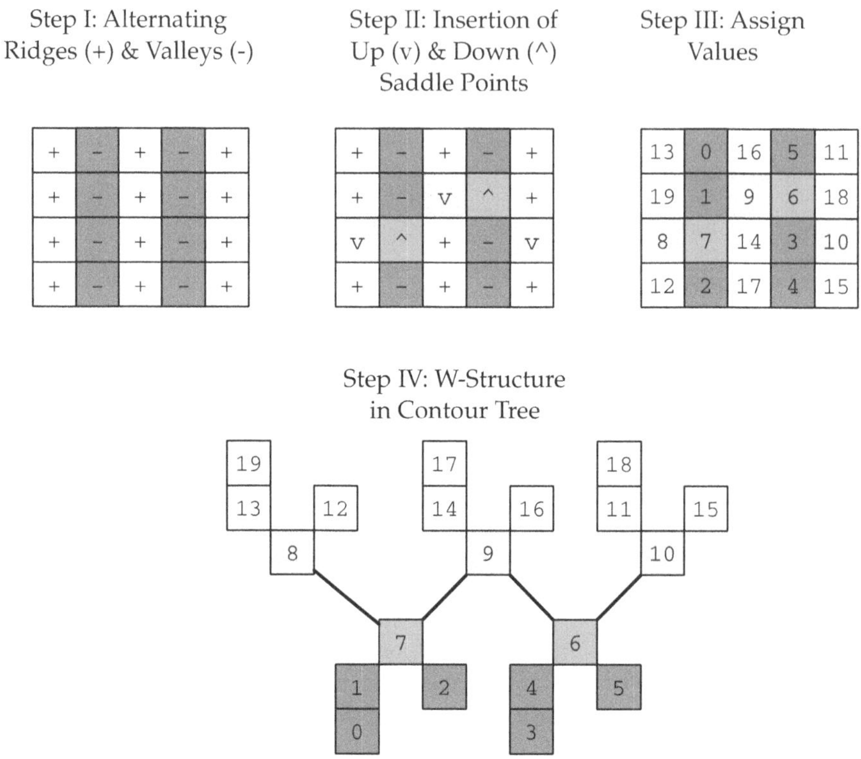

Fig. 3 An example of a "W structure" in a contour tree, with construction

6 Concrete Examples

In addition to the specific examples of flat regions and W structures, we have historically used a range of data sets for testing. While the following list is not exhaustive, it covers a range of data sets we have found useful in practice for test purposes.

The 5b Dataset This dataset was built around 2000 for testing contour tree construction. The intent was to pack the maximum number of topological features into the smallest possible space. As a result, it has shown up in a number of papers. It was constructed by electing to have two minima, one interior and one exterior, and four maxima arranged in pairs with toroidal isosurfaces nested around them. Initially, this data set was constructed as a $3 \times 3 \times 3$ grid, but was embedded in a layer of 0s—as a result, it is a $5 \times 5 \times 5$ grid, as shown in Fig. 4.

Volvis Data We also use the volvis data repository for testing, in particular the fuel dataset, which has around 100 critical points in a $64 \times 64 \times 64$ data set, although with more maxima than minima. The hydrogen dataset has also proved useful, as several of the features of interest form flat regions in the data, which is a useful test of the simulation of simplicity used to guarantee topological properties.

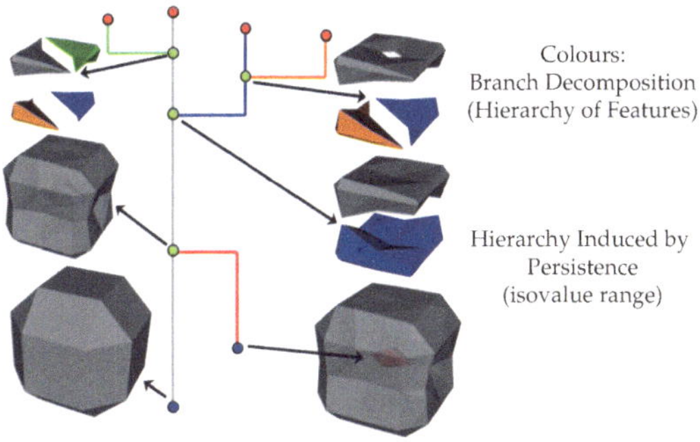

Plane z=0

0	0	0	0	0
0	0	0	0	0
0	0	0	0	0
0	0	0	0	0
0	0	0	0	0

Plane z=1

0	0	0	0	0
0	99	90	85	0
0	95	80	95	0
0	85	95	99	0
0	0	0	0	0

Plane z=2

0	0	0	0	0
0	75	50	65	0
0	55	15	45	0
0	60	40	70	0
0	0	0	0	0

Plane z=3

0	0	0	0	0
0	97	87	82	0
0	92	77	92	0
0	82	87	97	0
0	0	0	0	0

Plane z=4

0	0	0	0	0
0	0	0	0	0
0	0	0	0	0
0	0	0	0	0
0	0	0	0	0

Fig. 4 5b: a synthetic $5 \times 5 \times 5$ dataset, constructed to have four maxima, two minima, three connectivity-critical points and five additional Morse critical points in the central $3 \times 3 \times 3$ block

Protein Data Base Another source of test data is the PDB protein data base (www.wwpdb.org). One of the utilities from this project generates sampled electrostatic potential fields at any desired resolution, and these tend to have many topological events occurring at bonds between atoms. One dataset from this source was therefore instrumental in trapping a typographical error in a lookup table that caused the same surface to be extracted over a hundred times instead of once.

GTOPO30 Data We have found terrain data to be valuable for testing for several reasons. First, terrain data is defined in two dimensions, not three, and therefore tends to be more useful for small scale testing. Secondly, it is naturally self-similar, so interesting topology occurs throughout much of the world. Thirdly, thanks to US government policy, it is freely available from the US Geographical Survey, in particular the 30 arc-s GTOPO30 dataset (https://lta.cr.usgs.gov/GTOPO30).

We have found that a small section of terrain around Vancouver captures 40 odd topological events in about 400 data values, which is still feasible to verify manually. For scaling studies, sections of the Canadian Rockies proved exceptionally useful: they naturally give rise to W structures due to the parallel nature of mountain ranges. Finally, a section of low-relief terrain from the Canadian Shield tests both the handling of W structures and of clumped data values. This is not to say that these are ideal choices: merely that they have been good tests in the past.

Nasty W During the development of the most recent contour tree algorithm, W structures caused particular problems, and we therefore built a number of small examples, which led us to the construction described above. These were crucial in constructing valid parallel algorithms, and had the side effect of developing our understanding of this pathological case. Eventually, we constructed an extreme case to track down a particularly nasty bug, simplifying the construction to a single triangle strip for ease of manual debugging. We show a developed version of this, called the "nasty W" in Fig. 5, in which every vertex of the mesh is a supernode in the contour tree.

One of the particular values of this example is that the W structure ensures that only two branches are candidates for simplification at any given time, with the lower priority one chosen for removal. Here, our priority measure is the "height" of the branch, which is not in fact the persistence of the extremum [15]. The right-hand branch (80 – 60) in the illustration has a priority of 20, lower than the priority (110) of the left-hand branch (230 – 120), and is removed first, revealing branch (20 – 160) with priority 140. The decomposition then proceeds left to right since none of the

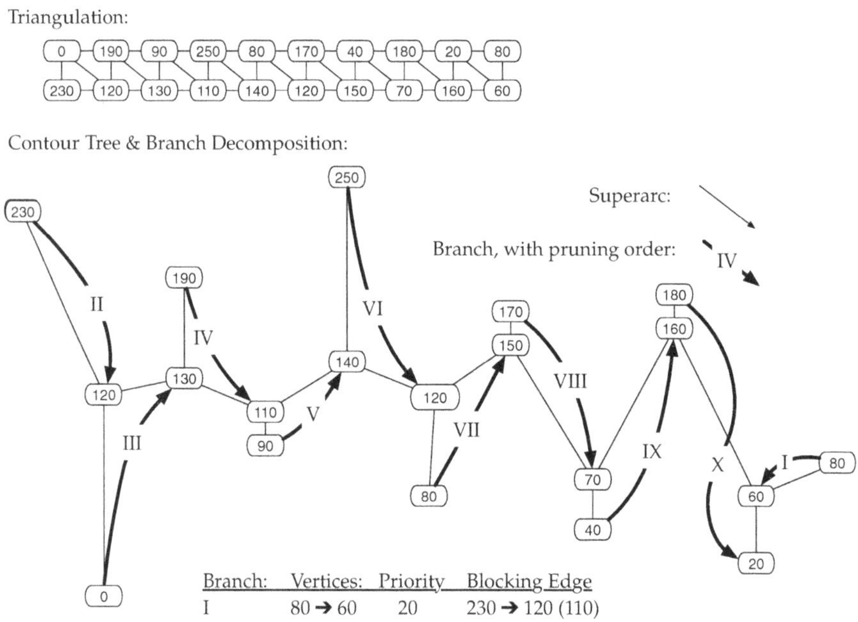

Branch:	Vertices:	Priority	Blocking Edge
I	80 ➔ 60	20	230 ➔ 120 (110)
II	230 ➔ 120	110	20 ➔ 160 (140)
III	0 ➔ 130	130	20 ➔ 160 (140)
IV	190 ➔ 110	80	20 ➔ 160 (140)
V	90 ➔ 140	50	20 ➔ 160 (140)
VI	250 ➔ 120	130	20 ➔ 160 (140)
VII	80 ➔ 150	70	20 ➔ 160 (140)
VIII	170 ➔ 70	130	40 ➔ 160 (140)
IX	180 ➔ 20	160	None - Master Branch

Fig. 5 The Nasty W test example. The contour tree was constructed first, then the saddles arranged along one row of the mesh, the extrema along the other row

remaining branches have height greater than 140. In the result, the master branch is (180 – 20).

The Nasty W has several interesting properties. First, the master branch includes neither the global minimum nor the global maximum. Second, the global minimum and global maximum do not pair with each other, indirectly demonstrating that branch decomposition and contour tree simplification are not equivalent to persistence. And third, the master branch isn't even the longest monotone path in the tree! As a result, this (and several other W structures) are now part of our standard test suite for working on contour tree algorithms (Fig. 6).

One Cube and One Cube Forking While less work has been done on Reeb spaces than on contour trees and Reeb graphs, the same pattern of test case construction was visible. Real data sets were too complex, while the existing mathematical examples were difficult to construct in practical data. As a result, we constructed three small examples of volumetric bivariate meshes, all employing six tetrahedra packed into a single cube with a shared major diagonal (i.e. a Freudenthal subdivision).

Here, the goal was not to maximize the complexity (at least for our first example), but rather to develop a small example with non-trivial Reeb space in order to assist in our own understanding of fiber topology. After several years of limited progress based on combinations of polynomial functions, this example was developed with a small test harness and immediately led to fiber surfaces, then played a role in algorithm development for Reeb spaces [7, 21]. It is therefore recommended both as a first tiny data set for testing correctness, but also as an example for the process of learning and reasoning about Reeb spaces and fiber topology.

Figure 6 shows an example following a standard approach in fiber topology, by using a linear ramp for one of the functions, and choosing a second so that the contour tree on one face is an upwards fork, while the contour tree on the opposite face is a downwards fork. In its earliest incarnation in 2002, it was used by the first author to reason about time-varying contour trees. It then became an example used to explain Reeb Spaces to students: when the time came to construct a small data set for Reeb Space computation, it was natural to embody it as a small tetrahedral mesh.

Due to the tetrahedralization chosen, the result is slightly different. However, the downward fork is recognizable as two flaps (the white regions on the left), while the upward fork became another two (the white regions on the right). In the middle gray region, only one fiber exists, represented by gluing the two partial Reeb spaces together. We constructed paper models of this (and our other examples), and it was these models that led us to construct fiber surfaces [7].

We then constructed the second example, in Fig. 7, by replacing the linear ramp with a rotated copy of the first function. Here there are up to three fiber components for any given value. Moreover, two "vertices" of the central square are not vertices of the mesh, but intersections of the projections of the mesh edges. This property showed us that certain algorithmic lines of attack would be unfruitful.

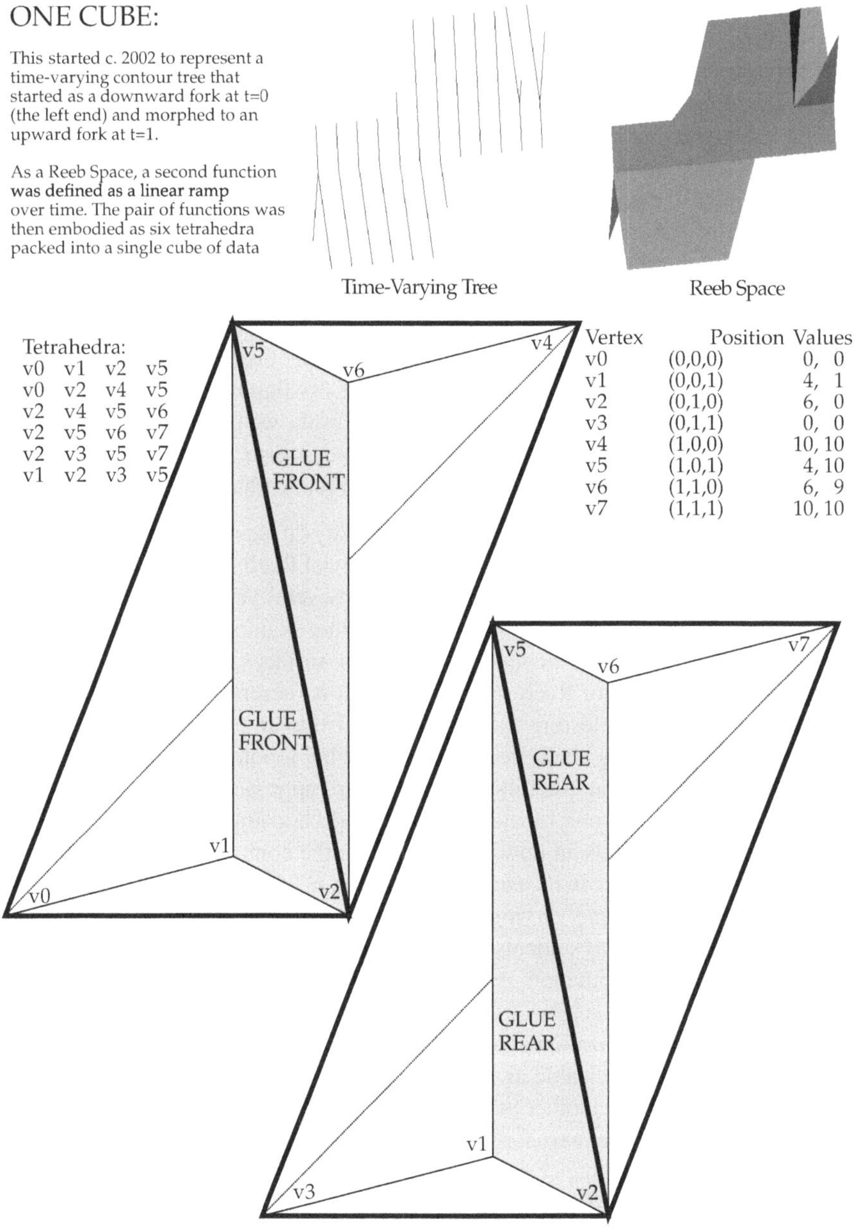

Fig. 6 A small Reeb space constructed from six tetrahedra packed into a single cube in the domain with shared edge v2v5

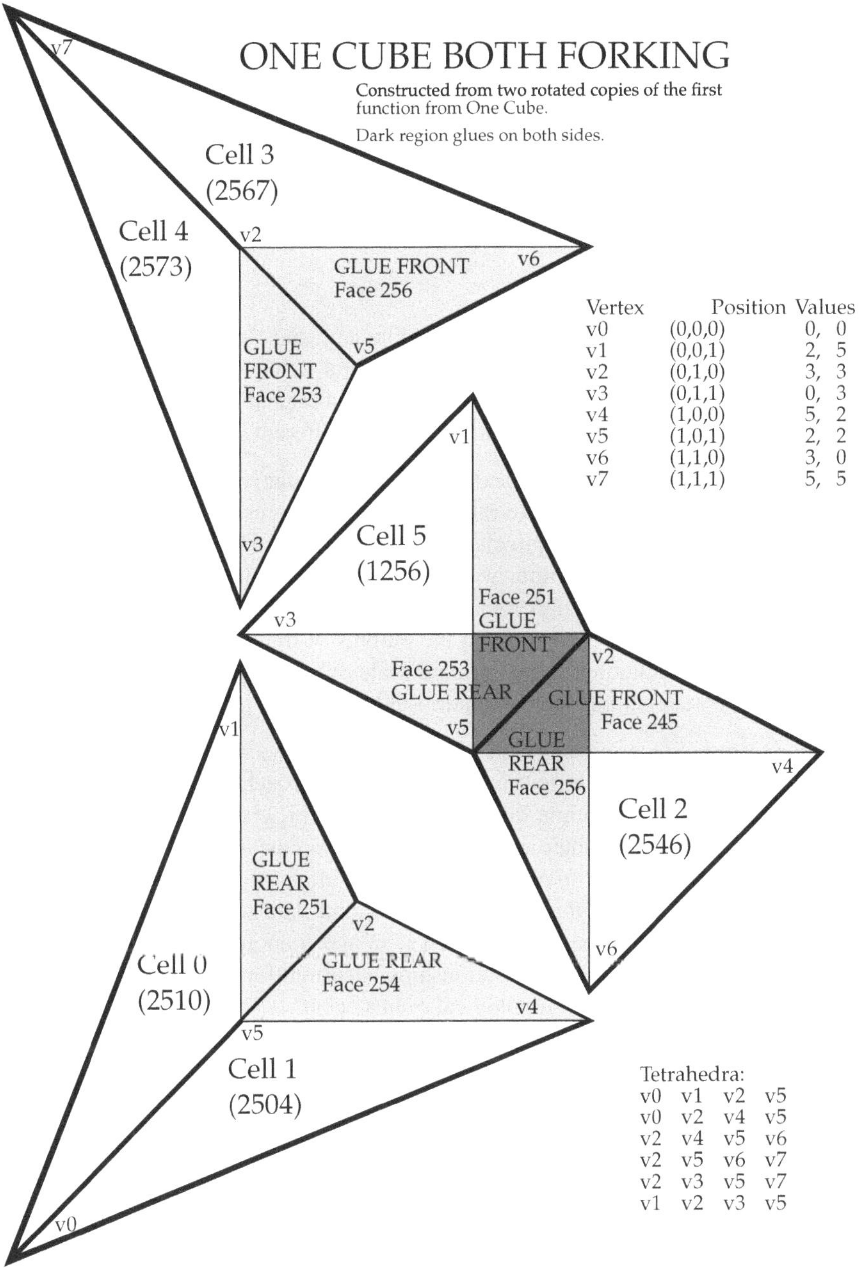

Fig. 7 A second Reeb space constructed from six tetrahedra packed into a single cube. Here, the first field is the same as in Fig. 6, while the second as a copy of the first field, rotated by 90°

Finally, since all of the fibers in these examples were open at the boundary, we constructed a third to have closed loop fibers around the main diagonal of the cube. This was less useful than we had hoped, but still helped us to think about vertices and tetrahedra that overlapped in projection. We therefore omit this example.

These examples are small enough to print out and assemble manually, and we have found them very useful for comprehension and for debugging.

7 Debug Tricks

Once suitable test data is available, development and debugging can proceed. And here, too, experience indicates that a general debug procedure needs some modification to accommodate the demands of topological computation. We have noticed three basic tricks that we tend to repeat in different contexts:

Text Output As with all code, text output is particularly useful. Part of this is because the internal structures are rarely intuitive. As a result, having a consistent well-formatted text output is priceless for tracking down bugs. Moreover, when improving an existing algorithm, we have found that sharing a common output format between versions makes it easier to identify where bugs occur, simply by using the command line tool 'diff' on the outputs of the versions. This technique is particularly valuable in practice as data scales: for example, this allowed us to validate new parallel algorithms [8] against old serial code [6].

Visual Output Since our target is to visualize data, we normally operate in an environment where visual output is feasible. One approach is to export meshes at various stages of an algorithm, then use an external program such as paraview or TTK to inspect them. Another approach is to build a custom application, either to support the debug process, to play with particular ideas, or to illustrate the process for others. One example of this with two-dimensional data is to render a terrain with the contour tree superimposed: since the (x, y) and h coordinates are known, this means that visual inspection of relationships is straightforward at smaller scales, although difficult when many topological events occur.

Over time, the desire to have intermediate visual output was part of the motivation for the development of the Topology Tool Kit (TTK) by the second author [23], and readers may find its features valuable.

Graph Output Since scalar Reeb analysis results in graph-like structures, one of the most useful debug tricks is to export the internal data structures to a graph format such as graphviz [12], then to invoke external programs such as dot to generate PDFs and display them. This is a variant of the stepping method described by Laramee [17] which we have found useful, especially for small test sets.

At one stage of developing a new parallel algorithm [8], debug involved manual cross-checking of a contour tree with over 1000 nodes, shown in Fig. 8, which involved using a GUI-based graph editor on the dot format output. Most recently, improvements to internal data structures have been simplified considerably by graph

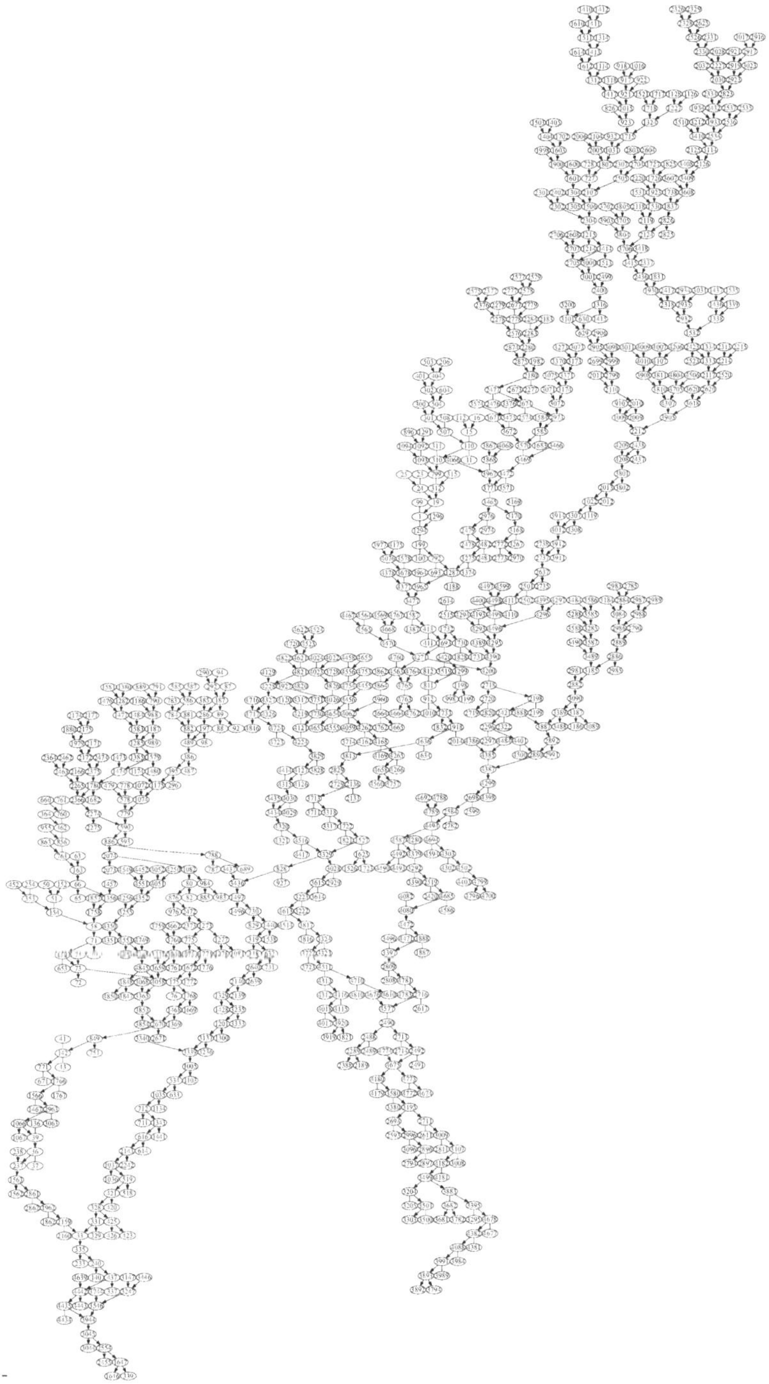

Fig. 8 A large contour tree visualized using graphviz for debug purposes. One edge had been mis-computed, and had to be identified manually

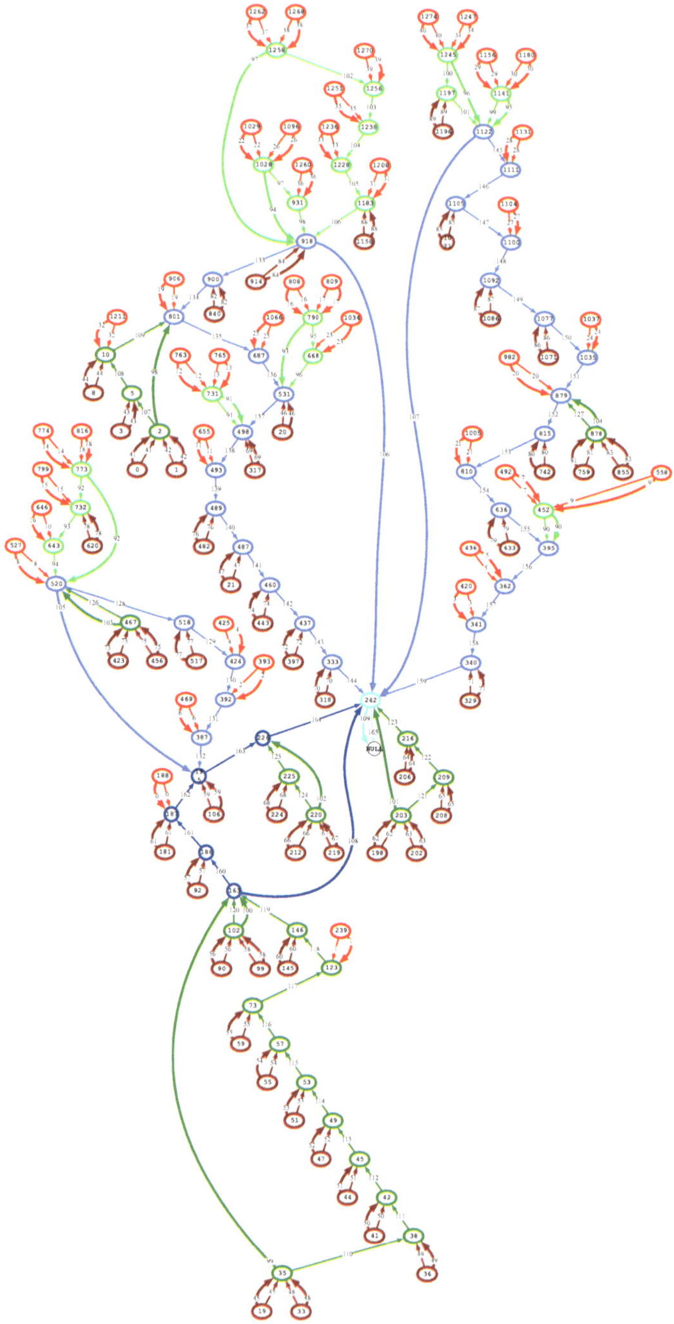

Fig. 9 Contour tree of a subset from Fig. 8, with additional pointers, and colour-coding for the iterations of an algorithm

outputs that show all of the internal cross-linked pointers (Fig. 9), using colour-coding to show which vertices are processed in which iteration.

None of these techniques is unprecedented in general algorithmic procedure. However, simple inspection of data structures in memory is particularly difficult with topological code, so secondary routines such as those described are strongly recommended to accelerate the debug process, and we now ask at an early stage what debug visualizations we will need.

8 Conclusions

In this paper, we have attempted to report on a crucial phase in algorithmic development for computational topology: selection of suitable test cases and debug procedure. As is apparent above, we have found that the skills of counter-example construction, and the consideration of pathological cases, have given the greatest insight into the mathematics, into our algorithms, and into the debug process.

We have already started work on the theoretical implications of the W-structure, and intend to report on in due course [15]. In the ideal case, we would also use this understanding to resolve the algorithmic implications, but these are non-trivial, and must wait until the current parallel work is fully reported.

Equally, we would encourage our colleagues to report on their test data and strategies, as these are crucial to developing modern topological algorithms, but are currently communicated by word of mouth, if at all. We note that, with the possible exception of the 5b data set, no standard benchmarks yet exist for topological algorithms, and suggest that this may be a fruitful direction for the community.

Acknowledgments In the UK, this work was supported by the Engineering and Physical Sciences Research Council (EPSRC) project EP/J013072/1. In the US, this work was supported by the Exascale Computing Project (17-SC-20-SC), a collaborative effort of the U.S. Department of Energy Office of Science and the National Nuclear Security Administration and by the Director, Office of Science, Office of Advanced Scientific Computing Research, of the U.S. Department of Energy under Contract No. DE- 392 AC02-05CH11231. In France, this work was supported by the BPI grant "AVIDO" (Programme 393 d'Investissements d'Avenir, reference P112017-2661376/DOS0021427).

References

1. Allili, M., Corriveau, D., Derivière, S., Kaczynski, T., Trahan, A.: Discrete dynamical system framework for construction of connections between critical regions in lattice height data. J. Math. Imaging Vis. **28**(2), 99–111 (2007)
2. Boyell, R.L., Ruston, H.: Hybrid techniques for real-time radar simulation. In: IEEE 1963 Fall Joint Computer Conference, pp. 445–458 (1963)
3. Carr, H., Duke, D.: Joint contour nets. IEEE Trans. Vis. Comput. Graph. **20**(8), 1100–1113 (2014)

4. Carr, H., Snoeyink, J., Axen, U.: Computing contour trees in all dimensions. Comput. Geom. Theory Appl. **24**(2), 75–94 (2003)
5. Carr, H., Snoeyink, J., van de Panne, M.: Simplifying flexible isosurfaces with local geometric measures. In: IEEE Visualization, pp. 497–504 (2004)
6. Carr, H., Snoeyink, J., van de Panne, M.: Flexible isosurfaces: simplifying and displaying scalar topology using the contour tree. Comput. Geom. Theory Appl. **43**(1), 42–58 (2010)
7. Carr, H., Geng, Z., Tierny, J., Chattopadhyay, A., Knoll, A.: Fiber surfaces: generalizing isosurfaces to bivariate data. Comput. Graph. Forum **34**(3), 241–250 (2015)
8. Carr, H., Weber, G., Sewell, C., Ahrens, J.: Parallel peak pruning for scalable SMP contour tree computation. In: IEEE Large Data Analysis and Visualization (LDAV) (2016)
9. Cox, J., Karron, D., Ferdous, N.: Topological zone organization of scalar volume data. J. Math. Imaging Vis. **18**(2), 95–117 (2003)
10. Edelsbrunner, H., Mücke, E.P.: Simulation of simplicity: a technique to cope with degenerate cases in geometric algorithms. ACM Trans. Graph. **9**(1), 66–104 (1990)
11. Edelsbrunner, H., Harer, J., Patel, A.K.: Reeb spaces of piecewise linear mappings. In: ACM Symposium on Computational Geometry, pp. 242–250 (2008)
12. Ellson, J., Gansner, E., Koutsofios, L., North, S.C., Woodhull, G.: Graphviz – open source graph drawing tools. In: International Symposium on Graph Drawing, pp. 483–484. Springer, Berlin (2001)
13. Gold, C., Cormack, S.: Spatially ordered networks and topographic reconstruction. In: ACM Symposium on Spatial Data Handling, pp. 74–85 (1986)
14. Helman, J., Hesselink, L.: Representation and display of vector field topology in fluid flow data sets. Computer **1**, 27–36 (1989)
15. Hristov, P., Carr, H.: W-Structures in contour trees. Accepted for publication in Topological Methods in Data Analysis and Visualization VI, Springer (2021)
16. Kaczynski, T.: Multivalued maps as a tool in modeling and rigorous numerics. J. Fixed Point Theory Appl. **4**(2), 151–176 (2008)
17. Laramee, R.: Using visualization to debug visualization software. IEEE Comput. Graph. Appl. **6**, 67–73 (2009)
18. Reeb, G.: Sur les points singuliers d'une forme de Pfaff complètement intégrable ou d'une fonction numérique. C. R. Acad. Sci. Paris **222**, 847–849 (1946)
19. SFB 382 of the German Research Council (DFG): Hydrogen atom. Available at http://schorsch.efi.fh-nuernberg.de/data/volume/
20. Theisel, H.: Designing 2D vector fields of arbitrary topology. Comput. Graph. Forum **21**(3), 595–604 (2002)
21. Tierny, J., Carr, H.: Jacobi fiber surfaces for bivariate Reeb space computation. IEEE Trans. Vis. Comput. Graph. **1**, 960–969 (2017)
22. Tierny, J., Gyulassy, A., Simon, E., Pascucci, V.: Loop surgery for volumetric meshes: Reeb graphs reduced to contour trees. IEEE Trans. Vis. Comput. Graph. **15**(6), 1177–1184 (2010)
23. Tierny, J., Favelier, G., Levine, J.A., Gueunet, C., Michaux, M.: The topology toolkit. IEEE Trans. Vis. Comput. Graph. **24**(1), 832–842 (2018)
24. Weber, G.H., Scheuermann, G., Hamann, B.: Detecting critical regions in scalar fields. In Visualization Symposium (VisSym), EUROGRAPHICS and IEEE TCVG (2003)
25. Zhang, E., Mischaikow, K., Turk, G.: Vector field design on surfaces. ACM Trans. Graph. **25**(4), 1294–1326 (2006)

Part III
Time-Varying Topology

Abstracted Visualization of Halo Topologies in Dark Matter Simulations

Karsten Schatz, Jens Schneider, Christoph Müller, Michael Krone, Guido Reina, and Thomas Ertl

Abstract This work focuses on particle-based ΛCDM (cold dark matter) simulations. The features of interest are clusters of dark matter particles, called halos. Halos are governed by the laws of motion and gravitation, and they may, consequently, merge over time. In this paper, we present visualization methods for the topology of the resulting tree-like accumulation history of the halos, as well as for the underlying halo data. We combine direct visualization methods of merger trees, in which trajectories over time are depicted in 3D space, with novel visual topological abstracts that are obtained by mapping time to one spatial axis while projecting halo positions on the remaining two axes. The user can explore and analyze both halos and merger trees through our unified visualization interface, which uses linked views complementing each other. All of our methods pay special attention to the periodic boundary conditions that are typically used during the underlying physical simulation.

1 Introduction

In cosmology, large-scale particle-based ΛCDM (cold dark matter) simulations are a popular way to gain knowledge about the structure formation in the early universe. The simulated dark matter movements allow the scientists to understand the formation of galaxies and other major stellar structures [31]. To deduce galaxy

K. Schatz (✉) · C. Müller · G. Reina · T. Ertl
Visualization Research Center (VISUS), University of Stuttgart, Stuttgart, Germany
e-mail: karsten.schatz@visus.uni-stuttgart.de; christoph.mueller@visus.uni-stuttgart.de; guido.reina@visus.uni-stuttgart.de; thomas.ertl@visus.uni-stuttgart.de

J. Schneider
College of Science and Engineering, Hamad Bin Khalifa University, Doha, Qatar
e-mail: jeschneider@hbku.edu.qa

M. Krone
Big Data Visual Analytics (BDVA), University of Tübingen, Tübingen, Germany
e-mail: michael.krone@uni-tuebingen.de

H. Carr et al. (eds.), *Topological Methods in Data Analysis and Visualization V*, Mathematics and Visualization, https://doi.org/10.1007/978-3-030-43036-8_8

formation models from these simulations, it is necessary to comprehend how the simulated dark matter particles attract each other and form clusters, called halos. These halos, again, exert gravitational forces on each other and merge to even larger structures. Using the history of these merging processes it is possible to build so-called *merger trees* that describe the mass assembly of the halos in a topological manner [10].

Visualization of the halos and their assembly histories can be a useful tool for understanding the simulated phenomena. To this end, visualizing merger trees and the accompanying halos were the subject of the IEEE SciVis Contest in 2015 [8]. The data published for the contest originate from the Dark Sky simulations [27], which are a series of some of the largest cosmological N-body simulations ever performed.

Contributions We present visualization methods both for dark matter halos as well as merger trees. In particular, we present a direct visualization of the halo positions in the merger trees. This visualization highlights each halo's trajectory and the individual merge events. To better understand the history of the halo trajectories in the merger tree, we furthermore present a novel method to generate visual *topological abstracts*. Inspired by the clarity of two-dimensional graph drawings, the topological abstracts map the time axis to one of the three spatial dimensions. In contrast to 2D drawings, however, our method is able to communicate the spatial relationship between a primary halo trajectory and secondary trajectories that merge with the primary. This is achieved by projecting the 3D halo positions onto the remaining two coordinate axes in a consistent fashion. Using the data provided for the aforementioned contest, we demonstrate that our visual topological abstracts highlight the mutual gravitational interactions between halo trajectories effectively. By rendering the resulting structures as colored tubes, our method further allows us to communicate the virial radius of halos as well as one additional attribute, for example velocity dispersion, spin, mass, etc. All of our methods specifically resolve periodic boundary conditions. Such boundary conditions are common for ΛCDM simulations and, if unaccounted for, will result in visual artifacts as halo trajectories leave through one face of the domain and re-enter at the opposite side.

2 Related Work

Large N-body simulations have become more and more popular over the last years. Reasons are the ever increasing computational power of the supercomputers used to run the simulation, coupled with algorithmic advances. As a result, the data output has grown tremendously in size and fidelity. In 2005, the Millennium Run simulated about ten billion particles [28], whereas the Dark Sky simulations of 2014 comprise around one trillion particles in the largest run [27]. This amounts to an increase of two orders of magnitude in less than 9 years.

The resulting data can be visualized showing the raw particles [7, 11, 23]. Rendering large point-based data sets is a well-known problem in scientific visualization and has been addressed many times, for example by Hopf and Ertl [9] or Rizzi et al. [22]. More related to this specific area, Ahrens et al. [1] presented an approach for comparative visualizations of cosmological data sets, where the results of different simulations can be compared. For more information on visualizing data from physical sciences, we would like to refer to the survey by Lipşa et al. [13].

Alternatively or to support the visualization of the raw particles, halos and other structures can be extracted from the data to gain insight into the data's underlying topology. To this end, a wide variety of halo identification tools have been proposed (e.g., Knebe et al. [12] or Onions et al. [18]). Most of these tools try to apply the density profile proposed by Navarro et al. [17] to the particle data sets to detect the halos. In the context of the Dark Sky simulations, halos have been extracted using a modified version of the ROCKSTAR algorithm [3] running in largely distributed processing environments. The modification allowed the direct generation of merger trees, which are normally computed separately using other algorithms [10, 31]. Especially the so-called *major mergers*, which are merge events of halos of similar size, are of interest to the community [4].

While the visualization of raw particles is rather common, visualizations of halo data or their merger trees are rare [29, 30], although the assembly history of dark matter halos is of keen interest, as stated by Wechsler et al. [33], for example. These may have been the main motivations for the task of the 2015 IEEE SciVis Contest [8], which actively requested the visualization of the merger trees. Scherzinger et al. [24], who won the contest, addressed this topic with a direct graph drawing of the merger tree graph. Merger trees of the accumulation topology can also be understood as a special case of contour trees, which themselves are special cases of Reeb graphs [20]. In contrast to merger trees, the construction [5] and visualization [19, 32] of contour trees is well-studied. Unlike contour trees, however, merger trees typically allow joining of nodes only, as splitting would indicate erroneous cluster assignment of a given halo.

3 Visualization Methods

Our data was released during the 2015 IEEE SciVis Contest. It comprises 89 time steps of raw dark matter particles (128^3), halos ($\sim$550,000), and spatio-temporal halo merger trees ($\sim$7500). In this work, we deliberately ignore the raw particle data and focus on the halos. Each halo comprises a wide variety of physical quantities, including position, radius, angular momentum, and other, more specialized variables. As depicted in Fig. 1, merger trees are constructed bottom-up, starting with the oldest halos. A halo is assigned a new ID for each time step, and old halos store the ID of the younger halo they evolve into. If halos merge, each of the older halos stores the ID of the younger, merged halo. Furthermore, halos store their virial radius and, additionally, eccentricity. The virial radius R_{vir} of a

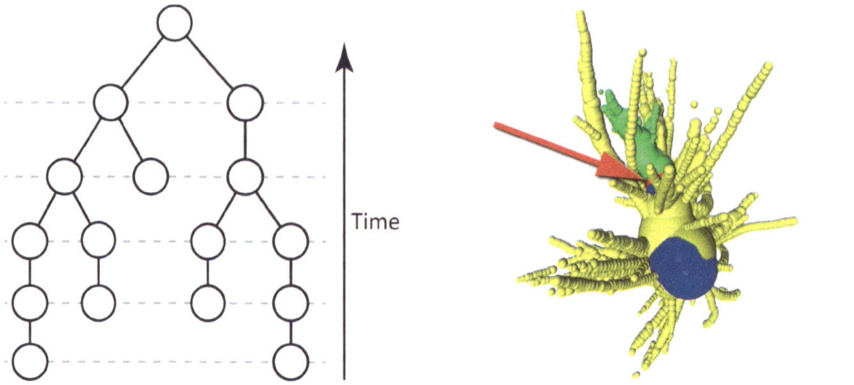

(a) Planar merger tree. (b) Merger tree with preserved halo positions.

Fig. 1 (**a**) Possible merger tree in two-dimensional representation, with the youngest halo as root node at the top. (**b**) Merger tree with halo positions, colored according to merging relationships. Direct temporal predecessors of the selected halo (red) are green, temporal successors are blue

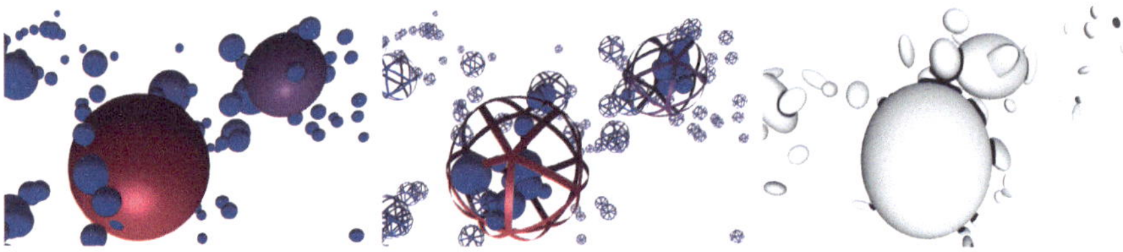

(a) Solid halo spheres (b) Host halos with cutouts (c) Depth darkening

Fig. 2 Halos rendered as spheres (**a**–**b**) or ellipsoids (**c**). (**a**) and (**b**) Colored according to velocity dispersion (blue=low, red=high). Cutouts in (**b**) allow exploring sub-halos. In (**c**), depth darkening depicts a large distance between the shadowing and the shadowed object

halo is the radius where the dark matter density inside the described sphere exceeds a certain threshold that depends on the chosen properties of the simulated universe (analogously, M_{vir} is the mass of dark matter inside the sphere).

The spatial locations of the halos and their size can be directly visualized in the 3D view (see Fig. 2a) of our application, which combines different linked views on the halo data. In this view, the user can also select a halo for inspecting its history (see Fig. 1b). This adds the predecessors and successors of the selected halo, showing the merger tree while preserving spatial positions and sizes of the halos. Furthermore, a line representation of the merger trees (see Fig. 3) can be selected to counter occlusion caused by large halos. However, halos can have thousands of predecessors, which still results in a very complex visual representation of limited utility. We therefore propose an additional view showing a topological abstract of the merger tree, which enables the user to explore the development of the selected halo over time while preserving the virial radii of the halos involved as well as limited information about the spatial relations between them.

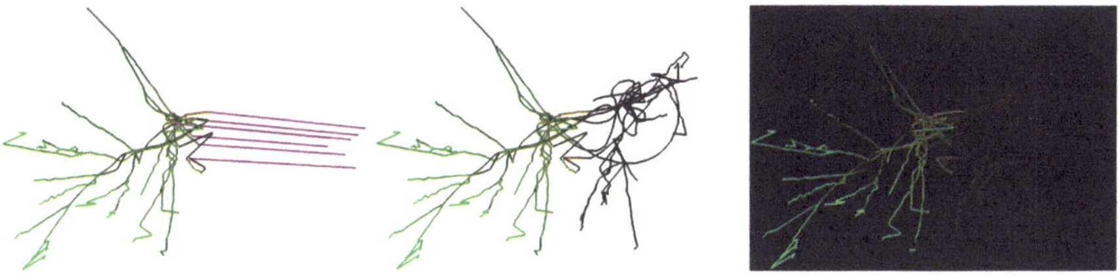

(a) Periodic boundary hints (b) Duplicated tree half (c) Emphasized desaturation

Fig. 3 Tree visualization behavior at bounding box borders. Positions of early time steps are green, positions of late time steps are red. In (**a**), the periodic boundary condition hints are depicted in magenta. In (**b**), the missing part (desaturated) is shown directly at the side of the available part. The desaturation is best visible with dark backgrounds (**c**)

3.1 Direct Halo Visualization

Our direct visualisation of the halo data is based on ellipsoids, which are tessellated and scaled on the GPU from an icosahedron created for each halo. Using geometry instead of impostors allows for displaying sub-halos by cutting out occluding geometry in a shader program (see Fig. 2b). Sub-halos are halos which are already bound by gravitation to a bigger host halo, but which have not yet merged. As a consequence, sub-halos typically lie inside their corresponding host halo. In addition to Blinn-Phong shading, we also use depth darkening by Luft et al. [15] to add an ambient occlusion-like shadow effect behind every object. The effect increases with the distance between the front objects and objects behind them (see Fig. 2c). The resulting additional depth cue eases depth perception of overlapping halos, while maintaining the overall appearance of non-occluded objects in the background.

The ellipsoidal representation of the halos can be used to construct a spatial representation of a merger tree by just showing the predecessors and successors of a selected halo (see Fig. 1b). Using color coding, we can, on the one hand, show the temporal relation of the elements in the tree to the selected halo. On the other hand, the user may also choose any of the halo properties from the simulation and map them to the color of both, merger tree and halos.

Besides preserving important 3D spatial relationships, this representation also allows for showing the virial radius in a natural way, namely as size of the ellipsoids. However, as can be seen in Fig. 1b, halos with a large virial radius may hide merge events when using solid ellipsoids. Sacrificing the size as attribute for mapping data, a line representation can solve this issue. It can be used to obtain an image containing the trajectories of all halos of the entire simulation run. To improve the perception of line orientation, we applied the line lighting technique by Mallo et al. [16] (see also Fig. 3).

This line representation nevertheless suffers from artifacts induced by periodic boundary conditions, which are common in cosmological simulations. The periodic boundary conditions result in one half of a tree appearing on one side of the bounding box of the simulation domain, while the other half appears on the opposite side. We provide two ways to remedy this situation. First, the user can enable a visual cue of where straight edges leaving or entering the bounding box appear also on the other side (see Fig. 3). This allows for a quick diagnosis of boundary effects. Second, to fully eliminate this kind of distraction, we also offer the option to duplicate the part of the tree that crosses the boundary. The duplicated part of the tree is shifted such that the merger tree is rendered as a whole. This shifted part of the tree, which is outside the simulation bounding box, is rendered with desaturated colors to indicate the duplication. Consequently, this allows the user to explore the tree as a whole, that is, as if there was no boundary between the two halves. The rendering of the duplicated version of the trees is performed in a geometry shader. For each tree that has to be duplicated, a second instance of each rendered line is spawned on the other side of the bounding box.

While rendering merger trees directly as depicted in Fig. 1 conveys the spatial geometry of each trajectory well, it is not free of problems. Firstly, since this is essentially a projection along the time axis, the history of each trajectory is hard to grasp. Secondly, additional visual clutter may be produced by instancing a primitive per halo and per time step, given that some merger trees comprise more than 4000 halos. While the first issue can be resolved by color-coding time, this solution may be inadequate if additional attributes are to be visualized on top of the geometry.

3.2 Topological Abstracts of Merger Trees

For the aforementioned reasons, we also propose visual topological abstracts of merger trees. The key idea of these visual abstracts is similar to the tx transform video technique popularized by Reinhart [21]. In essence, we map time to one of the spatial axes while projecting 3D positions to the remaining two spatial axes.

Given a primary trajectory $\mathbf{P}$ consisting of samples $\mathbf{p}_i$ in space-time $\mathbb{R}^3 \times \mathbb{R}_0^+$,

$$\mathbf{P} = \{\mathbf{p}_i\}_{i=1}^{N}, \quad \mathbf{p}_i \in \mathbb{R}^3 \times \mathbb{R}_0^+, \; N \in \mathbb{N}, \tag{1}$$

we would like to project samples of a second trajectory

$$\mathbf{S} = \{\mathbf{s}_i\}_{i=1}^{M}, \quad \mathbf{s}_i \in \mathbb{R}^3 \times \mathbb{R}_0^+, \; M \in \mathbb{N} \tag{2}$$

onto the XY plane. For this, we postulate the following goals:

(1) Convey the 3D distance $\|\mathbf{s}_i - \mathbf{p}_j\|_2$, where $\mathbf{p}_j$ is the sample on $\mathbf{P}$ closest in time to $\mathbf{s}_i$.
(2) Preserve the altitudinal angle between $\mathbf{s}_i$ and the primary trajectory $\mathbf{P}$.
(3) Show both the virial radius of the halo at samples $\mathbf{s}_i, \mathbf{p}_j$ and one additional, user-selectable attribute.

In addition, we want to straighten **P** such that we can map time onto the remaining Z axis. Goal (1) ensures that the user can judge the relative distance between **P** and **S**, while goal (2) provides visual information on the relative orientation of **S** with respect to **P**. Finally, goal (3) provides the user with a sense of the relative scale of each halo and conveys additional attributes like velocity dispersion, spin, mass, etc.

To achieve these goals, we begin by computing the orthonormal Frenet-Serret frame using only the three spatial components of each point $\mathbf{p}_i$:

$$\mathbf{T}_i = \frac{\partial}{\partial t}\mathbf{P}\bigg|_{\mathbf{p}_i}, \qquad \mathbf{N}_i = \mathbf{T}_i^{\perp}\left(\frac{\partial^2}{\partial t^2}\mathbf{P}\bigg|_{\mathbf{p}_i}\right), \qquad \mathbf{B}_i = \mathbf{T}_i \otimes \mathbf{N}_i,$$

followed by normalization,

$$\mathbf{T}_i \leftarrow \frac{\mathbf{T}_i}{\|\mathbf{T}_i\|_2}, \qquad \mathbf{N}_i \leftarrow \frac{\mathbf{N}_i}{\|\mathbf{N}_i\|_2}, \qquad \mathbf{B}_i \leftarrow \frac{\mathbf{B}_i}{\|\mathbf{B}_i\|_2}. \tag{3}$$

Here, $\mathbf{T}_i^{\perp}$ denotes projection into the orthogonal complement of $\mathbf{T}_i$ and $\otimes$ denotes the cross-product. $\mathbf{T}_i, \mathbf{N}_i, \mathbf{B}_i$ denote, respectively, tangent, normal, and binormal at sample $\mathbf{p}_i$ of the trajectory **P** with respect to time t (also see Fig. 4, left).

We then proceed by projecting the samples of the secondary trajectory. For each sample $\mathbf{s}_i$, we first find two samples $\mathbf{p}_j, \mathbf{p}_{j+1}$ such that the time of $\mathbf{p}_j$ is smaller and the time of $\mathbf{p}_{j+1}$ is greater or equal to the time of $\mathbf{s}_i$. Using linear interpolation, we then compute the point $\mathbf{p}'$ on **P** that is closest in time to $\mathbf{s}_i$. We also use linear interpolation followed by re-orthonormalization to compute a Frenet-Serret frame $\mathbf{T}, \mathbf{N}, \mathbf{B}$ at $\mathbf{p}'$. Using this fame, we project the spatial part of $\mathbf{s}_i$ along **T** onto the **NB**

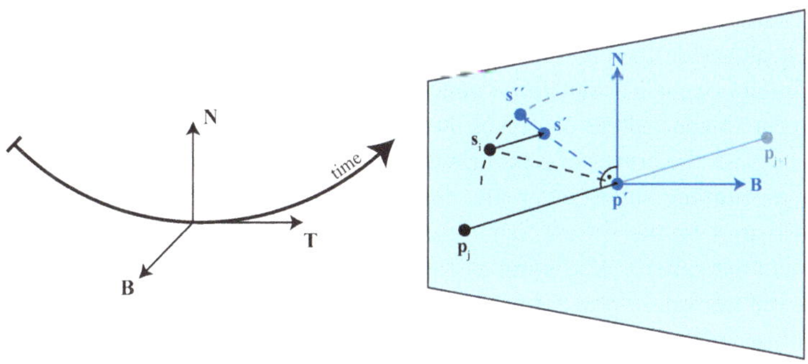

Fig. 4 Quantities arising in the projection of samples $\mathbf{s}_i$ onto the primary halo trajectory. **Left**: Frenet-Serret frame along a trajectory, parametrized by time. **Right**: Projecting a sample $\mathbf{s}_i$ onto $\mathbf{s}''$, on the **NB** plane through position $\mathbf{p}'$ of the primary trajectory (blue quantities on the **NB** plane)

plane. To simplify the following exposition, we again consider only the three spatial components of the involved vectors (refer also to Fig. 4, right).

$$\mathbf{s}' := \mathbf{p}' + \mathbf{T}^{\perp}\left(\mathbf{s}_i - \mathbf{p}'\right), \tag{4}$$

where $\mathbf{T}^{\perp}$ again denotes projection into the orthogonal complement of $\mathbf{T}$. While this ensures that goal (2) is achieved, a re-normalization establishes goal (1):

$$\mathbf{s}'' := \mathbf{p}' + \frac{\mathbf{s}' - \mathbf{p}'}{\|\mathbf{s}' - \mathbf{p}'\|_2} \left\|\mathbf{s}_i - \mathbf{p}'\right\|_2 . \tag{5}$$

This allows us to express $\mathbf{s}''$ solely in terms of $\mathbf{N}$ and $\mathbf{B}$. Straightening the trajectories $\mathbf{P}, \mathbf{S}$ after this projection thus becomes a simple matter of mapping the local coordinate axes: $\mathbf{B} \mapsto$ X, $\mathbf{N} \mapsto$ Y and $\mathbf{T} \mapsto$ Z. For samples of the primary and secondary trajectories, 3D positions are thus obtained as

$$\mathbf{p}_j \mapsto \begin{bmatrix} 0 \\ 0 \\ \text{time}\left(\mathbf{p}_j\right) \end{bmatrix}, \quad \mathbf{s}_i \mapsto \begin{bmatrix} \langle \mathbf{s}'', \mathbf{B} \rangle \\ \langle \mathbf{s}'', \mathbf{N} \rangle \\ \text{time}\left(\mathbf{s}_i\right) \end{bmatrix}. \tag{6}$$

Finally, to achieve goal (3), we extrude the trajectories obtained by the above embedding using a logarithmic mapping of the virial radius. The logarithmic mapping is necessary in order to cope with the high dynamic range of this attribute. We first assign radii to samples using a homeomorphic mapping,

$$r = \alpha \frac{\log\left(1 + R_{\text{vir}}\right)}{\log\left(1 + \max R_{\text{vir}}\right)}, \tag{7}$$

which results in $r \in [0, \alpha]$. R_{vir} is the so-called virial radius of a halo, whereas $\alpha \in \mathbb{R}^+$ is a user-defined parameter to control the overall thickness. To compute the final geometry, we use a transfinite generalization of the Power Diagram [2]. Power Diagrams are a generalization of Voronoi diagrams, in which the Euclidean distance is replaced with the power of a point with respect to a sphere. Given a point $\mathbf{p}$ and a sphere with center $\mathbf{c}$ and radius r, the power is defined as $\|\mathbf{p} - \mathbf{c}\|_2^2 - r^2$. The power vanishes if $\mathbf{p}$ is on the sphere, it is positive for $\mathbf{p}$ outside the sphere and negative for $\mathbf{p}$ inside the sphere. Assigning to each sample on $\mathbf{P}$ and $\mathbf{S}$ a radius allows us to compute the minimum power value on a regular 3D grid. The 0-isocontour, which can be obtained by contouring algorithms such as Marching Cubes [14], is then the desired surface fulfilling all of our goals. In our case, the generalization to the transfinite case is achieved by considering linear interpolations between two circles. Figure 5 depicts selected iso-contours of a single segment in 2D.

In addition to the actual scalar power field, we also keep track of the nearest site on the trajectories for each 3D grid position. This allows us to propagate attributes from the trajectory to the actual surface. Since time and virial radius are already

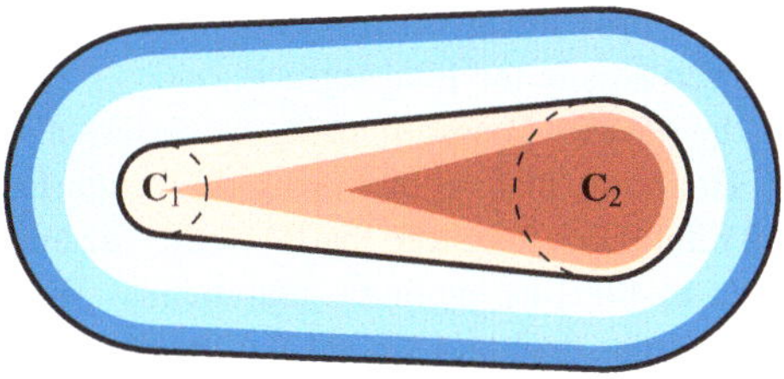

Fig. 5 Power isocontours of two circles $\mathbf{C}_1$ at (0, 0) with radius 0.5, and $\mathbf{C}_2$ at (5, 0) with radius 1. We plot isocontours at -0.5, -0.25, 0, 1, 2, 3. Blue hues correspond to positive values, whereas red hues correspond to negative values

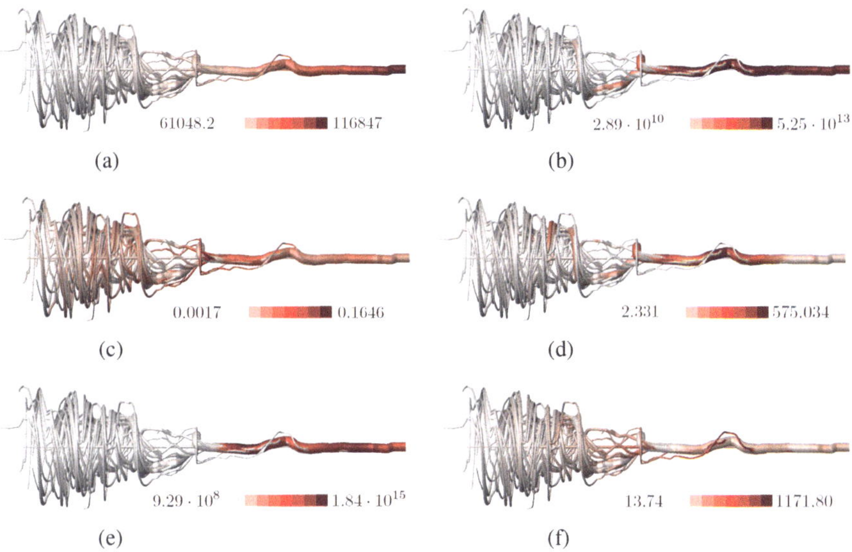

Fig. 6 A topological abstract with six selected, color-coded attributes (units in brackets). (**a**) Virial density (M_{sun}/h). (**b**) Virial mass (M_{sun}/h). (**c**) Spin. (**d**) Scale radius (kpc/h). (**e**) Angular momentum magnitude (($M_{sun} \cdot Mpc \cdot$km)/($h^2 \cdot$s)). (**f**) Velocity magnitude (km/s)

encoded in the geometry, an additional attribute can then be mapped to a color to satisfy the second half of goal (3) as depicted in Fig. 6.

3.3 Implementation Details

To ensure that the Frenet-Serret frame is smooth, we first compute $\mathbf{T}_i$ using central differences, followed by a smoothing step and then re-normalization prior to computing $\mathbf{N}_i$, $\mathbf{B}_i$ as described in Eq. (3). In order to obtain smooth trajectories, we subdivide segments $\mathbf{p}_i$, $\mathbf{p}_{i+1}$ and $\mathbf{s}_i$, $\mathbf{s}_{i+1}$. Due to the scaling of Eq. (5), which preserves 3D distances, simple linear interpolation between the end-points of line segments results in curved arcs in the final surface.

We implemented the Power Diagram using a variation of a parallel vector propagation [25, 26]. However, instead of storing the vector to the closest point as in the aforementioned references, we store the ID of the site and compute the power of a point with respect to the line segment analytically. We compute the Power Diagrams on a $1\,024^3$ regular grid and we set the maximum radius α to about 15 voxel lengths.

In order to obtain consistent topological abstracts, we resolve periodic boundary conditions as follows: We choose the first sample of the primary trajectory as anchor point. For every remaining sample of the primary trajectory, we consider the Euclidean distance between the previous sample and the current sample. Of the six wrap-around cases, we choose the one resulting in the minimum distance. The periodicity is resolved analogously for the secondary trajectories, except for the first sample of each trajectory where we minimize the distance to the anchor point.

4 Discussion

The visualization methods described in Sect. 3 are combined in an interactive, unified visualization framework (see also Fig. 7). Our framework uses linked views in which selections are propagated across all different visualization methods. This allows users to quickly move between halo and merger tree visualizations. In the halo view, the user has the option to select any of the visualizations depicted in Fig. 2. In particular, cutouts of tessellated host halo clusters allows to explore the interior of halos (Fig. 2b). Cutouts were chosen over transparency since they do not require depth sorting, which can quickly become prohibitive for large simulation data. Furthermore, depth darkening provides the user with additional depth cues (Fig. 2c). Unlike using fogging to provide depth cues, depth darkening does not change the appearance of non-occluded objects.

For visualizing halo trees, we provide two visualization approaches. The first approach is the direct visualization of halos and merger trees, explained in Sect. 3.1. To provide users with intuitive illustrations, special attention has been paid to the periodic boundary conditions commonly used in ΛCDM simulations. In our framework, edges of the trajectory crossing boundaries can either be highlighted (Fig. 3a) or be desaturated (Fig. 3c) to aid the user in understanding the domain extent. Alternatively, they can be fully resolved to help in understanding the underlying spatial structure (Fig. 3b). Our direct visualization approach provides a good spatial overview, but it may be difficult to assess a single halo trajectory's history. To address this issue, we provide as second visualization approach, namely our novel visual topological abstracts explained detailed in Sect. 3.2. While our topological abstracts maintain some spatial information (distance and altitudinal angle), they do not match the spatial clarity of the direct method. In contrast to the direct method, our topological abstracts provide the user with clear cues

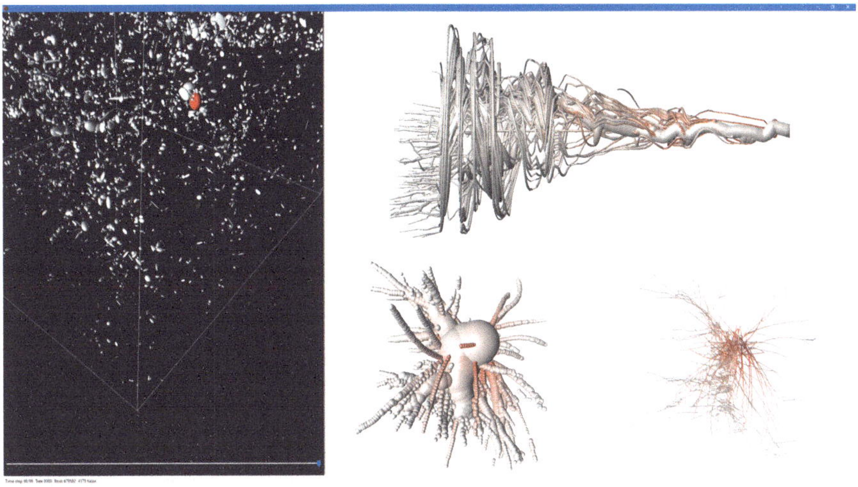

Fig. 7 Overview of our unified visualization framework. **Left**: Halo view with slider to select timestep. The user can pick a halo to display the corresponding merger tree's topology in the linked views to the right. **Right, top**: Topological abstract. **Right, bottom**: Direct merger tree visualization using spheres (left) and lines (right). Periodic boundary conditions are fully resolved. All topology visualizations show the color-coded velocity magnitude. The color-encoded value range is: 6.59 km/s 2009.36 km/s

about the history of each trajectory, mutual gravitational interactions, and geometric deformations of the halos. This is illustrated in Fig. 8, which shows three different visual topological abstracts generated with our method. The straight line in the center represents the primary halo trajectory with multiple secondary trajectories projected into the primary's Frenet-Serret space. As can be seen, merging secondary halos enter the first halo's gravitational field. The mutual gravitational pull coupled with the inertia of the involved halos frequently results in oscillations prior to halos merging that can be clearly seen in the visualization. Furthermore, as halos are absorbed in the primary halo, their virial radius commonly expands, which is also obvious from our visual abstract. For these reasons, our visualization framework combines complementing visualization methods using linked views. While the camera is freely movable in the halo visualization (Fig. 7 left), the two direct merger tree views (Fig. 7 bottom right) allow a linked rotation around the center. The rendering of the topological abstract (Fig. 7 top right) can be rotated around the main axis, that is, around the straightened primary halo trajectory. Panning and zooming is supported by all views.

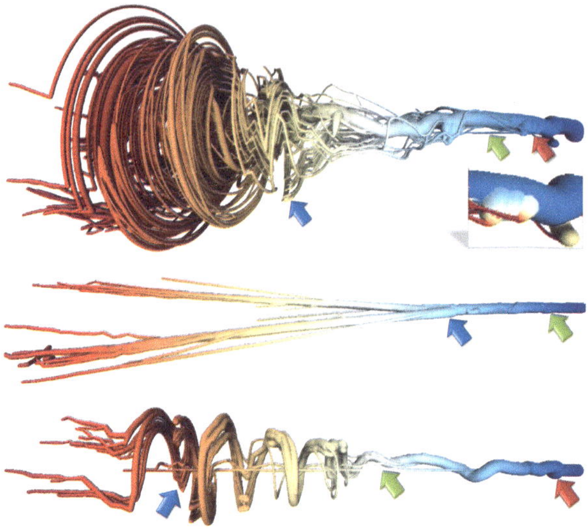

Fig. 8 Topological abstracts of three merger trees. **Top**: A halo merger tree comprising a total of 4175 halo samples. The primary halo trajectory (green arrow) is aligned in the center, with time increasing from left to right and mapped to the following color scale: min max. Mutual gravitational pull between primary and secondary trajectories coupled with the inertia of the halos result in clearly visible oscillations in the altitudinal angle (blue arrow). As secondary halos merge with the primary halo, their radius increases (red arrow) as a result of the gravitational pull of the primary trajectory. This can also be seen in the inset showing the velocity dispersion in this area, using the same coloring scheme in the value range [0.0, 1069.9] km/s. The part of the halo closer to the future host halo is accelerated faster than the part further away, resulting in an increase of the dispersion. **Middle**: The primary halo (green arrow) is the result of two large secondary halos merging (blue arrow). **Bottom**: Multiple secondary halos approach the primary halo trajectory (green arrow) and start oscillating (blue arrow) until they finally merge (red arrow)

5 Conclusion and Future Work

We presented a unified interactive visualization framework for rendering merger trees and their accompanying halo data. Our framework uses linked views with complementing visualization methods to comprehensively communicate the complex topology arising in merger trees effectively and efficiently.

The richness of cosmological data sets leaves plenty of tasks for the future: We would like to incorporate our halo topology visualization into existing visualization frameworks geared towards the visual exploration of large particle simulations, e.g., [22, 23]. Our direct visualization method could be improved further by the use of methods specialized to the rendering of dense line ensembles, such as the method of Everts et al. [6]. Furthermore, we also want to incorporate non-spatial visualizations of merger trees similar to the tracking graphs of Widanagamaachchi et al. [34].

While our framework is geared towards the understanding of the topology arising in merger trees, we would further like to grow our visualization into a semi-

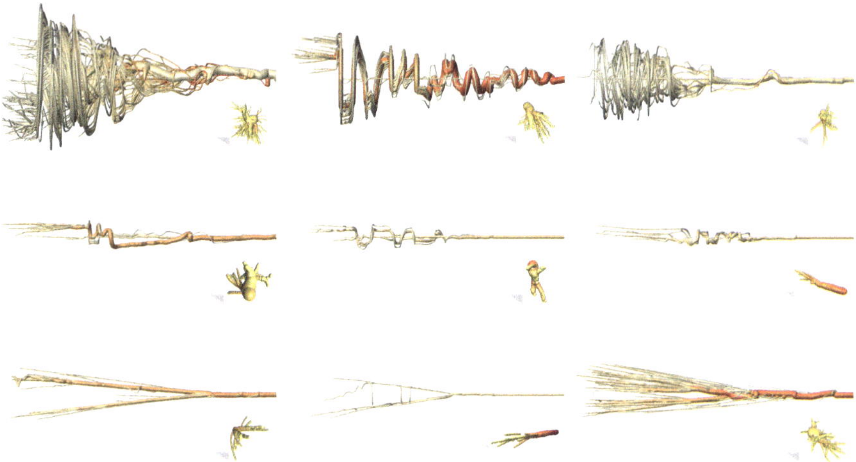

Fig. 9 Representative examples for visually identified topological classes, with direct rendering of the merger tree as inset. Shown value: Max. velocity (39.09 km/s 1164.39 km/s) **Top row**: Complex interactions of multiple secondary trajectories swirling around a primary halo forming early in time. **Middle row**: Smaller halo merger trees result in significantly less complex structures. **Bottom row**: If the primary halo forms later in time, secondary trajectories are straightened prior to the start of the primary trajectory

automated analysis tool. In collaboration with domain scientists, we would like to provide users with automatically extracted and scientifically significant statistics in regions of interest around the halos. In particular, we believe that our topological abstracts are helpful in visually classifying different types of halo formation, as depicted in Fig. 9. However, a comparative study of such classes is left for future work.

Acknowledgments This work was partially funded by Deutsche Forschungsgemeinschaft (DFG) as part of SFB 716 projects D.3 and D.4. J. Schneider is funded by the College of Science and Engineering, Hamad Bin Khalifa University, Qatar. M. Krone is funded by Carl-Zeiss-Stiftung.

References

1. Ahrens, J., Heitmann, K., Habib, S., Ankeny, L., McCormick, P., Inman, J., Armstrong, R., Ma, K.L.: Quantitative and comparative visualization applied to cosmological simulations. J. Phys. Conf. Ser. **46**, 526–534 (2006)
2. Aurenhammer, F.: Power diagrams: properties, algorithms and applications. SIAM J. Comput. **16**(1), 78–96 (1987)
3. Behroozi, P.S., Wechsler, R.H., Wu, H.Y.: The ROCKSTAR phase-space temporal halo finder and the velocity offsets of cluster cores. ApJ **762**(2), 109–129 (2013)

4. Behroozi, P., Knebe, A., Pearce, F.R., Elahi, P., Han, J., Lux, H., Mao, Y.Y., Muldrew, S.I., Potter, D., Srisawat, C.: Major mergers going Notts: challenges for modern halo finders. Mon. Not. R. Astron. Soc. **454**, 3020–3029 (2015)
5. Carr, H., Snoeyink, J., Van De Panne, M.: Flexible isosurfaces: simplifying and displaying scalar topology using the contour tree. Comput. Geom. **43**, 42–58 (2010)
6. Everts, M.H., Bekker, H., Roerdink, J., Isenberg, T.: Depth-dependent halos: illustrative rendering of dense line data. IEEE Trans. Vis. Comput. Graph. **15**(6), 1299–1306 (2009)
7. Fraedrich, R., Schneider, J., Westermann, R.: Exploring the millennium run – scalable rendering of large-scale cosmological datasets. IEEE Trans. Vis. Comput. Graph. **15**(6), 1251–1258 (2009)
8. Hentschel, B., Geveci, B., Turk, M., Skillman, S.: Scivis-contest 2015: visualize the universe (2015). http://darksky.slac.stanford.edu/scivis2015/
9. Hopf, M., Ertl, T.: Hierarchical splatting of scattered data. In: Proceedings of the IEEE VIS, p. 57 (2003)
10. Jiang, A.F., van den Bosch, F.C.: Generating merger trees for dark matter haloes: a comparison of methods. Mon. Not. R. Astron. Soc. **440**, 193–207 (2014)
11. Kähler, R., Hahn, O., Abel, T.: A novel approach to visualizing dark matter simulations. IEEE Trans. Vis. Comput. Graph. **18**(12), 2078–2087 (2012)
12. Knebe, A., Knollmann, S.R., Muldrew, S.I., et al.: Haloes gone mad: the halo-finder comparison project. Mon. Not. R. Astron. Soc. **415**(3), 2293–2318 (2011)
13. Lipşa, D.R., Laramee, R.S., Cox, S.J., Roberts, J.C., Walker, R., Borkin, M.A., Pfister, H.: Visualization for the physical sciences. In: Computer Graphics Forum, vol. 31, pp. 2317–2347 (2012)
14. Lorensen, W., Cline, H.: Marching cubes: a high resolution 3d surface reconstruction algorithm. ACM Comput. Graph. **21**(4), 163–169 (1987)
15. Luft, T., Colditz, C., Deussen, O.: Image enhancement by unsharp masking the depth buffer. In: ACM Trans. Graph., pp. 1206–1213 (2006)
16. Mallo, O., Peikert, R., Sigg, C., Sadlo, F.: Illuminated lines revisited. In: Proceedings of the IEEE VIS, pp. 19–26 (2005)
17. Navarro, J.F., Frenk, C.S., White, S.D.M.: A universal density profile from hierarchical clustering. ApJ **490**, 493–508 (1997)
18. Onions, J., Ascasibar, Y., Behroozi, P., et al.: Subhaloes gone Notts: spin across subhaloes and finders. Mon. Not. R. Astron. Soc. **429**(3), 2739–2747 (2013)
19. Pascucci, V., Cole-McLaughlin, K., Scorzelli, G.: Multi-resolution computation and presentation of contour trees. In: Proceedings of the IASTED VIIP, pp. 452–290 (2004)
20. Reeb, G.: Sur les points singuliers d'une forme de Pfaff completement intégrable ou d'une fonction numérique. CR Acad. Sci. Paris **222**, 847–849 (1946)
21. Reinhart, M.: Verfahren zur Transformation von Filmaufnahmen. Patent EP 0967572 A2 (1999)
22. Rizzi, S., Hereld, M., Insley, J., Papka, M.E., Uram, T., Vishwanath, V.: Large-scale parallel visualization of particle-based simulations using point sprites and level-of-detail. In: Proceeding of the EGPGV (2015). https://doi.org/10.2312/pgv.20151149
23. Schatz, K., Müller, C., Krone, M., Schneider, J., Reina, G., Ertl, T.: Interactive visual exploration of a trillion particles. In: Proceedings of the LDAV (2016). https://doi.org/10.1109/LDAV.2016.7874310
24. Scherzinger, A., Brix, T., Drees, D., Völker, A., Radkov, K., Santalidis, N., Fieguth, A., Hinrichs, K.H.: Interactive exploration of cosmological dark-matter simulation data. IEEE Comput. Graph. Appl. **37**(2), 80–89 (2017)
25. Schneider, J., Kraus, M., Westermann, R.: GPU-based real-time discrete Euclidean distance transforms with precise error bounds. In: Proceedings of the VISAPP, pp. 435–442 (2009)
26. Schneider, J., Kraus, M., Westermann, R.: GPU-based Euclidean distance transforms and their application to volume rendering. In: CCIS, pp. 215–228. Springer, New York (2010)
27. Skillman, S.W., Warren, M.S., Turk, M.J., Wechsler, R.H., Holz, D.E., Sutter, P.M.: Dark sky simulations: early data release (2014). arXiv:1407.2600

28. Springel, V., Evrard, A., Thomas, P., et al.: Simulating the joint evolution of quasars, galaxies and their large-scale distribution. Nature **435**(7042), 629–636 (2005)
29. Takle, J., Silver, D., Heitmann, K.: A case study: tracking and visualizing the evolution of dark matter halos and groups of satellite halos in cosmology simulations. In: Proceedings of the VAST, pp. 243–244 (2012)
30. Takle, J., Silver, D., Kovacs, E., Heitmann, K.: Visualization of multivariate dark matter halos in cosmology simulations. In: Proceedings of the LDAV, pp. 131–132 (2013)
31. Tweed, D., Devriendt, J., Blaizot, J., et al.: Building merger trees from cosmological n-body simulations. Astron. Astrophys. **506**, 647–660 (2009)
32. Weber, G.H., Bremer, P.T., Pascucci, V.: Topological cacti: visualizing contour-based statistics. In: Topological Methods in Data Analysis and Visualization, pp. 63–76. Springer, Cham (2012)
33. Wechsler, R.H., Bullock, J.S., Primack, J.R., Kravtsov, A.V., Dekel, A.: Concentrations of dark halos from their assembly histories. ApJ **568**, 52–70 (2002)
34. Widanagamaachchi, W., Christensen, C., Pascucci, V., Bremer, P.T.: Interactive exploration of large-scale time-varying data using dynamic tracking graphs. In: Proceedings of the LDAV, pp. 9 –17 (2012)

Persistence Concepts for 2D Skeleton Evolution Analysis

Bastian Rieck, Filip Sadlo, and Heike Leitte

Abstract In this work, we present concepts for the analysis of the evolution of two-dimensional skeletons. By introducing novel persistence concepts, we are able to reduce typical temporal incoherence, and provide insight in skeleton dynamics. We exemplify our approach by means of a simulation of viscous fingering—a highly dynamic process whose analysis is a hot topic in porous media research.

1 Introduction

There are many research problems that express themselves more in terms of topological structure than morphology. Typical examples of such processes include electrical discharge, the growth of crystals, and signal transport in networks. In this paper, we address *viscous fingering*, where the interface between two fluids is unstable and develops highly-dynamic "finger-like" structures. A prominent cause for such structures are setups where a fluid with lower viscosity (Fig. 1a–c, left) is injected into a fluid with higher viscosity (Fig. 1a–c, right). To analyze these processes, a straightforward approach employs traditional skeletonization techniques for extracting the topology of each time step independently. Here, we employ iterative thinning [11]. However, like all skeletonization techniques, the resulting skeletons tend to be to temporally incoherent because the extraction is susceptible to small variations and noise. We present persistence concepts to address these issues and provide insight into the underlying processes.

B. Rieck (✉) · H. Leitte (✉)
TU Kaiserslautern, Kaiserslautern, Germany
e-mail: rieck@cs.uni-kl.de; leitte@cs.uni-kl.de

F. Sadlo
Heidelberg University, Heidelberg, Germany
e-mail: sadlo@uni-heidelberg.de

H. Carr et al. (eds.), *Topological Methods in Data Analysis and Visualization V*, Mathematics and Visualization, https://doi.org/10.1007/978-3-030-43036-8_9

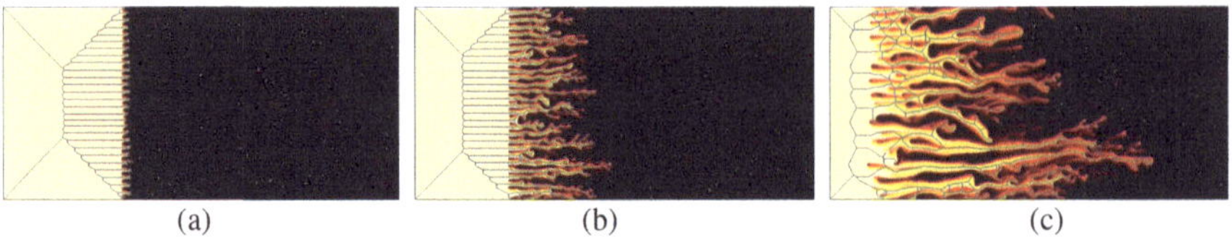

Fig. 1 Selected time steps (**a**)–(**c**) of a 2D viscous fingering simulation [15], with extracted skeleton (overlay). We used a conservative threshold for segmentation to suppress dark-red parts. (**a**) $t = 8$. (**b**) $t = 30$. (**c**) $t = 70$

2 Viscous Fingering

Even though the methods described in this paper are generically applicable to time-varying skeletons, we focus our analysis on skeletons that we extracted from *viscous fingering* processes. The term viscous fingering refers to the formation of structural patterns that appear when liquids of different viscosity are mixed. Under the right conditions, e.g., when water is being injected into glycerine, branch-like structures—the eponymous *viscous fingers*—begin to appear and permeate through the liquid of higher viscosity. Understanding the formation of these patterns is a prerequisite for the description of many natural processes, such as groundwater flows. Consequently, researchers are interested in setting up simulations that closely match the observations of their experiments.

Since each simulation uses a different set of parameters, summary statistics and comparative visualizations are required in order to assess how well a simulation describes an experiment. As a first step towards analyzing these highly-complex dynamics, we extract skeletons for each time step of a simulation or an experiment. In this paper, we introduce several concepts for assessing the inherent dynamics of these skeletons, permitting a comparative analysis.

2.1 Other Methods

In the context of analyzing viscous fingering, several other techniques exist. An approach by Lukasczyk et al. [12], for example, uses tracking graphs to visualize the spatio-temporal behavior of such processes. In a more general context, discrete Morse theory could be applied to detect persistent structures in gray-scale images [5]. The applicability of these approaches hinges on the data quality, however. Our experimental data suffers from a high noise level in which many smaller fingers cannot be easily identified by the other approaches. This is why we decided to focus on conceptually simpler skeletonization techniques for now.

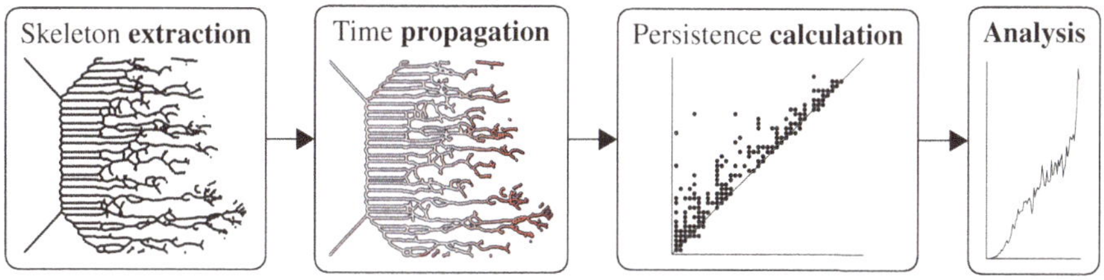

Fig. 2 The basic pipeline of our approach. The first step, i.e., skeleton extraction, strongly depends on the desired application. Likewise, the analysis step can comprise different diagrams, summary statistics, and goals. Individual parts of the pipeline are replaceable, making our approach highly generic. Our current implementation uses an algorithm by Zhang and Suen [16] for skeleton extraction (Sect. 3.1.3). The subsequent propagation of creation times between time steps along all branches of the skeleton uses the methods described in the same section. From this extended skeleton, Sect. 3.3 describes how to derive numerous persistence diagrams. Following this, we define multiple activity indicators based on these diagrams in Sect. 3.4. Finally, Sect. 4 presents an analysis of different data sets under different aspects

3 Overview and Methods

In this paper, we implement a pipeline that comprises the whole range of the analysis process of a series of time-varying skeletons. Figure 2 shows a schematic illustration and points to the corresponding sections in which individual parts are described. We provide an open-source implementation (in `Python`) of the pipeline on GitHub.[1] The repository includes all examples, data, and instructions on how to reproduce our experiments. For the analysis of our persistence diagrams, we implemented tools that build upon `Aleph`,[2] an open-source library for persistent homology calculations. We stress that our implementation is a proof of concept. Its computational bottleneck is the brute-force matching (which could be improved by using an approximate matching algorithm) that is required as a precursor to creation time propagation. More precisely, calculating all matches over all time steps takes between 2 h and 6 h, while the subsequent propagation of creation times takes 82 s (example data, 839 px $\times$ 396 px, 84 time steps), 384 s (measured data, 722 px $\times$ 1304 px, 58 time steps), and 524 s (simulation data, 1500 px $\times$ 1000 px, 37 time steps). Finally, persistence diagram creation requires 100 s (example data), 183 s (simulation data), and 926 s (measured data), respectively. The time for calculating activity indicators (Sect. 3.4), e.g., total persistence, is negligible, as the persistence diagrams only contain a few hundred points. Please refer to Sect. 4 for more information about the individual data sets.

Subsequently, we will first briefly discuss skeleton extraction—both in terms of sets of pixels as well as in terms of graphs. Next, we explain the necessary steps for obtaining information about the "creation time" of pixels and how to propagate said

[1] https://github.com/Submanifold/Skeleton_Persistence.

[2] https://github.com/Submanifold/Aleph.

information over all time steps in order to obtain evolution information. Based on this, we derive and exemplify several concepts motivated by topological persistence.

3.1 Skeleton Extraction and Propagation of Pixel Creation Time

Iterative thinning provides skeletons from binary images in a pixel-based format. A sequence of skeletons thus gives rise to a sequence of pixel sets $\mathscr{P}_0$, $\mathscr{P}_1$, ..., $\mathscr{P}_k$, each corresponding to a time step $t_0, t_1, \ldots, t_k$. We employ 8-neighborhood connectivity around each pixel, i.e., the set of all neighbors including the diagonal ones, to convert each pixel set $\mathscr{P}_i$ into a graph $\mathscr{G}_i$. Depending on the degree d of each vertex in $\mathscr{G}_i$, we can classify each pixel as being either a regular point ($d = 2$), a start/end point ($d = 1$), or a branch point ($d \geq 3$). This also permits us to define *segments* formed by connected subsets of regular pixels.

3.1.1 Pixel Matching

Since the skeleton changes over time, we need to characterize the creation time of each pixel, i.e., the time step t_i in which it initially appears. Moreover, we want to permit that a pixel "moves" slightly between two consecutive time steps in order to ensure that drifts of the skeleton can be compensated. Our experiments indicate that it is possible to obtain consistent creation times for the pixels based on their nearest neighbors, regardless of whether the simulation suffers from a coarse time resolution or not. Given two time steps t_i, t_{i+1}, we assign every pixel $p \in \mathscr{P}_i$ the pixel $p' \in \mathscr{P}_{i+1}$ that satisfies

$$p' := \arg \min_{q \in \mathscr{P}_{i+1}} \operatorname{dist}(p, q), \tag{1}$$

where dist$(\cdot)$ is the Euclidean distance. Likewise, we assign every pixel in $\mathscr{P}_{i+1}$ its nearest neighbor in $\mathscr{P}_i$, which represents a match from $\mathscr{P}_{i+1}$ to $\mathscr{P}_i$. This yields a set of directed matches between $\mathscr{P}_i$ and $\mathscr{P}_{i+1}$. Each pixel is guaranteed to occur at least once in the set. We refer to matches from $\mathscr{P}_i$ to $\mathscr{P}_{i+1}$ as *forward* matches, while we refer to matches in the other direction as *backward* matches. A match is *unique* if the forward and backward match connect the same pair of pixels. Figure 3 depicts matches for selected time steps and illustrates the movement of pixels.

3.1.2 Pixel Classification

We now classify each pixel in time step t_{i+1} according to the forward matches between $\mathscr{P}_i$ and $\mathscr{P}_{i+1}$, as well as the backward matches between $\mathscr{P}_i$ and $\mathscr{P}_{i+1}$.

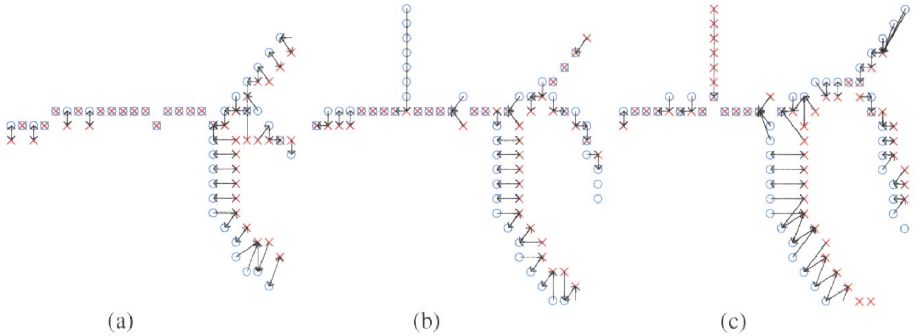

Fig. 3 An excerpt demonstrating matches between two time steps. Some of the pixels of the current time step (blue circles) overlap with pixels from the previous time step (red crosses). We use arrows to indicate forward and backward matches. (**a**) $t = 72$. (**b**) $t = 73$. (**c**) $t = 74$

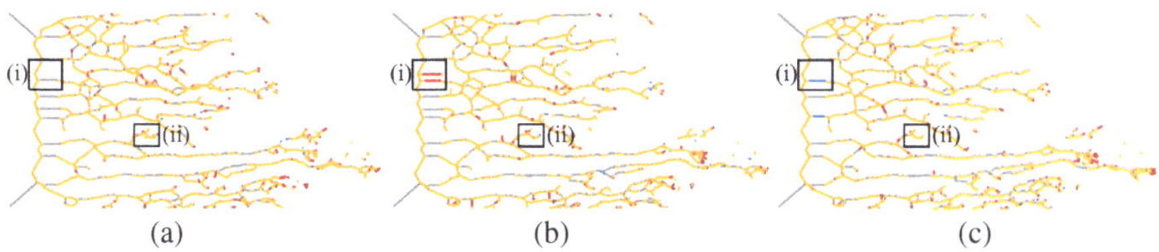

Fig. 4 Classification of all pixels into growth pixels (red filled circle), decay pixels (blue filled circle), known pixels (gray filled circle), and irregular pixels (yellow filled circle). The abrupt appearance (**b**) or disappearance (**c**) of segments is a challenge for skeleton extraction and tracking. (**a**) $t = 68$. (**b**) $t = 69$. (**c**) $t = 70$

We call a pixel *known* if their match is unique, i.e., there is exactly one forward and one backward match that relate the same pixels with each other. Known pixels are pixels that are already present in a previous time step with a unique counterpart in time step t_{i+1}. Similarly, we refer to a pixel in $\mathscr{P}_{i+1}$ as a *growth* pixel if there is a unique match in $\mathscr{P}_i$ and at most one forward match from some other pixel in $\mathscr{P}_i$. Growth pixels indicate that new structures have been created in time step t_{i+1}, or that existing structures have been subject to a deformation. The counterpart to a growth pixel is a *decay* pixel in $\mathscr{P}_{i+1}$, which is defined by a unique match in $\mathscr{P}_i$ and at most one backward match to the same pixel in $\mathscr{P}_i$ from another pixel in $\mathscr{P}_{i+1}$. Decay pixels indicate that a skeleton region has been lost in time step t_{i+1}. We refer to all other pixels as *irregular*. In our experiments, irregular pixels, which are caused by small shifts between consecutive time steps, comprise about 60% of all pixels. As we subsequently demonstrate, we are able to assign consistent creation times despite the prevalence of irregular pixels. Figure 4 depicts classified pixels for consecutive time steps. It also demonstrates that skeletons may be temporally incoherent: pixels in region (i) only exist for a single time step, forming long but short-lived segments. Pixels in region (ii), by contrast, form short but long-lived segments. We want to filter out segments in region (i), while keeping segments in region (ii) intact. This requires knowledge about pixel creation times.

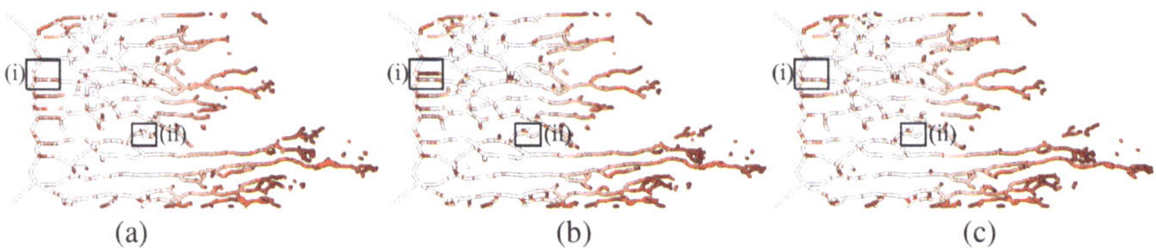

Fig. 5 Propagated age per pixel, using a white–red color map. The skeleton inconsistencies in region (i) impede the temporal coherence of neighboring pixels. (**a**) $t = 68$. (**b**) $t = 69$. (**c**) $t = 70$

3.1.3 Propagating Creation Times

Initially, each pixel in $\mathscr{P}_0$ is assigned a creation time of 0. Next, we classify the pixels in each pair of consecutive time steps t_i and t_{i+1} as described above. For known pixels, we re-use the creation time of t_i. For growth pixels, we distinguish two different cases: (1) If a growth pixel in time step t_{i+1} is not the target of a forward match from time step t_i, we consider it to be a new pixel and hence assign it a creation time of t_{i+1}. (2) Else, we re-use the creation time just as for known pixels. This procedure ensures that we are conservative with assigning "new" creation times; it turns out that a small number of growth pixels with increased creation times is sufficient for propagating time information throughout the data. For all other types of pixels, we assign them the minimum of all creation times of their respective matches from $\mathscr{P}_i$, ignoring the direction of the matching. Again, this is a conservative choice that reduces the impact of noise in the data.

Thus, every pixel in every time step has been assigned a creation time. This time refers to the first time step in which the pixel was unambiguously identified and appeared. By propagating the creation time, we ensure that skeletons are allowed to exhibit some movement between consecutive time steps. Figure 5 depicts the creation times for several time steps. For temporally coherent skeletons, recent creation times (shown in red) should only appear at the end of new "fingers". We can see that the brief appearance of segments causes inconsistencies. Ideally, the creation time of pixels should vary continuously among a segment.

3.2 Improving Temporal Coherence

To improve temporal coherence, i.e, creation times of adjacent pixels, we observe that inconsistencies are mainly caused by a small number of growth pixels along a segment. These are a consequence of a "drift" in pixel positions over subsequent time steps, which our naive matching algorithm cannot compensate for. A simple neighborhood-based strategy is capable of increasing coherence, though: for each growth pixel, we evaluate the creation times in its 8-neighborhood. If more than 50% of the neighbors have a different creation time than the current pixel, we replace its creation time by the *mode* of its neighbors' creation times. This strategy is remi-

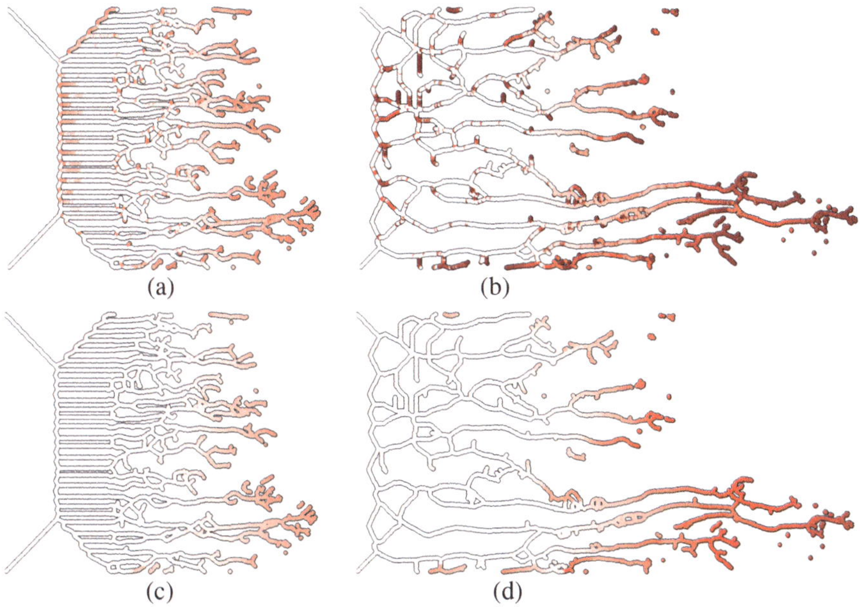

Fig. 6 Pixel creation times at two selected time steps. Recent creation times are shown in shades of red. We can see that the "front" of the fingers is always recent, while the oldest structures have been created at the very beginning. This example also demonstrates how the temporal coherence of creation times can be improved. (**a**) $t = 42$ (no coherence). (**b**) $t = 84$ (no coherence). (**c**) $t = 42$ (coherence). (**d**) $t = 84$ (coherence)

niscent of *mean shift smoothing* [4]. Figure 6 compares the original and improved creation times for two time steps. Ideally, all segments should exhibit a gradient-like behavior, indicating that their structures have been expanded continuously. We see that this is only true for the longest segments. Erroneous creation times are an inevitable byproduct of instabilities in skeleton extraction, which can be mitigated through persistence-based concepts.

3.3 Persistence Concepts

Persistence is a concept introduced by Edelsbrunner et al. [6–8]. It yields a measure of the range (or scale) at which topological features occur in data and is commonly employed to filter or simplify complex multivariate data sets [14]. For skeletons, i.e., *graphs*, the standard topological features are well known, comprising connected components and cycles. While these features are useful in classifying complex networks [2], for example, they do not provide sufficient information about skeleton evolution processes because they cannot capture the growth of segments. Hence, instead of adopting this viewpoint, we derive several concepts that are inspired by the notion of persistence. A crucial ingredient for this purpose is the availability of creation times for every pixel in every time step.

3.3.1 Branch Inconsistency

Using the graph $\mathcal{G}_i$ for a time step t_i, we know which pixels are branch points, i.e., points where multiple segments meet. Let c_b be the creation time of such a branch point, and let $c_1, c_2, \ldots$ refer to the creation times of the first adjacent point along each of the segments meeting at the branch point. We define the *branch inconsistency* for each branch–segment pair as $|c_i - c_b|$, and we refer to the diagram formed by the points (c_b, c_i) as the branch persistence diagram. The number of points in the branch inconsistency diagram indicates how many new branches are being created in one time step. Moreover, it can be used to prune away undesired segments in a skeleton: if the branch inconsistency of a given segment is large, the segment is likely an artifact of the skeletonization process—thinning algorithms often create segments that only exist for a single time step. Overall, those segments thus have a late creation time. In contrast to the persistence diagrams in topological data analysis, where closeness to the diagonal indicates noise, here, points that are *away* from the diagonal correspond to erroneous segments in the data. Points *below* the diagonal are the result of inconsistent creation times for some segments—a branch cannot be created *before* its branch point.

Figure 7 shows the branch inconsistency diagram and colored skeletons for $t = 69$. It also depicts how to filter segments with a large branch inconsistency, which already decreases the number of noisy segments. Please refer to the accompanying video for all branch inconsistency values.

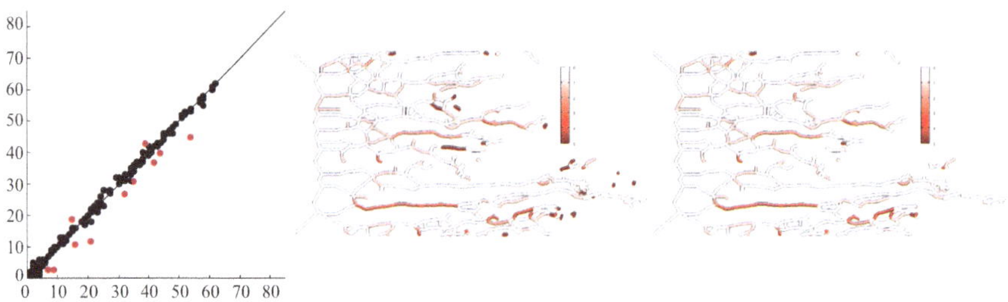

(a) (b) (c)

Fig. 7 Branch inconsistency diagram and branch inconsistency values on the skeleton for $t = 69$. The diagram indicates that most branches are temporally coherent. Some of them are removed from the diagonal (or below the diagonal), which may either indicate inconsistencies in skeleton tracking or cycles. Removing segments with a branch inconsistency ≥ 5 (red dots in the diagram, dark red segments in the skeleton) can be used to filter the skeleton. (**a**) Branch inconsistency. (**b**) Skeleton. (**c**) Skeleton, filtered

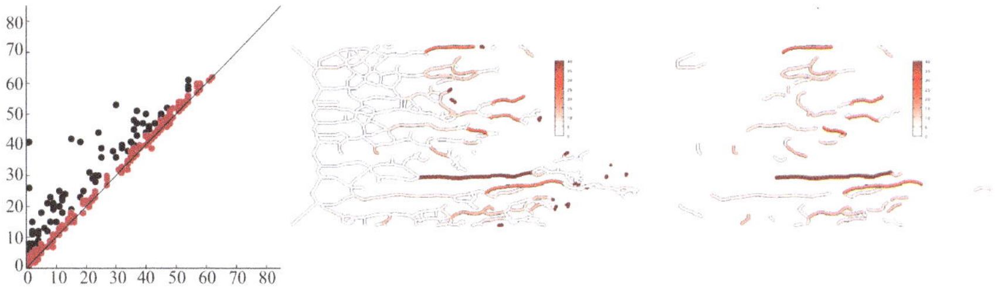

Fig. 8 Age persistence diagram and age persistence values on the skeleton for $t = 69$. Numerous segments towards the "front" of the fingers appear to be active here. Removing all segments whose age persistence is ≤ 5 (red points in the diagram) leaves us with the most active segments. (**a**) Age persistence. (**b**) Skeleton. (**c**) Skeleton, filtered

3.3.2 Age Persistence

Analogously to branch inconsistency, we obtain an *age persistence* diagram for each branch–segment pair when we use the maximum creation time of points along each segment. Age persistence is capable of measuring whether a segment is young or old with respect to its branch point. Here, the "persistence" of each point is an indicator of how much the skeleton grows over multiple time steps: if segments stagnate, their points remain at the same distance from the diagonal. If segments continue to grow, however, their points will move away from the diagonal.

Figure 8 shows the age persistence diagram and the age persistence values on the skeleton for $t = 69$. The filtered skeleton only contains the most active segments, which facilitates tracking. We can combine branch inconsistency and age persistence to remove fewer segments than in Fig. 7c. For example, we could remove segments that correspond to points below the diagonal of the branch inconsistency diagram and keep those for which *both* branch inconsistency and age persistence are high. These segments commonly correspond to cycles that were formed during the evolution of the skeleton. An isolated analysis of branch inconsistency is unable to detect them. Figure 9 depicts the results of such a combined filtering operation.

3.3.3 Growth Persistence

We define the *growth persistence* of a segment in $\mathcal{G}_i$ as the difference between the maximum creation time $t_{\max}$ of its pixels and the current time step t_i. Intuitively, this can be thought of performing "time filtration" of a simplicial complex, in which simplices may be created *and* destroyed (notice that such a description would require zigzag persistence for general simplicial complexes). A small value in this quantity indicates that the segment is still growing, while larger values refer to segments that stagnate. Growth persistence is useful to highlight segments that are relevant for tracking in viscous fingering processes. In contrast to the previously-

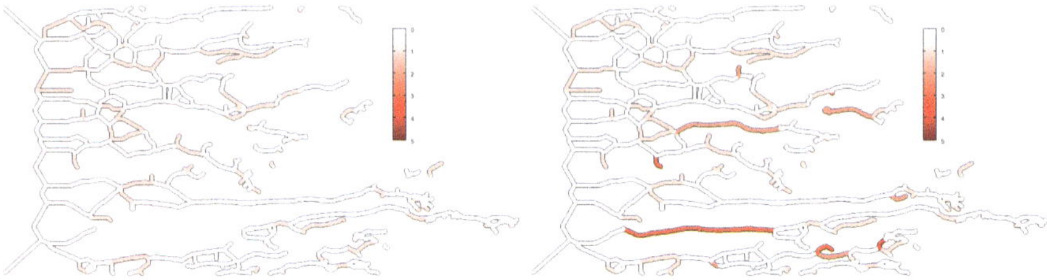

(a) (b)

Fig. 9 (**a**) Filtering segments using branch inconsistency may destroy longer segments. (**b**) If we combine both branch inconsistency and age persistence, keeping only those segments whose age persistence is high or whose branch inconsistency is low, we can improve the filter results by removing noisy segments while keeping more cycles intact

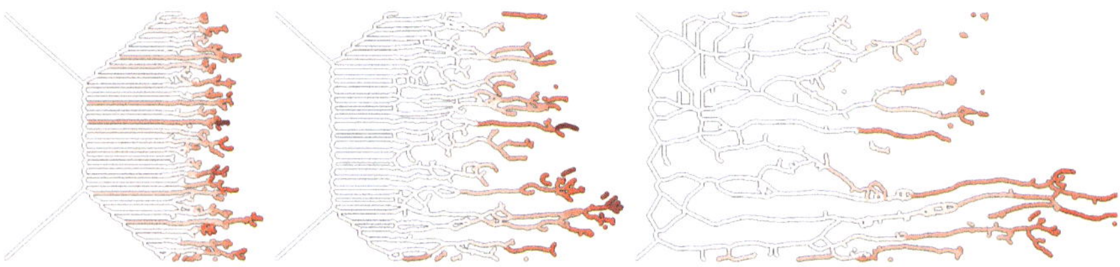

(a) (b) (c)

Fig. 10 Growth persistence values. Red segments are highly active in the evolution of the skeleton. In this example, red segments are mostly those that are at the tips of individual "fingers". (**a**) $t = 21$. (**b**) $t = 42$. (**c**) $t = 84$

defined persistence concepts, growth persistence is only defined per segment and does not afford a description in terms of a persistence diagram. Figure 10 depicts the growth persistence of several time steps. Red segments are growing fast or have undergone recent changes, such as the creation of cycles. A low branch persistence in segments, coupled with a low growth persistence corresponds to features that are "active" during skeleton evolution. Please refer to the accompanying video for the evolution of growth persistence.

3.4 Activity Indicators

In order to capture the dynamics of skeleton evolution, we require a set of activity indicators. They are based on the previously-defined concepts and can be used to quickly summarize a time series of evolving skeletons.

3.4.1 Total Persistence

There are already various summary statistics for persistence diagrams. The *second-order total persistence* $\text{pers}(\mathscr{D})$ [3] of a persistence diagram $\mathscr{D}$ is defined as

$$\text{pers}(\mathscr{D})_2 := \left(\sum_{(c,d)\in\mathscr{D}} \text{pers}^2(c,d) \right)^{\frac{1}{2}}, \tag{2}$$

i.e., the sum of powers of the individual persistence values (i.e. coordinate differences) of the diagram. Total persistence was already successfully used to assess topological activity in multivariate clustering algorithms [13].

Here, the interpretation of total persistence depends on the diagram for which we compute it. Recall that in a branch inconsistency diagram, points of high "persistence" indicate inconsistencies in branching behavior. Total persistence thus helps detect anomalies in the data; see Fig. 12 for a comparison of total branch persistence in different data sets. For age persistence, by contrast, high persistence values show that a skeleton segment is still actively changing. The total age persistence hence characterizes the dynamics of the data, e.g., whether many or few segments are active at each time step. Figure 14 depicts a comparison of total age persistence in different data sets.

3.4.2 Vivacity

We also want to measure the "vivacity" of a viscous fingering process. To this end, we employ the growth persistence values. Given a growth threshold t_G, we count all growth pixels with $\text{pers}_G \leq t_G$ and divide them by the total number of pixels in the given time step. This yields a measure of how much "mass" is being created at every time step of the process. Similarly, we can calculate vivacity based on *segments* in the data. However, we found that this does not have a significant effect on the results, so we refrain from showing the resulting curves. Figure 15 depicts vivacity curves for different data sets with $t_G = 10$.

4 Analysis

Having defined a variety of persistence-based concepts, we now briefly discuss their utility in analyzing time-varying skeleton evolution. In the following, we analyze three different data sets: (1) the *example* data set that we used to illustrate all concepts, (2) a *measured* data set, corresponding to a slowly-evolving viscous fingering process, (3) and a *simulation* of the example data set. Figure 11 depicts individual frames of the latter two data sets. The measured data set is characterized

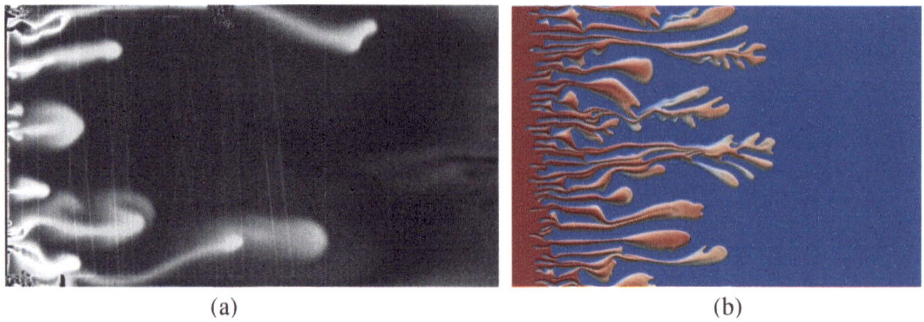

Fig. 11 Selected still images from the two remaining data sets. The measured data in (**a**) exhibits artifacts (parallel lines) that are caused by the experimental setup. The simulation data (**b**), by contrast, does not contain any noise. (**a**) Measured data set, $t = 33$. (**b**) Simulation data set, $t = 26$

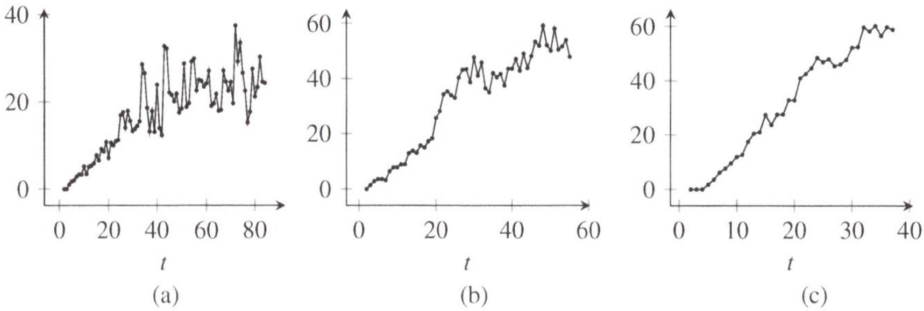

Fig. 12 A comparison of total persistence of the branch inconsistency diagram for three different data sets. The first data set (**a**) exhibits more anomalies; these are indicated by "jumps" in the total persistence curve. (**a**) Example. (**b**) Measured. (**c**) Simulation

by a viscous fingering process whose fingers evolve rather slowly over time. Moreover, this experiment, which was performed over several days, does not exhibit many fingers. The simulation data, by contrast, aims to reproduce the dynamics found in the example data set; hence, it contains numerous fast-growing fingers. Please refer to the accompanying videos for more details.

4.1 Anomaly Detection

To detect anomalies in skeleton extraction and tracking, we calculate the total persistence of the branch inconsistency diagram. Figure 12 compares the values for all data sets. We observe that the example data set, Fig. 12a, exhibits many "jumps" in branch inconsistency. These are time steps at which the skeleton (briefly) becomes inconsistent, e.g., because a large number of segments disappears, or many small cycles are created. At $t = 43$ and $t = 72$ (both local maxima in the diagram), for

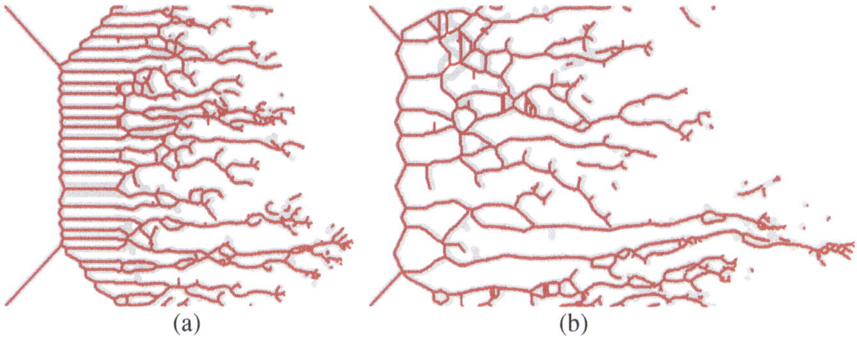

Fig. 13 Comparing the previous (gray) and the current time step (red) based on total persistence of the branch inconsistency diagram helps uncover problems with skeleton extraction. (**a**) $t = 43$. (**b**) $t = 72$

example, we observe changes in the number of cycles as well as the appearance of numerous segments of various lengths, which makes it harder to assign consistent creation times according to Sect. 3.1. Figure 13 depicts the changes in skeleton topology at these time steps. The other two data sets contain fewer anomalies. For the measured data, this is caused by lower propagation velocities and fewer "fingers" in the data. For the simulation data, this is due to a better separation of individual fingers, caused by the synthetic origin of the data.

4.2 Active Branches

We use total age persistence to assess the rate at which existing branches move. Figure 14 compares the data sets, showing both the original total age persistence values as well as a smooth estimate, obtained by fitting Bézier curves [9] to the sample points. In Fig. 14c, the simulated origin of the data is evident: while the other data sets exhibit changes in the growth rate of total age persistence, the simulation data clearly exhibits almost constant growth. Moreover, we observe that the measured data in Fig. 14b has a period of constant growth for $t \in [20, 40]$, while the example data displays a slightly diminished growth rate for $t \in [25, 65]$, only to pick up at the end. Age persistence may thus be used to compare the characteristics of different skeleton evolution processes.

4.3 Quantifying Dissimilarity

To quickly quantify the dissimilarity between different curves, e.g., the vivacity curves that we defined in Sect. 3.4.2, we can use *dynamic time warping* [1], a

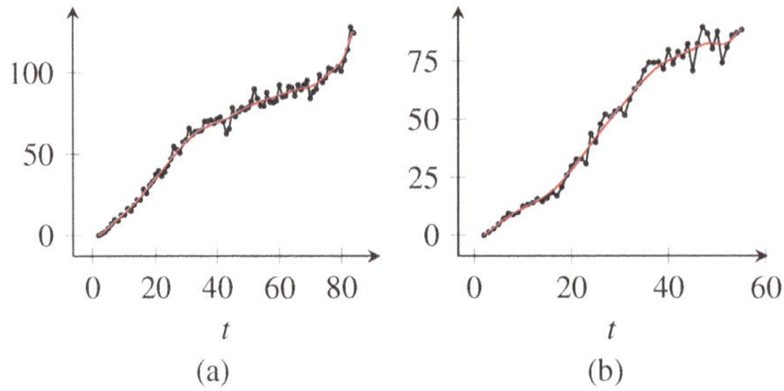

Fig. 14 A comparison of total age persistence for the three different data sets, along with a smooth estimate for showing trends. (**a**) Example. (**b**) Measured. (**c**) Simulation

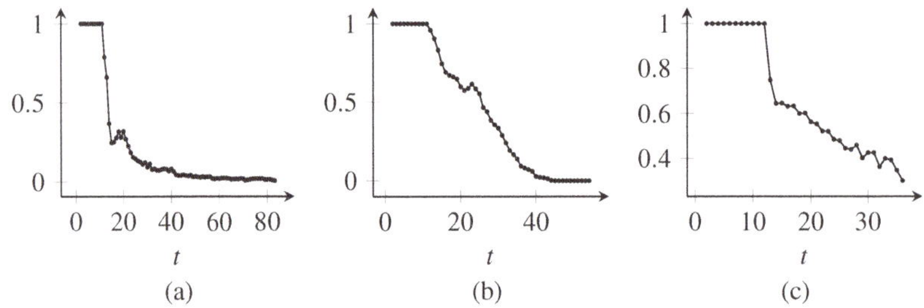

Fig. 15 Vivacity curves (pixel-based) for the three different data sets. At a glance, the curves permit comparing the dynamics of each process. The sampling frequencies are different, necessitating the use of dynamic time warping. (**a**) Example. (**b**) Measured. (**c**) Simulation

technique from dynamic programming that is able to compensate for different sampling frequencies and different simulation lengths. Figure 15 depicts the vivacity curves of the data sets. We can see that the measured data in Fig. 15b is characterized by a slower process in which new mass is continuously being injected to the system. Hence, its vivacity does not decrease steeply as that of the example data in Fig. 15a. The vivacity curve for the simulation, shown in Fig. 15c, appears to differ from the remaining curves. As a consequence, we can use these curves in visual comparison tasks and distinguish between different (measured) experiments and simulations. The dynamic time warping distance helps quantify this assumption. We have $\mathrm{dist}(a, b) \approx 442$, $\mathrm{dist}(a, c) \approx 1135$, and $\mathrm{dist}(b, c) \approx 173$. This indicates that the characteristics of the simulation in Fig. 15c differ from those found in a real-world viscous fingering process, shown in Fig. 15a, while being reasonably close to another measured experiment, which is depicted by Fig. 15b. Vivacity curves may thus be used for parameter tuning of simulations in order to obtain better approximations to measured data.

5 Conclusion

Driven by the need for a coherent analysis of time-varying skeletons, we developed different concepts inspired by topological persistence in this paper. We showed how to improve the consistency of tracking algorithms between consecutive time steps. Moreover, we demonstrated the utility of our novel concepts for different purposes, including the persistence-based filtering of skeletons, anomaly detection, and characterization of dynamic processes.

Nonetheless, we envision numerous other avenues for future research. For example, the propagation velocity of structures in the data may be of interest in many applications. We also plan to provide a detailed analysis of viscous fingering, including domain expert feedback, and extend persistence to physical concepts within this context. More generally, our novel persistence-inspired concepts can also be used in other domains, such as the analysis of motion capture data (which heavily relies on skeletonization techniques) or time-varying point geometrical point clouds, for which novel skeletonization techniques were recently developed [10].

References

1. Berndt, D.J., Clifford, J.: Using dynamic time warping to find patterns in time series. Technical Report WS-94-03, AAAI (1994)
2. Carstens, C., Horadam, K.: Persistent homology of collaboration networks. Math. Prob. Eng. **2013**, 815,035:1–815,035:7 (2013)
3. Cohen-Steiner, D., Edelsbrunner, H., Harer, J., Mileyko, Y.: Lipschitz functions have L_p-stable persistence. Found. Comput. Math. **10**(2), 127–139 (2010)
4. Comaniciu, D., Meer, P.: Mean shift: a robust approach toward feature space analysis. IEEE Trans. Pattern Anal. Mach. Intell. **24**(5), 603–619 (2002)
5. Delgado-Friedrichs, O., Robins, V., Sheppard, A.: Skeletonization and partitioning of digital images using discrete Morse theory. IEEE Trans. Pattern Anal. Mach. Intell. **37**(3), 654–666 (2015)
6. Edelsbrunner, H., Harer, J.: Computational Topology: An Introduction. AMS, Providence (2010)
7. Edelsbrunner, H., Letscher, D., Zomorodian, A.J.: Topological persistence and simplification. Discrete Comput. Geom. **28**(4), 511–533 (2002)
8. Edelsbrunner, H., Morozov, D., Pascucci, V.: Persistence-sensitive simplification of functions on 2-manifolds. In: Proceedings of the 22nd Annual Symposium on Computational Geometry, pp. 127–134. ACM Press, New York (2006)
9. Farin, G.: Curves and Surfaces for Computer-aided Geometric Design: A Practical Guide, 3rd edn. Elsevier, New York (1993)
10. Kurlin, V.: A one-dimensional homologically persistent skeleton of an unstructured point cloud in any metric space. Comput. Graph. Forum **34**(5), 253–262 (2015)
11. Lam, L., Lee, S.W., Suen, C.Y.: Thinning methodologies – a comprehensive survey. IEEE Trans. Pattern Anal. Mach. Intell. **14**(9), 869–885 (1992)
12. Lukasczyk, J., Aldrich, G., Steptoe, M., Favelier, G., Gueunet, C., Tierny, J., Maciejewski, R., Hamann, B., Leitte, H.: Viscous fingering: a topological visual analytic approach. In: Physical Modeling for Virtual Manufacturing Systems and Processes. Applied Mechanics and Materials, vol. 869, pp. 9–19 (2017)

13. Rieck, B., Leitte, H.: Exploring and comparing clusterings of multivariate data sets using persistent homology. Comput. Graph. Forum **35**(3), 81–90 (2016)
14. Rieck, B., Mara, H., Leitte, H.: Multivariate data analysis using persistence-based filtering and topological signatures. IEEE Trans. Vis. Comput. Graph. **18**(12), 2382–2391 (2012)
15. Viscous fingering displacement. https://www.youtube.com/watch?v=NZEB8tQ3eOM
16. Zhang, T.Y., Suen, C.Y.: A fast parallel algorithm for thinning digital patterns. Commun. ACM **27**(3), 236–239 (1984)

Fast Topology-Based Feature Tracking using a Directed Acyclic Graph

Himangshu Saikia and Tino Weinkauf

Abstract We present a method for tracking regions defined by Merge trees in time-dependent scalar fields. We build upon a recently published method that computes a directed acyclic graph (DAG) from local tracking information such as overlap and similarity, and tracks a single region by solving a *shortest path* problem in the DAG. However, the existing method is only able to track one selected region. Tracking *all* regions is not straightforward: the naïve version, tracking regions one by one, is very slow. We present a fast method here that tracks all regions at once. We showcase speedups of up to two orders of magnitude.

1 Introduction

We are concerned with the tracking of regions defined by merge trees. In [12], we devised a method that tracks the superlevel or sublevel sets of a scalar field as defined by the subtrees of the merge tree. However, once these regions have been extracted in each time step, we neglect their origin 1 and record tracking information such as overlap and histogram similarity in a directed acyclic graph (DAG). Its nodes are the regions (Figs. 1 and 2). Overlapping and similar regions in consecutive time steps are connected by an edge, weighted by the amount of overlap and similarity. We solve a *shortest path* problem to track a region over time. This global approach to tracking prevents the issue with only local decisions as shown in Fig. 1.

In [12], we present, among other things, a method for tracking a single region using the DAG.This is done by computing shortest paths to all reachable sources and sinks from a given node and combining those two paths. This however, is not how one might define a shortest path via a node. In this paper, we define a shortest path as the path with the least objective function value out of all paths starting at a source, going through the given node and ending at a sink. An objective function can

H. Saikia (✉) · T. Weinkauf
KTH Royal Institute of Technology, Stockholm, Sweden
e-mail: saikia@kth.se; weinkauf@kth.se

H. Carr et al. (eds.), *Topological Methods in Data Analysis and Visualization V*, Mathematics and Visualization, https://doi.org/10.1007/978-3-030-43036-8_10

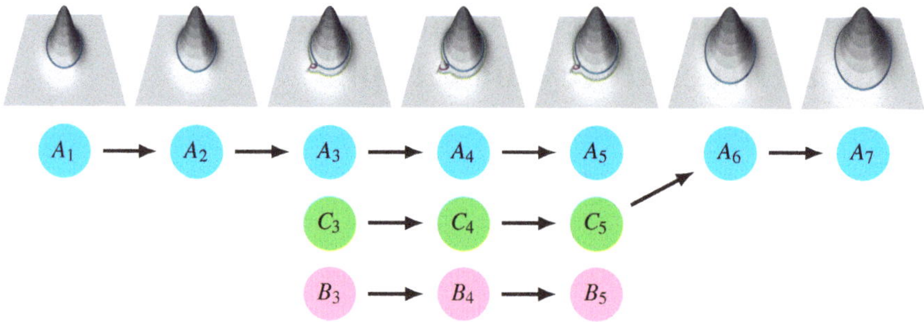

Fig. 1 Tracking regions solely based on local decisions leads to broken tracks. In this simple example, a small fluctuation between time steps t_3 and t_5 causes the creation of a region C that has significant overlap and similarity with region A. Assigning the locally best match neglects that there can be more than one suitable track between two time steps (e.g., between t_5 and t_6), and causes tracks to break. A graph structure as illustrated in Fig. 3a and b helps circumvent this problem

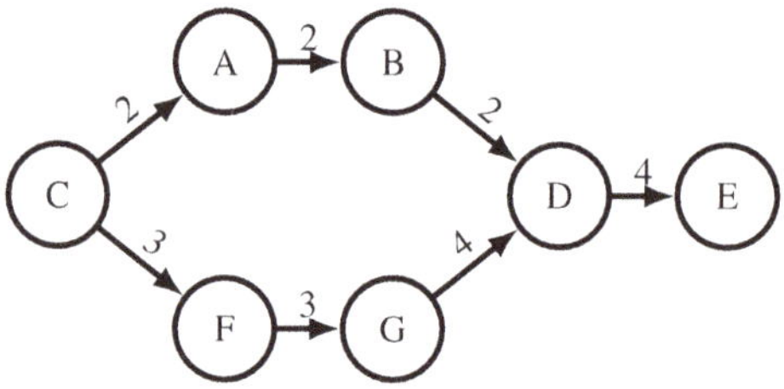

Fig. 2 Illustration of an objective function that does not satisfy the condition expressed in Eq. (4). If f signifies the standard deviation of the weights along a path, $f(CABD) < f(CFGD)$ but $f(CABD \cup DE) > f(CFGD \cup DE)$

be any function which assigns a score to a path based on how well it represents the evolution of a particular feature along that path. Using this definition of a shortest path, the previous method of combining backward and forward shortest paths may not work.

In this work, we extend the previous work and present a non-trivial solution to tracking all regions from all time steps, i.e., a method for extracting all feature tracks. The trivial solution is to iterate over all nodes of the DAG and execute the single region tracking algorithm from [12]. However, we will show in this paper how this leads to very long running times. Our approach is up to two orders of magnitude faster. Our method employs a shortest path algorithm but is quite different from the standard Djikstra's algorithm or Floyd-Warshall's algorithm to compute all pairs shortest paths. Our DAG being structured in a way that only temporal edges—edges between nodes of two successive time steps—exist, facilitates better runtime bounds than standard algorithms.

2 Related Work

The sheer size of time-dependent data sets often necessitates a data reduction step before an efficient analysis can take place. It is therefore a common approach to extract and track features.

Many methods track topological structures. Tricoche et al. [23] track critical points and other topological structures in 2D flows by exploiting the linearity of the underlying triangle grid. Garth et al. [4] extend this to 3D flows. Theisel and Weinkauf [18, 26] developed feature flow fields as a general concept to track many different features independent of the underlying grid. Reinighaus et al. [11] extended this idea to the discrete setting.

In the area of time-dependent scalar fields several methods exist to track and visualize topological changes over time. Samtaney et al. [15] provides one of the first algorithms to track regions in 3D data over time using overlap. Kettner et al. [7] presents a geometric basis for visualization of time-varying volume data of one or several variables. Szymczak [17] provides a method to query different attributes of contours as they merge and split over a certain time interval. Sohn and Bajaj [16] presents a tracking graph of contour components of the contour tree and use it to detect significant topological and geometric evolutions. Bremer et al. [2] provide an interactive framework to visualize temporal evolution of topological features.

Other methods for tracking the evolution of merge trees such as the method due to Oesterling [8] track changes to the hierarchy of the tree. This comes at the price of a very high computation time. Its runtime complexity is polynomial in the data size, more precisely, it is $O(n^3)$ with n being the number of voxels. However, the method tracks the unaugmented (full) merge tree instead of just critical points or super-arcs.

Vortex structures are another important class of features that can be tracked in time-dependent flows. Reinders et al. [10] track them by solving a correspondence problem between time steps based on the attributes of the vortices. Bauer and Peikert [1] and Theisel et al. [19] provide different methods for tracking vortices defined by swirling stream lines. This notion was extended later to include swirling path lines [25], swirling streak and time lines [27], swirling trajectories of inertial particles [5] and rotation invariant vortices [6].

Pattern matching has been originally developed in the computer vision community. A number of visualization methods have been inspired by that. Examples are pattern matching methods for vector fields based on moment invariants as proposed by Bujack et al.[3], or pattern matching for multi-fields based on the SIFT descriptor as proposed by Wang et al. [24].

A similar line of research, but technologically rather different, is the analysis of structural similarity in scalar fields, which gained popularity recently. Thomas and Natarajan detect symmetric structures in a scalar field using either the contour tree [20], the extremum graph [21], or by clustering contours [22]. Saikia et al. compared merge trees by means of their branch decompositions [13] or by means of histogram over parts of the merge tree [14]. Our method outputs a set of best tracks of topologically segmented structures in a spatio-temporal setting, and enables an all-to-all temporal pattern matching scheme using techniques like dynamic time warping.

3 Method

In the following, we will first briefly recapitulate the tracking method for single regions of [12], and then present our new and fast approach for tracking all regions.

3.1 *Tracking Merge Tree Regions using a Directed Acyclic Graph*

Given is a time-dependent scalar field. It could have any number of spatial dimensions, our implementation supports 2D and 3D. A merge tree is computed from each time step independently. After an optional simplification, all subtrees (as defined in [13]) are converted into a set of nodes to be used within the Directed Acyclic Graph (DAG). They represent the components of the superlevel or sublevel sets of the scalar field and are continuous regions in the domain.

All overlapping nodes from consecutive time steps are connected via edges in the DAG. Their weights represent local tracking information in the sense that a lower edge weight indicates a higher likelihood for the two connected regions to be part of the same track. We use a linear combination of *volume overlap* and a *histogram difference* to compute these weights.The *volume overlap* distance d_o between two non-empty regions $\mathcal{S}_a$ and $\mathcal{S}_b$ is determined from the number of voxels they have in common and the total number of voxels covered by both regions:

$$d_o(\mathcal{S}_a, \mathcal{S}_b) = 1 - \frac{|\mathcal{S}_a \cap \mathcal{S}_b|}{|\mathcal{S}_a \cup \mathcal{S}_b|} \quad . \tag{1}$$

The Chi-Squared histogram distance (see e.g. [9]) between two regions is defined as

$$d_s(\mathcal{S}_a, \mathcal{S}_b) = \chi^2(\mathbf{h}_a, \mathbf{h}_b) = \frac{1}{2} \sum_i \frac{(h_{a,i} - h_{b,i})^2}{h_{a,i} + h_{b,i}} \,, \tag{2}$$

where $h_{a,i}$ and $h_{b,i}$ denote the bins of the histograms $\mathbf{h}_a$ and $\mathbf{h}_b$, respectively. Here the histograms represent the number of vertices encapsulated by a region as described in [14].

Our combined distance measure for an edge is given by $d = \lambda d_s + (1 - \lambda)d_o$ where $\lambda \in [0, 1]$ is a tunable parameter. It is now possible to use this DAG for the next step as is, or it can be further thresholded to weed out extremely large

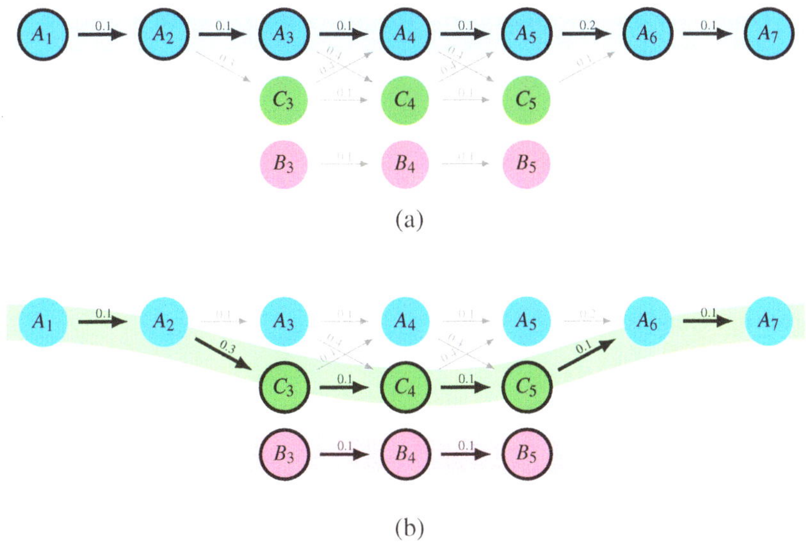

Fig. 3 Several nodes in the graph can have the same shortest path. Hence, running Djikstra's algorithm for every node independently will be expensive and redundant. (**a**) Dijkstra's algorithm to find the shortest path through the DAG represents the track of this region. Several regions may have the same source and sink, and result in the same shortest path. In this example, the shortest path starting from source A_1 and sink A_7, is common to all the nodes in bold outline. The path is shown as a blue band. (**b**) The shortest path through the green bold outlined nodes also has the same source-sink pair (A_1, A_7) as in (**a**). It is given by the green band. The shortest path through the red bold outlined nodes is given by the magenta band

weighted edges (For instance in Fig. 3a the edges between the green and pink nodes are removed).

We track a region by solving a shortest path problem with Dijkstra's algorithm on the DAG. The method in [12] does this for one region at a time. From the selected region, a shortest path is found to a source in an earlier time step, and another shortest path is found to a sink in a later time step. Combining these paths yields the track of the region. We describe a path to be a set of successive directed edges in the graph. Since there exists only a single directed edge between any two nodes in the graph, a path can also be described by all successive nodes that connect these edges. Source and sink refer in this context to nodes that have no incoming or outgoing edges, respectively. We discuss this in more detail in the next section.

3.2 Objective Function and Its Validity with Dijkstra's Algorithm

The classic Dijkstra algorithm finds the shortest path by summing up the edge weights along the path. Applying this directly to our setting would yield unsuitable tracks: instead of following a long path with many likely edges, the tracking would rather choose an unlikely edge to an immediate sink.

Hence, we use a measure assessing the average edge weight along a path. The goal is to find the path through a given node that has the smallest normalized squared sum of edge weights d_i:

$$f(\mathcal{P}) = \sqrt{\frac{\sum_{i \in \mathcal{P}} d_i^2}{|\mathcal{P}|}} \tag{3}$$

The purpose of this section is to demonstrate that the Dijkstra algorithm can be used to solve for this objective. To do so, let us define an objective function f which assigns a non-negative score to a path satisfying the following condition:

Condition 1 Consider two paths $\mathcal{P}_1$ and $\mathcal{P}_2$ with $f(\mathcal{P}_1) \leq f(\mathcal{P}_2)$. We require the objective function to maintain this relationship after adding an edge $\mathbf{e}$:

$$f(\mathcal{P}_1 \cup \mathbf{e}) \leq f(\mathcal{P}_2 \cup \mathbf{e}) \tag{4}$$

Dijkstra's algorithm can only be used to solve for an objective if this condition is fulfilled, since it allows to incrementally build a solution, which is the essential cornerstone of Dijkstra's algorithm.

The objective function used in the classic Dijkstra's shortest path algorithm is the sum of weights of all edges in a path $\sum_{i \in \mathcal{P}} d_i$. This function trivially satisfies the above condition. A non-satisfying objective function is the standard deviation of the weights as shown in Fig. 2. Thus we can see that not all objective functions that determine the quality of a path can be used with Dijkstra's algorithm.

Regarding the objective function (3) we note that it can be solved with Dijkstra's algorithm if $|\mathcal{P}_1| = |\mathcal{P}_2|$ holds, i.e., the two paths are of equal length. This keeps the denominator of (3) equal, and the numerators are just a sum of values consistent with Condition 1. The condition $|\mathcal{P}_1| = |\mathcal{P}_2|$ always holds true in our setting, since edges connect two consecutive time steps only and we start Dijkstra's algorithm at a particular source, which keeps all considered paths at equal length. Hence, Dijkstra's algorithm can be used to solve (3).

3.3 Algorithm for Finding All Paths

Tracking of a single node in the DAG is done by finding the shortest path $\mathcal{P}_{\min}$ through that node from any source to any sink of the DAG. It may be that the shortest path through other nodes coincides with $\mathcal{P}_{\min}$. This is illustrated in Fig. 3. Hence, to find the shortest paths through all nodes, running a naïve Dijkstra for every node independently will be expensive and redundant.

Instead, we run Dijkstra's algorithm for every source and sink (in a joint fashion in two passes, see below), record the gathered information at every node, and stitch this information together to obtain the shortest path for every node.

To facilitate this, we define a function to incrementally compute the objective function in (3). We denote this new function by the symbol $\oplus$ and call it the *incremental path operator*. The incremental path operator takes as input the objective value for path $\mathcal{P}$ and a connecting node $\mathbf{n}$ and computes the global measure for path $\mathcal{P} \cup \mathbf{n}$. If the weight of the connecting edge between $\mathcal{P}$ and $\mathbf{n}$ is given by d, $\oplus$ is defined as follows

$$f(\mathcal{P} \cup \mathbf{n}) = f(\mathcal{P}) \oplus d = \sqrt{\frac{f(\mathcal{P})^2 \cdot |\mathcal{P}| + d^2}{|\mathcal{P}| + 1}} \tag{5}$$

Furthermore, all nodes are topologically sorted. That is, for a node $n_{p,i}$ at timestep $t = p$ and another node $n_{q,j}$ at timestep $t = q$, node $n_{p,i}$ occurs before node $n_{q,j}$ in the sorted order if $p < q$.

Our algorithm works as follows. We make two passes through this list of sorted nodes. One in the sorted order (past time steps to future time steps) and one in the reverse sorted order. During the first pass, at every node, the best path from every reachable source to that given node is recorded. This is done by checking all incoming edges to that node and incrementally calculating the best path from all incoming edges from a single source. This becomes possible because all nodes connected to the incoming edges have already been processed earlier (they live at the previous time step). Consider a node n_i with some incoming edges as illustrated in Fig. 4. The best score from any given source to n_i is calculated using:

$$f(\mathcal{P}_{src \to i}) = \min_{n_j \in \mathbf{I}_i \wedge bestSource_j = src} (f(\mathcal{P}_{src \to j}) \oplus d_{j,i}). \tag{6}$$

Algorithm 1 shows the pseudo-code for the first pass described above. The second pass is equivalent to the first, but operates on the DAG with edges reversed. We record the best path to every reachable sink now. This is done by checking for outgoing edges and the sinks that they lead to. The best score to any given sink from n_i is calculated using:

$$f(\mathcal{P}_{i \to sink}) = \min_{n_j \in \mathbf{O}_i \wedge bestSink_j = sink} (f(\mathcal{P}_{j \to sink}) \oplus d_{i,j}) \tag{7}$$

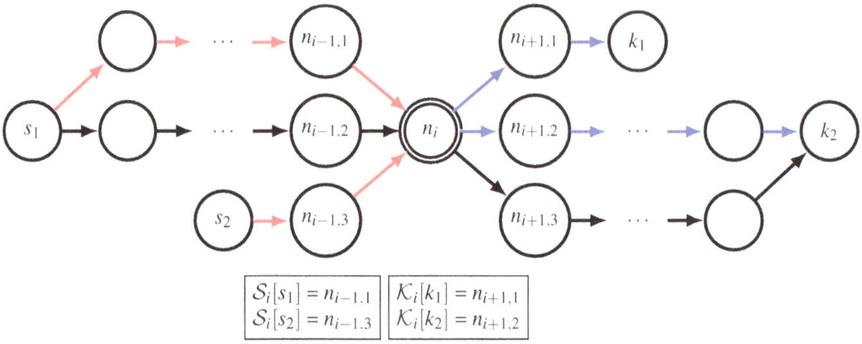

Fig. 4 Illustration of Algorithm 1. For every node $\mathbf{n}_i$ in the DAG, the lowest cost (and the corresponding best neighbor) to every reachable source is computed iteratively and stored in the associative map $\mathcal{S}_i$. In this figure, for example, the best path from n_i to source s_1 is via its neighbor $n_{i-1,1}$. Similarly lowest costs to all reachable sinks are stored in the map $\mathcal{K}_i$. After these values are computed, the best source-sink pair (s, k) is computed with the lowest cost using Eq. (8) and the best path $\mathcal{P}_{\min}$ from s to k passing through n_i is traced out. All nodes lying on $\mathcal{P}_{\min}$ which have the same best source-sink pair (s, k) need not be processed as the best path through any such node is $\mathcal{P}_{\min}$ itself. The final output is the set of all paths passing through every node in the DAG

Let the set of all reachable sources to node $\mathbf{n}_i$ be $\mathbf{S}_i$ and the set of all reachable sinks be called $\mathbf{K}_i$. After the two passes are complete, the combined best path for every node is calculated by choosing the paths $(\mathcal{P}^- \in \mathbf{S}_i, \mathcal{P}^+ \in \mathbf{K}_i)$ from the source-sink pair which minimizes the objective function on the combined path $\mathcal{P}^- \cup \mathcal{P}^+$ as follows:

$$\mathcal{P}_{\min} = \underset{\mathcal{P}^- \in \mathbf{S}_i, \mathcal{P}^+ \in \mathbf{K}_i}{\arg\min} \sqrt{\frac{|\mathcal{P}^-| \cdot f(\mathcal{P}^-)^2 + |\mathcal{P}^+| \cdot f(\mathcal{P}^+)^2}{|\mathcal{P}^-| + |\mathcal{P}^+|}}. \tag{8}$$

Algorithm 2 shows the pseudo-code to obtain all best paths. It can be observed that, if for any given node $\mathbf{n}_i$, the best source-sink pair is given by $(\mathbf{s}_i, \mathbf{k}_i)$ and the extracted best path is $\mathcal{P}_i$, all nodes lying on this path, having the same best source-sink pair $(\mathbf{s}_i, \mathbf{k}_i)$, will trace out the exact same path. Hence, while determining unique paths in our solution, we can avoid tracing paths from all such nodes. See Fig. 4 for an illustration.

After all nodes have been examined, we are left with the set of best paths passing through every single node in the DAG. An illustration of the output is shown in Fig. 5.

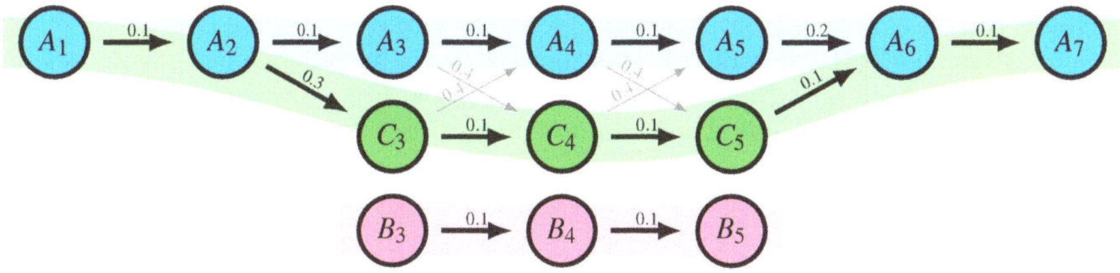

Fig. 5 The shortest paths through all nodes in the DAG combined represent our track graph structure. Our algorithm avoids computing the three shortest paths (given by the blue, green and magenta bands) for every single node naively, but instead traces a single shortest path only once. Nodes which lie on a shortest path and has the same source-sink pair, trivially trace the same path

Data: Set of all nodes $\mathbf{N} = \{\mathbf{n}_1, \mathbf{n}_2, \ldots, \mathbf{n}_m\}$ sorted topologically. Incoming nodes to n_i given by $\mathbf{I}_i$. The weight of an edge between nodes n_i and n_j is given by $d_{i,j}$.

Result: Set of associations of all reachable sources and the best path to them from every node n_i given by $\mathcal{S}_i$.

begin
  for $\forall n_i \in \mathbf{N}$ **do**
    for $\forall n_j \in \mathbf{N}$ **do**
      if $|\mathbf{I}_j| = 0$ **then**
        if $n_i = n_j$ **then**
          $\mathcal{S}_i[j] \leftarrow n_j$
          $\mathcal{S}_{i,cost}[j] \leftarrow 0$
        else
          $\mathcal{S}_i[j] \leftarrow -1$
          $\mathcal{S}_{i,cost}[j] \leftarrow \infty$
  for $\forall n_i \in \mathbf{N}$ **do**
    for $\forall n_j \in \mathbf{I}_i$ **do**
      for $\forall source \in \mathcal{S}_j$ **do**
        $cost_{new} \leftarrow \mathcal{S}_{j,cost}[source] \oplus d_{i,j}$
        if $cost_{new} < \mathcal{S}_{i,cost}[source]$ **then**
          $\mathcal{S}_i[source] \leftarrow n_j$
          $\mathcal{S}_{i,cost}[source] \leftarrow cost_{new}$

Algorithm 1: Algorithm to find the associations for the best routes to any node from all reachable sources. The best routes to all reachable sinks are determined by running the same algorithm with the nodes sorted in reverse order

3.4 Complexity Analysis

Let us assume, without loss of generality, that the average number of features in every timestep is n. For t timesteps, we would then have a total of tn nodes in the entire DAG. The number of edges is bounded by n^2 between every pair of successive timesteps, so the total number of edges would be bounded by tn^2. The naïve version of the algorithm is a combination of two simple Dijkstra runs from any given node to all reachable sources and sinks. As we know the runtime of this algorithm is $O(V + E)$, for V vertices and E edges in a graph, the runtime in the naïve case

Data: The sets of associated maps $\mathcal{S}_i$ and $\mathcal{K}_i$ for every node $\mathbf{n}_i$ as obtained from Algorithm 1. The function $tracePath(n_i, s, k)$ traces the path to source s and sink k from node n_i using the information present in $\mathcal{S}_i$ and $\mathcal{K}_i$.

Result: Set of all unique shortest paths **P** where there exists at least one path passing through every node n_i.

```
begin
    P ← ∅
    for ∀n_i ∈ N do
        bestScore ← ∞
        bestSource_i ← −1
        bestSink_i ← −1
        done_i ← false
    for ∀n_i ∈ N do
        for ∀s ∈ S_i do
            for ∀k ∈ K_i do
                score ← s_cost ⊕ k_cost
                if score < best then
                    bestScore ← score
                    bestSource_i ← s
                    bestSink_i ← k
    for ∀n_i ∈ N do
        if done_i ≠ true then
            𝒫 ← tracePath(n_i, bestSource_i, bestSink_i)
            done_i ← true
            P ← P ∪ 𝒫
            for ∀n_j ∈ 𝒫 do
                if bestSource_j = bestSource_i ∧ bestSink_j = bestSink_i then
                    done_j ← true
```

Algorithm 2: Algorithm to find shortest paths via every node

will be $O(tn + tn^2)$ or $O(tn^2)$ for every node. Hence, if we were to run the naïve algorithm for all nodes, the runtime would be given by $O(t^2n^3)$ in the worst case.

Now for our improved algorithm, assuming the number of sources/sinks is given by p, we can safely say that $p << tn$. The runtime of Algorithm 1 is then given by $O(tnp + tn^2p)$ or $O(tn^2p)$. For Algorithm 2, it is $O(tnp^2 + t^2n)$. So the total runtime of our algorithm would be given by $O(tn^2p + tnp^2 + t^2n)$ which in practice (as seen in Table 1) is far less than $O(t^2n^3)$.

The memory footprint for the naïve version is bounded by the normal Dijkstra runtime of $O(N)$ for N nodes. Thus, in our scenario, it is given by $O(tn)$ as the shortest path via every node is computed independently. For the improved algorithm however, we need to store the mappings of shortest paths from all incoming/outgoing edges to all reachable sources/sinks and hence the memory footprint is given by $O(tnp)$.

3.5 *Filtering Similar Paths for Visualization*

For visualization purposes we need to choose the best candidate paths which best represent a feature track at some spatio-temporal region.

In most cases, due to slight perturbations in the DAG, two unique paths may differ only at very few node positions with most of their nodes being identical. An example of this can be observed in Fig. 5, where the blue and green paths show in essence the same structure with only a slight perturbation.

We aim to show the path with the best objective score, while other similar paths falling within a specified threshold are filtered out. The similarity g between two paths is estimated using

$$g(\mathcal{P}_1, \mathcal{P}_2) = \frac{|\mathcal{P}_1 \cap \mathcal{P}_2|}{\max(|\mathcal{P}_1|, |\mathcal{P}_2|)} \tag{9}$$

where $|\mathcal{P}_1 \cap \mathcal{P}_2|$ represent the number of matching edges. The function g estimates the fraction of edges that are identical in both paths. The filtration using function g is applied as follows. All paths obtained by solving Eq. (8) for every node are sorted according to the best objective function score given by Eq. (3). Paths are then processed in this sorted order, from lowest score to highest. If a path falls above the similarity threshold with any other path encountered before, it is filtered out. All other paths are retained.

If the filter node is set to be 100%, we are left with the complete set of unique paths. In our experiments, a filter rate of 70% shows best results.

4 Results

The timing and memory consumption for our method are given in Table 1. Regarding the computation times, note how our algorithm improves over the naïve version by up to two orders of magnitude. Regarding the memory consumption, the naïve method has lower memory usage as it only processes one node at once, while our algorithm processes all nodes together. Hence, considering the number of nodes in each data set, our algorithm is quite efficient with regards to memory usage as well.

Figure 6 shows a rotating and translating benzene data set. Since the data is not truly time-dependent, but just transformed rigidly, this serves as a test case to show that we capture all expected tracks and that our method is invariant against rotations and translations.

Figure 7 shows the 2D time-dependent Streak Line Curvature dataset.

Figure 8 shows the tracks for the smallest super/sub level set regions in a 2D Checkerboard dataset. The checkerboard pattern starts off smoothly and becomes increasingly noisy with time.

Table 1 Computation runtimes and memory requirements of our algorithm versus the naïve one for several data sets

Dataset	Benzene	2D Checker-board	2D Streak line curvature	3D Square cylinder
Dimensions	127 × 127 × 255	128 × 128 × 1	750 × 136 × 1	192 × 64 × 48
Time steps	101	128	429	134
DAG nodes	2239	3413	18035	15818
DAG edges	2121	3198	85262	76739
Time to extract	10 ms	16 ms	1767 ms	3200 ms
Time to filter	5 ms	10 ms	80 ms	96 ms
Time naïve alg.	507 ms	936 ms	442050 ms	283448 ms
Memory	427 KB	758 KB	26322 KB	70814 KB
Memory naïve alg.	23 KB	12 KB	523 KB	372 KB

All our experiments were performed on a machine with a 2.3 GHz Intel i7 processor and 16 GB main memory. Timings are totaled over the entire dataset. The timings do not include computation and simplification of the merge trees. It is to be noted that runtimes and memory usage depend *only* on the size of the DAG and not on the size of the dataset

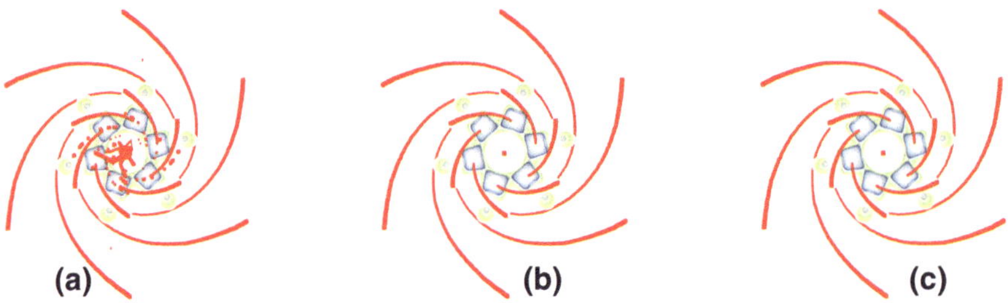

Fig. 6 Our method applied to the Benzene dataset. The paths indicate tracking of centers of mass of the regions signified by nodes in our DAG. (**a**) Paths of all lengths at filter rate 100%. (**b**) Paths of length 100 and above at filter rate 100%. (**c**) Paths of length 100 and above at filter rate 70%

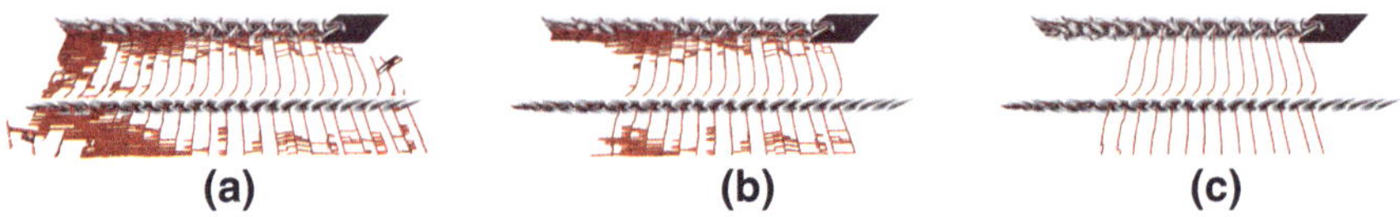

Fig. 7 (**a**) The 2D Streak Line Curvature dataset at filter rate 100% and showing paths of all lengths. (**b**) At 100% filter rate and full length paths only. (**c**) At 70% filter rate and full length paths only

Figure 9 shows all the tracks in a flow around a 3D Square Cylinder. The location of the center of mass of a region is used to visualize the paths in all result images.

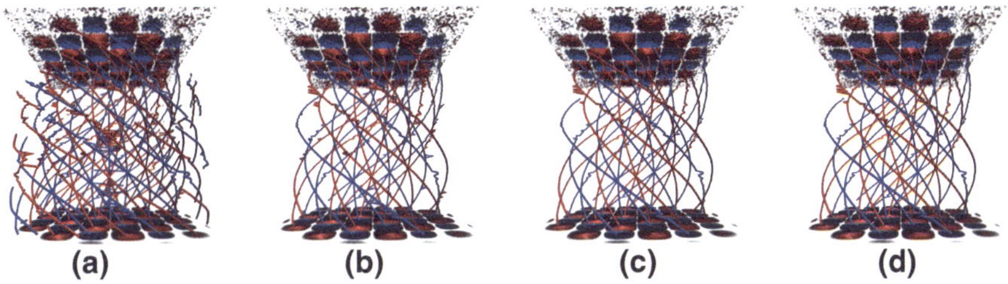

Fig. 8 Rotating 2D Checkerboard dataset. Tracks for the centers of mass of the smallest super/sub level sets are shown. (**a**) Filter rate 100% and paths of all lengths. (**b**) Filter rate 100% and long (100 length or more) paths only. (**c**) Filter rate 70% and long paths only. (**d**) Filter rate 70% with long paths obtained from the naive version of the algorithm

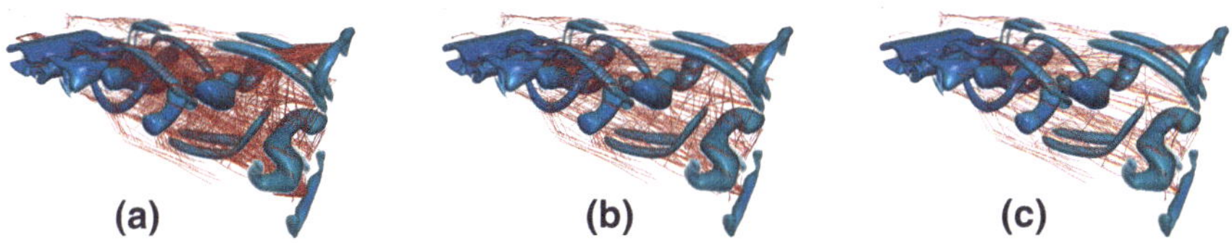

Fig. 9 Flow around a Square Cylinder dataset. (**a**) Paths of all lengths extracted at filter rate of 100%. (**b**) Paths of all lengths at filter rate 90%. (**c**) Paths of all lengths at filter rate 70%

5 Conclusion

We presented an extension of the method in [12] which was used to extract the best track through a chosen region at any given timestep in a time-dependent scalar field. These regions are based on topological segmentations in the spatial domain using merge trees and form the nodes in a Directed Acyclic Graph (DAG) structure in the spatio-temporal domain. Using the method in [12] to extract the best tracks through all nodes naïvely results in tracing the same paths multiple times. The algorithm presented in this paper makes use of the structure in the DAG to iteratively compute the best paths to every node from all reachable sources and sinks. This in turn allows us to compute the best paths through all nodes at orders of magnitude faster than the naïve approach. We also presented a filtering algorithm to filter out very similar paths for visualizing all paths together. Further work may include clustering these paths according to their similarity by using temporal similarity estimation techniques like dynamic time warping.

References

1. Bauer, D., Peikert, R.: Vortex tracking in scale-space. In: Proceedings of the Symposium on Data Visualisation 2002 (VISSYM '02), pp. 233–ff. Eurographics Association, Switzerland (2002)
2. Bremer, P.T., Weber, G., Tierny, J., Pascucci, V., Day, M., Bell, J.: Interactive exploration and analysis of large-scale simulations using topology-based data segmentation. IEEE Trans. Vis. Comput. Graph. **17**(9), 1307–1324 (2011)
3. Bujack, R., Hotz, I., Scheuermann, G., Hitzer, E.: Moment invariants for 2d flow fields using normalization. In: Proceedings of 2014 IEEE Pacific Visualization Symposium, pp. 41–48 (2014)
4. Garth, C., Tricoche, X., Scheuermann, G.: Tracking of vector field singularities in unstructured 3d time-dependent datasets. In: Proceedings of the Conference on Visualization '04 (VIS '04), pp. 329–336. IEEE Computer Society, Washington, (2004)
5. Günther, T., Theisel, H.: Vortex cores of inertial particles. IEEE Trans. Vis. Comput. Graph. **20**(12), 2535–2544 (2014)
6. Günther, T., Schulze, M., Theisel, H.: Rotation invariant vortices for flow visualization. IEEE Trans. Vis. Comput. Graph. **22**(1), 817–826 (2016)
7. Kettner, L., Rossignac, J., Snoeyink, J.: The safari interface for visualizing time-dependent volume data using iso-surfaces and contour spectra. Comput. Geom. **25**(1), 97–116 (2003). European Workshop on Computational Geometry—CG01
8. Oesterling, P., Heine, C., Weber, G.H., Morozov, D., Scheuermann, G.: Computing and visualizing time-varying merge trees for high-dimensional data. In: Carr, H., Garth, C., Weinkauf, T. (eds.) Topological Methods in Data Analysis and Visualization IV, pp. 87–101. Springer, Cham (2017)
9. Pele, O., Werman, M.: The quadratic-chi histogram distance family. In: Daniilidis, K., Maragos, P., Paragios, N. (eds.) Computer Vision—ECCV 2010, pp. 749–762. Springer, Berlin (2010)
10. Reinders, F., Sadarjoen, I.A., Vrolijk, B., Post, F.H.: Vortex tracking and visualisation in a flow past a tapered cylinder. Comput. Graphics Forum **21**(4), 675–682 (2002)
11. Reininghaus, J., Kasten, J., Weinkauf, T., Hotz, I.: Efficient computation of Combinatorial Feature Flow Fields. IEEE Trans. Vis. Comput. Graph. **18**(9), 1563–1573 (2012)
12. Saikia, H., Weinkauf, T.: Global feature tracking and similarity estimation in time-dependent scalar fields. Comput. Graphics Forum **36**(3), 1–11 (2017)
13. Saikia, H., Seidel, H.P., Weinkauf, T.: Extended branch decomposition graphs: structural comparison of scalar data. Comput. Graphics Forum (Proc. EuroVis) **33**(3), 41–50 (2014)
14. Saikia, H., Seidel, H.P., Weinkauf, T.: Fast similarity search in scalar fields using merging histograms. In: Carr, H., Garth, C., Weinkauf, T. (eds.) TopoInVis, pp. 1–14. Annweiler, Germany (2015)
15. Samtaney, R., Silver, D., Zabusky, N., Cao, J.: Visualizing features and tracking their evolution. Computer **27**(7), 20–27 (1994)
16. Sohn, B.S., Bajaj, C.: Time-varying contour topology. IEEE Trans. Vis. Comput. Graph. **12**(1), 14–25 (2006)
17. Szymczak, A.: Subdomain aware contour trees and contour evolution in time-dependent scalar fields. In: International Conference on Shape Modeling and Applications 2005 (SMI' 05), pp. 136–144 (2005)
18. Theisel, H., Seidel, H.P.: Feature flow fields. In: Proceedings of the Symposium on Data Visualisation 2003 (VISSYM '03), pp. 141–148. Eurographics Association, Switzerland, (2003)
19. Theisel, H., Sahner, J., Weinkauf, T., Hege, H., Seidel, H..: Extraction of parallel vector surfaces in 3d time-dependent fields and application to vortex core line tracking. In: IEEE Visualization, 2005 (VIS 05), pp. 631–638 (2005)

20. Thomas, D.M., Natarajan, V.: Symmetry in scalar field topology. IEEE Trans. Vis. Comput. Graph. **17**(12), 2035–2044 (2011)
21. Thomas, D.M., Natarajan, V.: Detecting symmetry in scalar fields using augmented extremum graphs. IEEE Trans. Vis. Comput. Graph. **19**(12), 2663–2672 (2013)
22. Thomas, D., Natarajan, V.: Multiscale symmetry detection in scalar fields by clustering contours. IEEE Trans. Vis. Comput. Graph. **20**(12), 2427–2436 (2014)
23. Tricoche, X., Wischgoll, T., Scheuermann, G., Hagen, H.: Topology tracking for the visualization of time-dependent two-dimensional flows. Comput. Graph. **26**(2), 249–257 (2002)
24. Wang, Z., Seidel, H.P., Weinkauf, T.: Multi-field pattern matching based on sparse feature sampling. IEEE Trans. Vis. Comput. Graph. (Proc. IEEE VIS) **22**(1), 807–816 (2016)
25. Weinkauf, T., Sahner, J., Theisel, H., Hege, H.C.: Cores of swirling particle motion in unsteady flows. IEEE Trans. Vis. Comput. Graph. (Proc. IEEE Visualization) **13**(6), 1759–1766 (2007)
26. Weinkauf, T., Theisel, H., Gelder, A.V., Pang, A.: Stable feature flow fields. IEEE Trans. Vis. Comput. Graph. **17**(6), 770–780 (2011)
27. Weinkauf, T., Hege, H.C., Theisel, H.: Advected tangent curves: a general scheme for characteristic curves of flow fields. Comput. Graph. Forum (Proc. Eurographics) **31**(2), 825–834 (2012). Eurographics 2012, Cagliari, Italy, May 13–18

Part IV
Multivariate Topology

The Approximation of Pareto Sets Using Directed Joint Contour Nets

Jan Bormann, Lars Huettenberger, and Christoph Garth

Abstract This paper presents the theoretical foundation to approximate the Pareto set and its affiliated reachability graph through the Joint Contour Net (JCN). The theory works for multivariate scalar fields with arbitrary numbers of domain dimensions and functions. This allows us to visualize critical regions, connected component inside the Pareto set, and their connections using the efficient and noise robust JCN algorithm. With visualization application in mind, we demonstrate the feasibility of our approach on $2D$ examples.

1 Introduction

Recent work introduced Pareto sets [9] as a method to visualize topology-based features in multivariate scalar fields. Together with the so-called reachability graph [10] it is possible to provide an abstract representation of the multifield data, critical regions and connections between those areas, with respect to common gradient directions.

Based on the previous work, the design of a fast algorithm to calculate the reachability graph is still an open task. One major obstacle is that the graph is based on continuous ascending paths between the critical regions. However in contrast to other related path-based concepts such as the Morse-Smale complex, the *steepest* gradient at a point does not exist in a multifield scenario but rather, the gradient is replaced as a range of ascending directions. Hence, instead of point-to-point connections, it is only possible to calculate a connection between a point and a set of points, making the reachability graph calculation exponentially more complicated.

J. Bormann (✉)
EXXETA AG, Karlsruhe, Germany
e-mail: jan.bormann@exxeta.com

L. Huettenberger · C. Garth
TU Kaiserslautern, Kaiserslautern, Germany
e-mail: l_huette@cs.uni-kl.de; garth@cs.uni-kl.de

H. Carr et al. (eds.), *Topological Methods in Data Analysis and Visualization V*, Mathematics and Visualization, https://doi.org/10.1007/978-3-030-43036-8_11

In addition, the connections have to be traced for every connected component in the Pareto set making it prone to small local structures. Note that in multifield scenarios, local structures in each individual field are cumulative, in that they might create large, significant structures in the combined data, but usually simply increase the number of small, insignificant structures in the multifield drastically.

In this work, to improve the calculation of the Pareto set and its reachability graph, we analyze the relation between this concept and an approach by Carr et al. [2, 3], the Joint Contour Net (JCN). Those comparisons are moreover important to place the concepts in context with each other and to improve the understanding and significance of multivariate topology in general. The significance of the two methods is also highlighted by the fact that different approaches and view points lead to similar results and visualizations. Previous work [11] already visually indicated some close relation, but want not able to prove those assumption.

The paper is structured as follows. After the related work and a brief summary of two major concepts, Pareto sets and JCNs, in Sects. 2 and 3, respectively, the main part of this work is the new definitions and mathematical proofs, in Sect. 4. These prove the relation between the mentioned concepts in general. Section 5 then analyzes these results, its implications and provides a visual proof of concept for $2D$ examples. While JCNs can also be calculated for higher dimensions, recent work also focused on efficiently building the Reeb space for those cases. We therefore discuss the usage of the continuous space instead of the approximation through the JCN. In the conclusion in Sect. 6, we summarize the previous sections and tasks for future work.

2 Related Work

As part of the introduction, the Pareto set concept was compared superficially with other multivariate approaches like Jacobi sets [6] and Joint Contour Nets [2, 3] in terms of domain restrictions as well as advantages and drawbacks.

Thus, Huettenberger et al. presented an algorithm to calculate Jacobi sets as a union of Pareto sets and conditions under which Pareto sets are subsets of the Jacobi sets [8]. Both concepts and their comparison are easy to visualize since both are defined in the domain of the data.

Chattopadhyay et al. [4] analyzed the projection of the Jacobi set in the Reeb space, introducing the Jacobi structure of a Reeb space. This structure separates the Reeb space into simple components which can be measured in a scale-invariant manner. This can directly be used as a multivariate topology simplification approach [5].

In the other direction, Tierny and Carr [14] recently used the Jacobi set to efficiently calculate the Reeb space [7]. Both Chattopadhyay et al. and Tierny and Carr present good examples of how the combination of two related concepts yields advancements, both algorithmically as well as in terms of understanding multivariate topology.

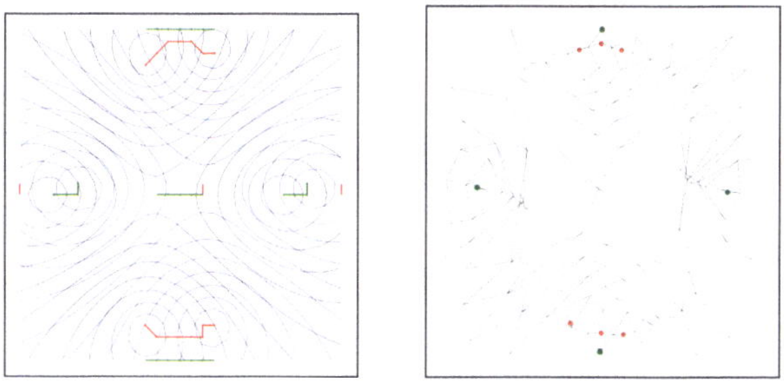

Fig. 1 Pareto set and dJCN for a bivariate scalar field from [11]. The field is outlined by blue and black contours in the left image. The Pareto set is colored red and green and the locations of critical slabs in the dJCN are indicated by red and green nodes. (**a**) Pareto set. (**b**) dJCN

To close the loop, i.e. the relation between Pareto sets and the Reeb space was investigated using the Joint Contour Net as an intermediate step. The JCN was proposed as an approximation of the Reeb space [2, 12] and, at least for $2D$ data, is easier to calculate than the continuous Reeb space itself.

In recent work, the relation between both concepts, JCN and Pareto set, was visually indicated in the introduction of directed JCNs (dJCN) [11]. The JCN was extended into a directed graph based on the ascending directions in the individual field. Slabs and their node representation in the dJCN were defined as critical based on the number of in- and outgoing edges (Fig. 1).

In this paper we present and extend main results from the master thesis by Jan Bormann [1] which investigated these observations mathematically. The thesis produced necessary modifications to the definition of critical slabs to allow a translation from dJCNs to Pareto sets. This enables us to combine the advantages of the Pareto concept, orientation-sensitive identification of critical regions, with the efficient calculation and noise robustness of the JCN. Also, we can use the graph structure of the dJCN to identify relations inside the Pareto set, namely the reachability graph, which so far was extremely costly to calculate [10].

3 Existing Concepts

For a brief summary on Pareto sets, we assume that the multivariate data is given as a set of n fields $f = (f_1, \ldots, f_n)$ over a common domain such that $f_i : \mathbb{S} \mapsto \mathbb{R}$ are continuous, piecewise-linear Morse functions with $\mathbb{S}$ a triangulated manifold of dimension d, i.e. a simplicial complex.

Two points $x, y \in \mathbb{S}$ are comparable if either $\forall_{i=1}^{n} f_i(x) \leq f_i(y)$ or $\forall_{i=1}^{n} f_i(x) \geq f_i(y)$ holds, denoted with $x \preceq y$ or $x \succeq y$, respectively. Otherwise, the points are incomparable, $x \prec\succ y$. We say for $x \succeq y$ that x dominates y, and for $x \preceq y$ that x

is dominated by y. For a given open neighborhood $U(x)$, $x \in U(x)$, the *ascending set* $H^+(x) := \{y \in U(x) \mid y \succeq x\}$ contains all ascending directions on which all separate fields can agree. The *descending set* $H^-(x)$ is defined analogously.

If $H^+(x) = \{x\}$, point x is defined as Pareto maximum, if $H^-(x) = \{x\}$ it is a Pareto minimum and if both sets only contain x itself, the point is a Pareto optimum. Otherwise, x is regular. The set of all Pareto minima, maxima, and optima is called the Pareto set $\mathbb{P}$ and all points in the set are generally described as Pareto extrema.

Furthermore, for two points $x, y \in \mathbb{S}$, we say that x *reaches* y $(x \rightarrow y)$ if a continuous ascending path $p : [0, 1] \mapsto \mathbb{S}$ exists such that $p(0) = x$, $p(1) = y$, and $0 \leq i < j \leq 1 \Rightarrow p(i) \preceq p(j)$. Connected components $X, Y \subseteq \mathbb{P}$ in the Pareto set, in terms of adjacency in the domain, are called *reachable* ($X \rightarrow Y$ or X *reaches* Y) if $x \in X$ and $y \in Y$ exist such that $x \rightarrow y$ holds.

With this, each connected component in $\mathbb{P}$ can be associated with a node in a graph structure and, the reachability between components can be associated to directed edges adjacent to the corresponding nodes, thus creating the *reachability graph*. More details, illustrations, and discussion can be found for example in [9, 10].

For the JCN, we assume the same multivariate data and an interval-based rounding function $r : \mathbb{R} \mapsto \mathbb{R}$ with $r(x) = \lfloor x/\delta \rfloor + \sigma$, where δ denotes the interval size and σ some offset, though, w.l.o.g. we assume $\sigma = 0$ throughout this paper. The discretized data $r \circ f = (r \circ f_1, \ldots, r \circ f_n)$ is separated into slabs such that two adjacent points $x, y \in \mathbb{S}$ are in the same slab if $\forall_{i=1}^n r \circ f_i(x) = r \circ f_i(y)$. Alternatively, x and y are in the same slab if there is a continuous path $p : [0, 1] \mapsto \mathbb{S}$ from x to y, i.e. $p(0) = x$, $p(1) = y$ such that $r \circ f(p(i))$ is constant for all $0 \leq i \leq 1$.

To create the JCN, each slab is associated with a node and two nodes are connected by an edge, if the corresponding slabs are adjacent. Note that in contrast to the reachability graph, the JCN uses undirected edges. A more detailed introduction on JCNs and their application can be found in [3, 13].

4 Calculating Pareto Sets with dJCNs

As mentioned above, the JCN does not have a notation of ascending or descending direction between adjacent slabs, though the Pareto set definition is based on it. Hence, it is unavoidable to introduce this concept into the JCN. Therefore, Huettenberger et al. [11] introduced the *directed Joint Contour Net* (dJCN) by replacing the undirected edges of the JCN with directed ones. In detail, for two adjacent slabs S_1, S_2 with points $s_1 \in S_1$ and $s_2 \in S_2$, we have an edge from the node corresponding to S_2 to the node corresponding to S_1, if $\forall_{i=0}^n r \circ f_i(s_1) \geq r \circ f_i(s_2)$. Thus, a notion of direction is introduced into the approximation of the Reeb space and we use this effect to compute an approximation of Pareto Sets. Note that the structure of the dJCNs depends strongly on the rounding function. By refining this function, i.e. decreasing the interval size, the size of the slabs decrease as well. We therefore define a series of rounding function with the identity function as limit, which leads to a series of dJCNs with a directed version of the Reeb space as limit. We will call such a series *refinement series* of dJCN.

4.1 The Limit of Refinement

A Pareto maximum is a point x with a neighborhood $U(x)$ around x such that each point $y \in U(x)$ is either $x \prec\succ y$ or $x \succeq y$. This implies, that x cannot reach any other point $y \in \mathbb{S}$, i.e. no monotone increasing path p to a dominating point y with $y \succeq x$ exists. Otherwise, the path has to intersect $U(x)$ somewhere, w.l.o.g. at point $p[i], 0 \leq i \leq 1$, and for each point $p[i]$ along the path $x = p[0] \preceq p[i]$ has to hold. Hence, either $x = p[i]$ or $f(x) = f(p[i])$ holds. This implies that either x is on the border of $U(x)$ or that $f(x) = f(p[j])$ for all $0 < j \leq i$, a contradiction to the assumption that f is a Morse function.

Analogously, Pareto minimum can therefore be redefined as points which cannot be reached via a monotone path from a dominated point y with $y \preceq x$. In both cases, y is not restricted to any neighborhood. Finally Pareto optimal points are surrounded by incomparable points, thus Pareto minimal and maximal simultaneously.

Obviously we can extend this notion to slabs quite easily. A slab is called Pareto maximal (minimal) if it cannot reach (not be reached from) dominating (dominated) slab. For dJCNs this simplifies to: A Pareto maximum slab has no outgoing edges and a minimum has no ingoing edges. As a reminder, in a dJCN there is only an edge if at least one of the rounded scalar field changes.

Theorem 1 *In the limit of a refinement series of dJCNs the union of all Pareto extremal slabs is equivalent to the Pareto set.*

Proof The limit of the series of the rounding functions is the identity. Therefore the limit of the slabs are just single points or areas with the same scalar value and the applied definitions are equivalent.

4.2 Recognition of Pareto Maxima and Minima

In coarse refinements, connected components of Pareto maxima and minima can be merged into one slab. This raises the question at which step in the refinement series all Pareto features are present in separated slabs in the dJCN. In order to decide when all connected components of Pareto maxima and minima are finally represented in their own critical slabs, we apply the concept of persistence, i.e. the scalar value difference between points.

Lemma 1 *After finite many refinement steps all Pareto maxima and minima slabs correspond to exactly one connected component of Pareto extrema.*

Proof A merge of two connected components of Pareto maxima or minima into the same slab is only possible if the refinement interval is larger enough. Namely, larger than the difference between any two points of these components with respect to any of the scalar fields. Otherwise, the slabs containing them would be separated by at least on edge. Hence, if the interval size is smaller than the minimal persistence over all scalar fields, all components of Pareto maxima and minima are separated from each other.

Note that, after this step is reached and under the assumption of a monotone refinement series, further refinement cannot merge existing slabs and thus undo this separation.

4.3 Recognition of Pareto Optima

Pareto optimal regions are maximal and minimal at the same time. By definition, there are no connection allowed to adjacent slabs. Hence at any slab border one scalar field has to be on the exact upper and one on the lower exact bound of the rounding function. In practical cases, this is only true in the limit of the refinement.

For example, assume a triangulation of a 2-manifold with three functions and three vertices: A, B and C. The values are $f(A) = (0.5, 0, 0)$, $f(B) = (0, 0.5, 0)$ and $f(C) = (0, 0, 1)$. Notice that all points in the triangle are incomparable to each other. Each function only increases in the direction of one vertex and decreases if we move towards the other two. So, all points in the triangle are Pareto optimal. But, as we show in Fig. 2, for all rounding functions the slabs have connections and therefore do not fulfill the condition for Pareto optimal slabs.

The above examples shows a general pattern. Scalar fields may divide the simplices at different points. As a consequence slabs will have connections although they contain only incomparable points. This can be avoided by generalizing the notion of Pareto extrema and define the changing rate.

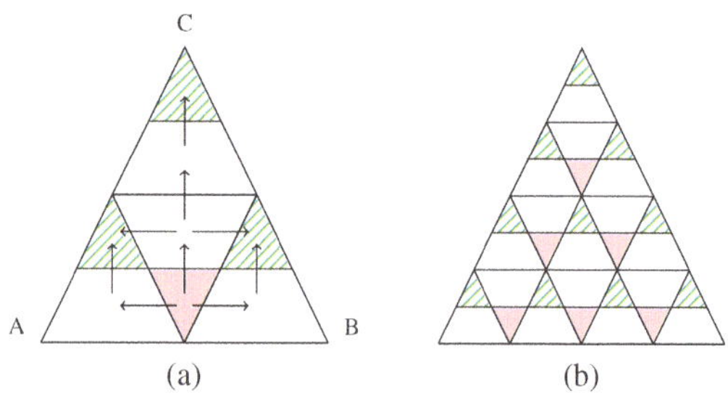

Fig. 2 An example with Pareto optimal points in slabs which are not marked as such. Pareto maximal slabs are marked in green hatch, the minimal in red, and regular slabs in white. (**a**) With one level of refinement. (**b**) With two levels of refinement

4.4 ϵ-Pareto Extrema

We call a point ϵ-Pareto maximum iff it cannot reach a point with Chebyshev (L_1) distance of more than ϵ via a monotone increasing path. Analogously, we define ϵ-Pareto minimum and optimum and again, extend this notion to ϵ-Pareto extremal slabs. By definition this notion is an over approximation of the Pareto set and if we chose $\epsilon = 0$ we get the original definition.

The idea is the Chebyshev distance creates a neighborhood around points x where ascending paths are allowed which otherwise would prohibit the classification of x as Pareto maxima. Hence, if no ascending path escapes this neighborhood, x is identified as such, i.e. a ϵ-Pareto maximum.

In addition, with the help of the gradient, we can identify the most increasing and most decreasing scalar field around each point. For each Pareto optimum x and each direction v, we call the quotient of the absolute values of the most increasing and most decreasing directional derivative over all separate fields f_i the *changing rate* of x towards v.

$$c_v(x) := \frac{|\max_i \nabla_v f_i(x)|}{|\min_i \nabla_v f_i(x)|}$$

We denote the maximal changing rate with $\mathrm{ch}_{\max}$. Note that a maximum $\mathrm{ch}_{\max}$ actually exists, since we assume that f is piece-wise linear, thus the gradients are piece-wise constant, and since otherwise, i.e. if no increasing or no decreasing gradient exist, x is not a Pareto optimum.

Lemma 2 *Let the interval size of the rounding function be smaller than the quotient of ϵ and $ch_{\max}$, then all Pareto optimal points are in ϵ-Pareto optimal slabs.*

Proof Given a Pareto optimal point $x \in \mathbb{S}$, some path p starting at x in direction v, the slab containing x can only be classified as ϵ-Pareto optimum, if along the path p at least one field increases and one field decreases, both with a value change of more than δ, the interval size of r. Then, the corresponding path in the dJCN could neither be classified as ascending nor descending path. Furthermore, these value changes need to appear within a Chebyshev distance of ϵ with respect of f around x. Otherwise, the ascending or descending subpath of p within the Chebyshev distance is sufficient to make x either not a ϵ-Pareto maximum or ϵ-Pareto minimum, respectively.

Hence, we approximate, from below, the number increasing and decreasing value changes along p within Chebyshev distance of ϵ that would result in slab changes along the corresponding path in the dJCN. If both numbers are equal or larger than 1, path p cannot be used to disqualify x as a ϵ-Pareto optimum.

Therefore, let y be the first point along p for which $\max_i(|f_i(x) - f_i(y)|) \geq \epsilon$. W.l.o.g. we can assume that the maximal argument of this Chebyshev distance is $i = 1$ such that $f_1(x) - f_1(y) = \epsilon$. For all other arguments, $i \neq 1$, it holds that $f_i(x) - f_i(y) < \epsilon$. Furthermore, since x is Pareto optimal, we find a minimal point y' between x and y and an arguments j, w.l.o.g. $j = 2$, with $f_2(x) - f_2(y') < 0$.

Otherwise, an ascending path from x to y exists such that x cannot be Pareto maximal and, therefore, neither Pareto optimal.

The number of slab changes c_1 based on f_1 is therefore approximated from below $\frac{f_1(x)-f_1(y)}{\delta} = \frac{\epsilon}{\delta}$. For an approximation of c_2, the slab changes based on f_2, we can assume that both f_1 and f_2 have the flattest ascend and descend, respectively, possible and, w.l.o.g. have $f_1(x) = f_2(x) = 0$. In other words, $f_1(t) = a \cdot t$ with $a = \frac{f_1(y)}{y}$ and $f_2(t) = b \cdot t$ with $b = \frac{f_2(y')}{y}$.

With this assumption we can define the location of y through ϵ and the gradient of f_1:

$$f_1(y) = \epsilon = a \cdot y \Rightarrow y = \frac{\epsilon}{a},$$

and calculate the number of slab changes c_2 based on f_2:

$$|\frac{f_2(y)}{\delta}| = c_2 \Rightarrow c_2 = |\frac{\epsilon \cdot b}{a \cdot \delta}| = \epsilon |\frac{b}{a}| \cdot \frac{1}{\delta}.$$

Since we assumed that $i = 1$ is the maximal argument for a point y with Chebyshev distance equals ϵ, we know that $|f_1(x) - f_1(y)| \geq |f_2(x) - f_2(y)|$ and thus, $|\frac{b}{a}| \leq 1$. Therefore, $c_2 \leq c_1$ is the lower boundary for direction v.

This argument, illustrated in Fig. 3, can be done analogously for every path originating from x to conclude the proof.

Since we have a lower bound for the interval size, the following theorem is a direct result.

Theorem 2 *Except for a bound interval, all refinement steps all connected components of Pareto extrema are contained in ϵ-Pareto extremal slabs.*

Note that by decreasing ϵ to zero we can reduce the over approximation for the price of more slabs to compute. On the other hand by choosing a rounding accuracy we can calculate a limit of the approximation rate of Pareto slabs.

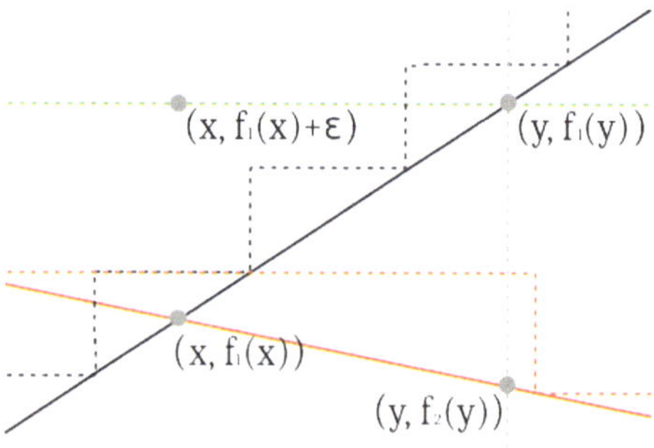

Fig. 3 Illustration of the number of slab changes between point x and y for two fields f_1, f_2

5 Discussion

We have shown that the error limit between the critical slabs and the Pareto set is reached after a finite amount of refinement steps. However, so far it is not possible to efficiently calculate $\mathrm{ch}_{\max}$ and thus the necessary resolution of the rounding function beforehand. Note that a naive calculation requires roughly the same computational effort as the calculation of the Pareto set itself, thus providing no faster runtime.

5.1 Implementation of JCN over Constant 2D-Grid

To support the changing of ϵ and the interval size, we use an alternative implementation of the slabs in contrast to existing literature [3, 13]. Assuming the grid is evenly triangulated, each triangle is recursively separated in equally sized subtriangles based on the interval size and the corresponding rounding function r. Each triangle is recursively subdivided until the distances to its adjacent triangles with respect to r is smaller than 1, i.e. $\forall_i |r(f_i(x)) - r(f_i(y))| \leq 1$ with x, y the centroids of the adjacent triangles. In a second step, if for any neighboring triangles their centroids are inside the same interval, i.e. $r(f_i(x)) = r(f_i(y))$ for all fields i, they are iteratively merged into triangle clusters representing the slabs.

This allows for fast refinement of the slabs if the interval size is changed.

5.2 Example

An simple example showing all types of ϵ-Pareto extrema is based on three shifted Gaussian functions, similar to the two used for Fig. 1. Figure 4 shows how the connected components of the Pareto set are over-approximated by ϵ-Pareto extremal slabs. Note how the shape of the extremal slabs in Fig. 4b is similar to the Pareto extrema in Fig. 4a as well as the correct classification into Pareto minimal, maximal, and optimal slabs. This effect can be improved with increased refinement at the cost of a higher number of slabs. The implementation is not restricted by the number of fields or the complexity of the data, see Fig. 5.

For the fields in Fig. 5, we used Gauss functions with slightly translated extrema. The movement was done parallel to the horizontal and vertical axis such that the components of the Pareto set are in cubic shapes. As in previous work for Pareto sets [9], we notice some irregularities (marked with blue circles) which we attribute to the coarseness of the data triangulation. Note how those irregularities start to show up in the critical slabs (Fig. 5a) depending on the interval size of the rounding function.

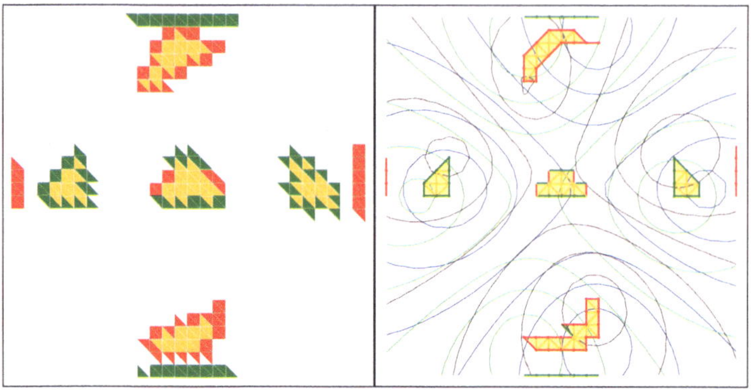

(a) (b)

Fig. 4 Pareto extremal points and slabs colored green, red, and yellow depending on their type. In (**b**) the three individual fields are indicated by a selection of contour lines. (**a**) Critical slabs. (**b**) Pareto set

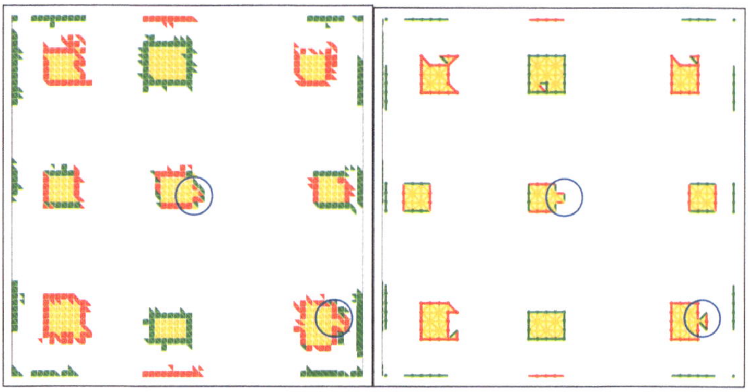

(a) (b)

Fig. 5 Pareto extremal points and slabs colored as in Fig. 4 but with an increased number of fields and univariate extrema within each field. (**a**) Critical slabs. (**b**) Pareto set

Furthermore, to our knowledge neither dJCNs nor any other approach can indicate where critical slabs appear after further refinement of the rounding function. This prohibits stepwise, local refinement, i.e. refinement only around critical slabs.

5.3 *Reachability Graph*

To visualize additional information about the Pareto set, a connected component in the Pareto set *reaches* another, if a monotone ascending path between any two points of those components exist. Obviously, this also implies that a path between

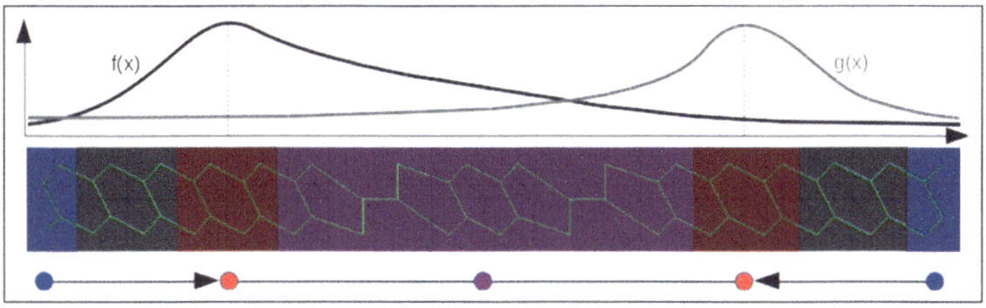

Fig. 6 Two $1D$ fields, a dJCN for a $2D$ extension of the fields, and the corresponding reachability graph

the slabs containing those points exist in the dJCN. Furthermore, since the ϵ-Pareto optimality is also defined through reachability, we can apply the same ideas to prove similar properties for the reachability graph, especially regarding its approximation by a dJCN.

The next example shows two $1D$ scalar fields $f_i : [0, 10] \mapsto \mathbb{R}$ in the upper part of Fig. 6, each with a distinct, obvious maximum in the center. The fields are extended to $2D$ with $f_i(x, y) = f(x)$ for $y \in [0, 1]$ and $i \in \{0, 1\}$. The grid is triangulated, the resulting slabs are shown in the middle part of the figure. Colors indicate the slab status while purple lines connect neighboring triangles if their centroids are comparable, i.e. dominating or dominated.

Using a Union-Find-algorithm to combine neighboring Pareto extremal slabs of the same type provides the nodes for the reachability graph. Edges are created by running a path-finding, for examples breadth-first-search, algorithm on the dJCN.

With this, we have the following algorithm.

- Create the dJCN
- Calculate the ϵ-Pareto set
- Identify clusters of Pareto extremal slabs through Union-Find
- Build the reachability graph through Breadth-First-Search on the dJCN

5.4 Reeb Space

The JCN is an approximation of the Reeb space, a generalization of the Reeb graph for multifield data. More precisely, given the equivalent class $x \sim y$ iff both points belong to the same path connected component of the preimage $f^{-1}(f(x)) = f^{-1}(f(y))$, the Reeb space is the space of all equivalent classes with the quotient topology inherited from $\mathbb{S}$. Thus it is reasonable to discuss the Reeb space as a means to calculate the Pareto sets and the reachability graph.

The Reeb space has the advantage that it is continuous and not an approximation as the JCN. Thus, we can apply the original definition of Pareto sets instead of the detour of ϵ-Pareto set. Assume that for two points $x, y \in \mathbb{S}, x \to y$ holds, i.e. a continuous ascending path from x to y exists. Then, there is also an ascending path

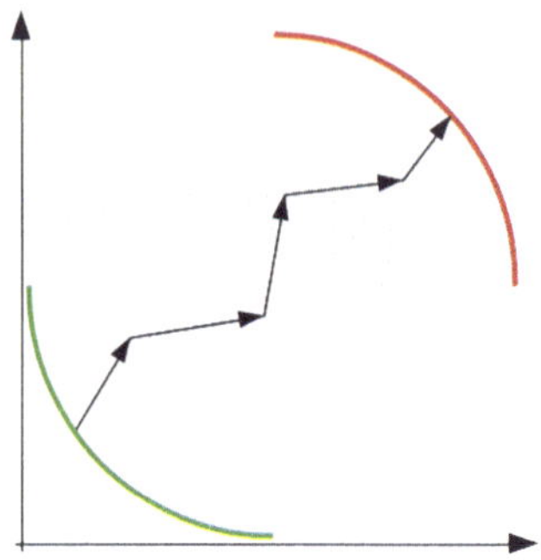

Fig. 7 The Reeb space for a disk with functions $f_1(x, y) = x$ and $f_2(x, y) = y$. The red and green lines indicate Pareto maxima and Pareto minima, respectively, and the arrows show one possible ascending path that connects the two Pareto extremal components

in the Reeb space from the equivalent class containing x to the class containing y. Therefore, corresponding to the definitions in Sect. 4.1, we can define points in the Reeb space as Pareto minimal, Pareto maximal, or Pareto optimal. Like Jacobi structures we could extract the Pareto set from the Reeb space, see [4]. The extraction might be improved by the fact that the Pareto set can be considered a subset of the border of the Reeb space, see Fig. 7.

However, the main advantage from the dJCN, namely the graph structure which allows a fast approximation of the reachability graph, is not given anymore. We hypothesize that the calculation of the ascending and descending paths is equally expensive in the domain $\mathbb{S}$ as well as in a triangulated Reeb space.

6 Conclusion

In contrast to previous work, we are not only able to show the connection between JCNs and Pareto set visually but prove and quantify the relation. In detail, we provide the following contributions to the study of multivariate topology-based analysis techniques:

- We disproved that the previous, naive definition of critical slabs is a correct approximation of the Pareto set.
- We therefore introduced the ϵ-based definition of Pareto extrema, both in context of simplicial complexes and the dJCN.
- We also proved that under the ϵ parameter and after a finite number of refinement steps, critical slabs are equivalent to Pareto extrema such that the dJCN can be used as an approximation of the Pareto set and its related structures.
- Furthermore, through the intermediate definition of ϵ-Pareto minima and maxima, we can limit the error between Pareto sets and dJCNs and the resolution of the rounding interval.

In summary, we can approximate the Pareto set through the dJCN and can limit its error to some degree. Furthermore, the dJCN can be used to approximate the reachability graph using simple path finding algorithms for graph structures in contrast to the existing, computationally expensive approaches on the piece-wise linear structure.

We implemented and applied the results on a set of artificial data sets with multiple fields and discussed the implications of our work for the relations between Pareto sets and the Reeb space. While the latter could provide a error-less identification of the Pareto set, we do not see a fast method to construct the reachability graph out of the Reeb space. With this, future work aims at utilizing simplification approaches based on JCN or the Reeb space to efficiently remove local structures from our multivariate visualization.

References

1. Bormann, J.: Joint contour nets revisited. Master's thesis, University of Kaiserslautern, Kaiserslautern (2016)
2. Carr, H., Duke, D.J.: Joint contour nets: computation and properties. In: IEEE Pacific Visualization Symposium, PacificVis 2013, February 27 2013–March 1, 2013, Sydney, NSW, pp. 161–168 (2013)
3. Carr, H., Duke, D.J.: Joint contour nets. IEEE Trans. Vis. Comput. Graph. **20**(8), 1100–1113 (2014)
4. Chattopadhyay, A., Carr, H., Duke, D., Geng, Z.: Extracting Jacobi structures in Reeb spaces. In: Elmqvist, N., Hlawitschka, M., Kennedy, J. (eds.) EuroVis – Short Papers. The Eurographics Association, Aire-la-Ville (2014)
5. Chattopadhyay, A., Carr, H., Duke, D., Geng, Z., Saeki, O.: Multivariate topology simplification. Comput. Geom. Theory Appl. **58**(C), 1–24 (2016)
6. Edelsbrunner, H., Harer, J.: Jacobi sets of multiple morse functions. In: Cucker, F., DeVore, R., Olver, P., Süli, E. (eds.) Foundations of Computational Mathematics: Minneapolis, 2002, pp. 37–57. Cambridge University Press, Cambridge (2004)
7. Edelsbrunner, H., Harer, J., Patel, A.K.: Reeb spaces of piecewise linear mappings. In: Proceedings of the 24th ACM Symposium on Computational Geometry, College Park, MD, June 9–11, 2008, pp. 242–250 (2008)
8. Huettenberger, L., Garth, C.: A Comparison of Pareto Sets and Jacobi Sets, pp. 125–141. Springer, Berlin (2015)
9. Huettenberger, L., Heine, C., Carr, H., Scheuermann, G., Garth, C.: Towards multifield scalar topology based on pareto optimality. Comput. Graph. Forum **32**(3), 341–350 (2013)
10. Huettenberger, L., Heine, C., Garth, C.: Decomposition and simplification of multivariate data using pareto sets. IEEE Vis. **20**(12), 2684–2693 (2014)
11. Huettenberger, L., Heine, C., Garth, C.: A comparison of joint contour nets and pareto sets. In: Carr, H., Garth, C., Weinkauf, T. (eds.) Topological Methods in Data Analysis and Visualization IV, pp. 51–65. Springer International Publishing, New York (2017)
12. Munch, E., Wang, B.: Convergence between categorical representations of Reeb space and mapper. In: 32nd International Symposium on Computational Geometry, pp. 53:1–53:16 (2016)
13. Nam, H.A., Staszczak, A., Schunck, N., Knoll, A., Carr, H., Duke, D.: Visualizing nuclear scission through a multifield extension of topological analysis. IEEE Trans. Vis. Comput. Graph. **18**(12), 2033–2040 (2012)
14. Tierny, J., Carr, H.: Jacobi fiber surfaces for bivariate Reeb space computation. IEEE Trans. Vis. Comput. Graph. **23**(1), 960–969 (2017)

Flexible Fiber Surfaces: A Reeb-Free Approach

Daisuke Sakurai, Kenji Ono, Hamish Carr, Jorji Nonaka, and Tomohiro Kawanabe

Abstract The *fiber surface* generalizes the popular isosurface to multi-fields, so that pre-images can be visualized as surfaces. As with the isosurface, however, the fiber surface suffers from visual occlusion. We propose to avoid such occlusion by restricting the components to only the relevant ones with a new component-wise flexing algorithm. The approach, *flexible fiber surface*, generalizes the manipulation idea found in the *flexible isosurface* for the fiber surface. The flexible isosurface in the original form, however, relies on the contour tree. For the fiber surface, this corresponds to the Reeb space, which is challenging for both the computation and user interaction. We thus take a *Reeb-free* approach, in which one does not compute the Reeb space. Under this constraint, we generalize a few selected interactions in the flexible isosurface and discuss the implication of the restriction.

1 Introduction

An isosurface is defined as the inverse image $f^{-1}(s)$ of some scalar value s in the scalar field $f : M(n) \to \mathbb{R}$, where $M(n)$ is an n-manifold. Though visualizing isosurfaces [19] is a routine task for scalar field analysis, it is often required to understand correlations between multiple quantities (e.g. temperature and pressure).

D. Sakurai (✉)
Pan-Omics Data-Driven Research Innovation Center Kyushu University, Fukuoka, Japan
e-mail: d.sakurai@computer.org

K. Ono
Research Institute for Information Technology, Kyushu University, Fukuoka, Japan

RIKEN Center for Computational Science, Kobe, Japan

H. Carr
School of Computing, University of Leeds, Leeds, UK

J. Nonaka · T. Kawanabe
RIKEN Center for Computational Science, Kobe, Japan

H. Carr et al. (eds.), *Topological Methods in Data Analysis and Visualization V*, Mathematics and Visualization, https://doi.org/10.1007/978-3-030-43036-8_12

In fact, scalar data analysis itself can be assisted by multi-field analysis when considering the gradient magnitude as the second field [17].

For 3-D multi-fields of the form $f : M(3) \rightarrow \mathbb{R}^m$, the range is extended to data tuples $(x_1, .., x_m) \in \mathbb{R}^m$. In particular, the *fiber surface* [8] generalizes the isosurface for $f : M(3) \rightarrow \mathbb{R}^2$, which is useful for extracting pre-images in the domain as a surface mesh. The fiber surface is the pre-image $f^{-1}(P)$ of a *control polygon* P, which is a polyline composed by l linear segments each having the endpoints (x_i, y_i) and (x_{i+1}, y_{i+1}) $(1 \leq i \leq l+1;\ i, l \in \mathbb{N})$. Each vertex (x_i, y_i) is called a *control point*. The user can specify the fiber surface by drawing the control polygon in the 2D range. The control point shall be dragged to see how the shape of the fiber surface changes. This allows fiber surfaces to be simple to compute, compact in size, and quantitative as isosurfaces in scalar fields. Since the fiber of a point in an $\mathbb{R}^2$ range is a 1-D structure almost everywhere in the 3-D domain, the fiber surfaces are indeed surfaces in general. We call a connected component of a fiber surface a *fiber surface component*. (A *component* refers to a fiber surface component unless specified otherwise.) However, interacting with the components remains still a challenging task. As with the isosurface, the fiber surface suffers from overlaps and may have too many features for the user to comprehend. In case of the isosurface, one has been able to utilize the *flexible isosurface* [7] to distinguish the components, hide irrelevant ones such as noise or uninteresting phenomena, and even vary the isovalue per component. Such interactions were shown to be useful for visualizing different objects in a CT scan dataset separately, for example.

We thus adapt the component-wise manipulation of the isovalue, as found in the flexible isosurface for multi-fields, as the *flexible fiber surface* (Fig. 1). Where an isosurface component is contracted as a point in the *contour tree* [5], a component of a fiber $f^{-1}(p)$ of some value $p = (x, y)$ is contracted as a point in the *Reeb space* [13]. Hence, each component of a fiber surface $f^{-1}(P)$ finds its contraction in the Reeb space as a connected component.

We generalize the flexible isosurface, or to be more precise, its concept of component-wise manipulation. Our flexible isosurface interface, as implemented [7], pre-computes the contour tree in order to obtain the seed of isosurface

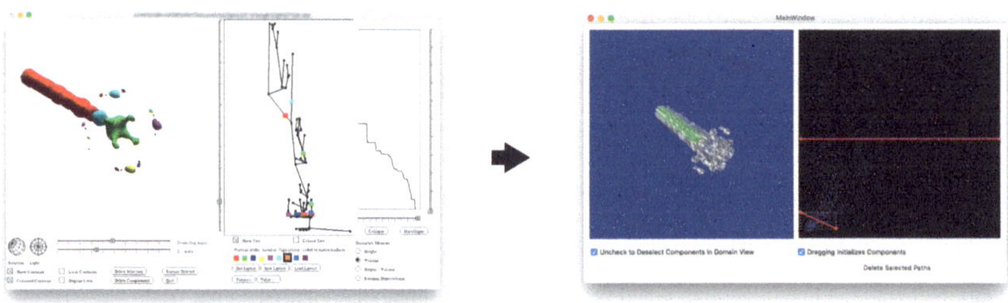

Fig. 1 Our flexible fiber surface generalizes the flexible isosurface to multi-fields without pre-computing the topology of pre-images. The domain is shown on the left and the range on the right

components and the connectivity of these components across different isovalues. While the contour tree is generalized into the Reeb space for multi-fields, in this work we avoid computating the Reeb space (we call such an approach to be *Reeb-free*), and for the following reasons. Firstly, the scalability of analysis is limited by the scale at which the Reeb space can be computed. Secondly, the current standard implementation [27, 28] of the Reeb space computation requires the user to subdivide the domain to a sufficiently fine resolution such that at least one tetrahedron is completely contained in each 3-sheet (which is challenging to achieve). Finally, the Reeb space is hard to be shown to, and to be utilized by, the user when the Reeb space is complicated [24], which is usual in real-world data. Interactions relying on the contour tree are not generalized in our approach (3 and 6) as we avoid obtaining the Reeb space.

Our contributions include (1) varying the input control polygon per fiber surface component, (2) freeing the operations of the flexible fiber surface from precomputing the Reeb space. For contribution (1), we devise a new flexing algorithm. Contribution (2) means that the flexible iso-surface also becomes Reeb-free as a special case. Our algorithm works for any 3-D tetrahedral grid, regardless of the homology of the domain. Our key idea is to follow the pre-image on-demand.

The remainder of this article is organized as follows. We introduce the related work in Sect. 2, and generalize the semantics of the original flexible isosurface to our Reeb-free flexible fiber surface in Sect. 3. We then revisit the existing algorithms of extracting a fiber surface in Sect. 4 with their implication to our computation. Section 4 explains how our algorithm identifies the connected components of fiber surfaces extracted with the algorithm by Klacansky et al., and further deform the fiber surface. Section 5 demonstrates the outcomes of our proof-of-concept implementation. Section 6 discusses the indication of generalizing flexible-isosurface to multi-fields in a Reeb-free manner including the limitations. Finally, Sect. 7 gives the conclusion and future work.

2 Related Work

The flexible fiber surface extends topological operations that are defined for isosurface. We thus introduce relevant work in the topological analysis for scalar fields and multi-fields.

2.1 Work in Scalar Field Analysis

Isosurface An isosurface can be visualized effectively with *marching cubes* [19]. Its basic idea is to rotate pre-defined cubes with triangular patches inside to reconstruct the isosurface. Since the original marching cubes algorithm had ambiguities of its topology, topological algorithms often tessellate the cubes into tetrahedra and apply *marching tetrahedra* [3]. In contrast to the marching cubes, the *continuation method* [14, 31] starts from seeds and proceeds to adjacent cells. This enables component-wise tracking of a pre-image, on which our algorithm is based.

Topological Analysis The Reeb graph [23] is a quotient space defined by contracting each of the connected components of isosurfaces to a point. For simple domains, their Reeb graph is guaranteed to be a tree—hence it has the special name *contour tree*. Reeb graphs and contour trees are useful for displaying the global configuration of the connected components of isosurfaces [2], simplifying the data [7, 10], indexing isosurfaces [16], extracting the topological changes in an isosurface [21, 25] and designing transfer functions [25, 30]. The standard algorithm for the contour tree [5] uses as input the isosurface topology along the edges of the simplicial mesh. On the other hand, van Kreveld et al. [29] tracked the pre-image explicitly, which is in fact the approach of our pre-image tracking as well. More generally, the Reeb graph must be computed [22] instead.

Parallelization of these algorithms has been, and is still, a challenge [11, 21] especially for distributed systems [9]. The dependency to pre-computed topology thus restricts the scalability of component-wise manipulations we aim to achieve. The manipulation, however, is in fact independent of the pre-computation as we show with our Reeb-free approach.

Interface One of the attributes of the contour tree is its ability to provide seeds for continuation. This forms the basis for the flexible isosurface [7]. This assumes that features are represented with connected components of isosurfaces, and lets the users analyze them while avoiding irrelevant surface components.

The original flexible isosurface interface (Fig. 1) consists of two views, one for the domain and another for the range. The interface shows the connected components in the domain. The components are assigned distinct colors so that the user can visually identify the connected components. These colors are simultaneously projected in the range view as color labels in order to indicate the component's isovalue. The user can see the pre-computed contour tree optionally in the 2D view on the right. This view shows the 1D range with an auxiliary dimension so that one can see the tree structure. In fact, labels are overlayed at the contraction points in the tree to indicate the position of the components. Though the contour tree was mandatory in this particular work, this was not necessary for data exploration as the user still could explore the global context by seeing the cumulative distribution.

We summarize several characteristic operations of the flexible isosurface below:

- Initialization: the user can *initialize* a flexible isosurface, i.e. show the entire pre-image or the *largest contour segmentations* [20].
- Selection: the user *selects* the components of interest in order to apply further operations. The user clicks on the components or on their labels mapped to the contour tree. Components can also be *deselected* with the mouse.

- Evolution: the user varies the isovalue for a subset of visible components by dragging the corresponding projection in the contour tree or all at once by manipulating the isovalue slider.
- Deletion: the user can *delete* selected components. Those are components such as artifacts and objects irrelevant to the analysis.
- Addition: the user *adds* a hidden surface component to the domain by clicking on its point contraction in the contour tree.
- Simplification: if the abundance of surface components makes navigating in the contour tree difficult, the user can *simplify* (i.e. hide) or *unsimplify* (show) the isosurface by thresholding surface statistics. This is achieved by cutting away or putting back the arcs of the contour tree, respectively.

2.2 Work in Multi-Fields Analysis

In multi-fields, the isosurface in scalar fields we have seen above generalizes as the fiber surface.

Fiber Surface Carr et al. [8] approximated the fiber surface as the isosurface of a scalar field, where the scalar value was the distance to the control polygon in the range. Later, Klacansky et al. [18] proposed an algorithm to compute the fiber surface without this approximation. We use the latter algorithm when extracting the fiber surface.

Topological Analysis Edelsbrunner et al. [13] generalized the Reeb graph to multi-fields as the Reeb space by contracting each connected component of the fiber f^{-1} to a point. The connectivity of connected components was not computed in their algorithm. This was later achieved by quantizing the range [4] into rectangular regions and connecting their pre-image in the domain. This approach found application to visualizing nuclear scission [12] and fiber topology [24].

Eventually, Tierny and Carr [27] computed the Reeb space without the quantization. The algorithm partitions the domain with *singular fibers*, at which topological events (such as mergers, splits, birth and death of the pre-image) happen. They estimate the steps of the algorithm to be $O(n_j \times n_T)$. n_j is the number of tetrahedral edges E; n_T is the number of tetrahedra. Though some optimization is possible [27], the fact remains that irrelevant topological events in the data may heavily increase the running time. As with the topological analysis for scalar fields, this can restrict the scalability of the analysis. Our *Reeb-free* computation, on the other hand scales linearly with the size of the feature that is actually of interest to the user. Last but not least, Pareto optimality gives an alternative generalization of scalar topology [15].

3 Generalizing the Semantics

Our flexible fiber surface generalizes the iso-surface evolution to multi-fields. The advantages and disadvantages of our generalization will be discussed in Sect. 6.

3.1 Generalizing the Interface

Figure 1 shows our fiber surfacing interface on the right. In the interface, the domain view on the left shows the fiber surface as the flexible isosurface interface does for the isosurface. In contrast, our range view on the right replaces the topological information with a scatterplot to navigate the user in the range. The prototype focuses on proving the concept.

3.2 Generalizing the Component-Wise Operations

Initialization When the user *initializes* the fiber surface of a control polygon, the domain view shows the user all the surface components to understand their distribution in the domain.

Selection and Deselection (De)selection can be achieved in the domain view. In contrast, the component-wise selection and deselection cannot be done in the range, since all the components of a fiber surface overlap in the view.

Evolution The user reshapes the control polygon for selected group of components by dragging the control points.

Deletion Selected components can be removed from the domain view.

Addition and Simplification As we free the flexible fiber surface from the Reeb space, these operations become infeasible. Indeed, addition requires displaying the components to the user, which is the Reeb space itself. Simplification, too, requires access to the global topology.

4 The Algorithm

We assume a tetrahedral grid and the barycentric interpolation. First we extract the fiber surface using the algorithm by Klacansky et al. [18]. In each segment of the control polygon, this algorithm first extracts a *base fiber surface*, which is the pre-image of the line that covers the segment (see Fig. 2). The surface is the set of triangular patches obtained by applying the marching tetrahedra to the scalar field

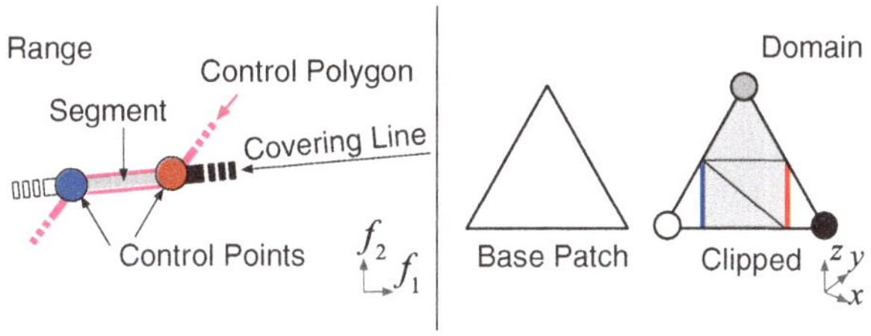

Fig. 2 Fiber surface extraction [18]. The white/grey/black color indicates the position in the covering line. A base fiber surface patch is clipped at the pre-image (blue and red) of control points

defined as the signed distance to the line. The pre-image of the complement of the segment is then *clipped*, i.e. cut away.

Next, we identify components, and the user selects interesting ones. The user shall then move the control point to specify the *target control polygon*. During the process, the flexing algorithm proposed in this article tracks the movement of the fiber surface components individually by tracking the *active tetrahedra*. (They cover the surface components being deformed (Algorithm 1).)

4.1 Identifying the Connected Components

Once the surface has been approximated, we can check its connectivity with the union-find algorithm [26]. The elements of union-find are the points in the mesh, and we connect the pairs of such points if they lie in adjacent patches. In the traditional marching tetrahedra, a point shared by adjacent mesh triangles resides in the same tetrahedral edge. Identifying the points is solved by identifying the edge [3] since no two different points reside in a single edge. However, a vertex may not lie in the edge for our fiber surface computation since the patches are clipped (Fig. 2).

A naive approach is to glue the triangular patch corner points when they have close coordinate values. However, a fiber surface can have intersecting patches, and thus two points in disconnected patches can share close coordinates. We instead find the connection between a base fiber's patches and clipped patch corners separately.

4.1.1 Patch Corners in Base Fiber Surface

To compute the location of corners in the base fiber surface is to extract the isosurface of the signed distance to a control polygon segment. The connectivity between the points can be obtained by recording the point ID at the tetrahedral edge. As a tetrahedral edge intersects with a base fiber surface only once at most, we keep record of the point IDs for each control polygon segment separately. By

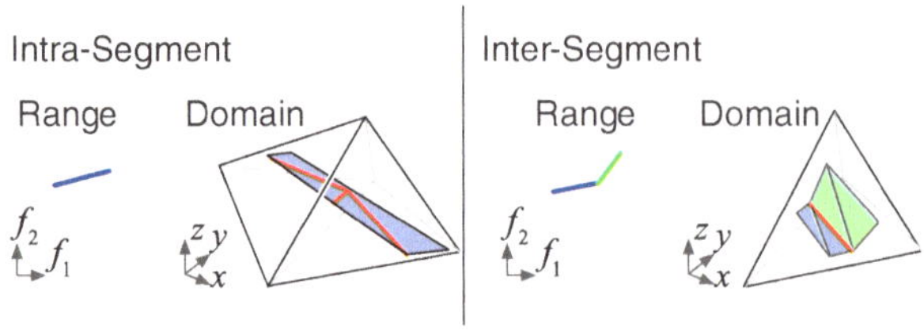

Fig. 3 Intra-segment connections and inter-segment connections (both in red) between clipped patches induced by the control polygons in the range

using these point IDs as the elements of the union-find data structure, we connect the patch corners in a base fiber surface as long as they are actually connected to each other.

4.1.2 Patch Corners Due to Clipping

As we can see in Fig. 3, two connected patches can belong to the same segment (*intra-segment* connection) or two neighboring ones (*inter-segment* connection).

Finding Intra-Segment Connections Figure 3 shows that an intra-segment connection can have a connection of corner points inside a single tetrahedron and across two tetrahedra adjacent to each other. To connect the points in the former manner, our algorithm joins two points with the union-find if they are from the same base fiber surface and inside the same tetrahedron. In addition, two points that are from the same base fiber surface and clipped by the identical control point are given the same point ID when they touch each other at a tetrahedral face. To do so, we record the point ID at the face for each control point separately just as we did for base fiber surfaces in Sect. 4.1.1. Notice that clipped patch corners are given a unique ID only when touching a tetrahedral face.

Finding Inter-Segment Connections Two clipped patches which are pre-image of different but adjacent segments belong to the same connected component as long as they are inside the same tetrahedron and were clipped due to the same control point. We loop through the segments, clip the patches, and connect every neighboring pairs. If the control polygon is closed, i.e. $(x_1, y_1) = (x_{l+1}, y_{l+1})$, we clip the first segment's patches but defer connecting them until the end of the loop.

4.2 Following the Connected Components

While the user drags a control point, the control polygon sweeps the range. As illustrated in Fig. 4, this starts from the *source segments* and ends at the

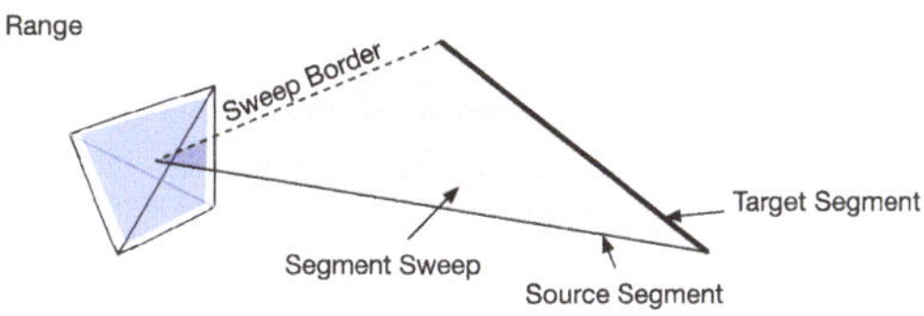

Fig. 4 When the user drags a control point, the segments sweep the range, starting from the source position and ending at the target position. Our algorithm walks through the *active tetrahedra*, which contain the pre-image of a moving segment

target segments. We model the control point to move along the *sweep border*, which is a line segment connecting the start and target positions. We follow the surface components being continuously deformed in the domain. If we observe a moving component from inside a tetrahedron, the component starts its motion from the original location and moves towards its destination. As the sweep proceeds, the pre-image penetrates into adjacent tetrahedra, and moves out from our example tetrahedron if the destination is outside.

We detail this procedure in Algorithm 1. We start by gathering the *active tetrahedra*, i.e. the tetrahedra holding the user-selected fiber surface patches. We pair each tetrahedron with the segments that define the component, and put all such pairs in a $queue$. If a tetrahedron overlaps with multiple segments, every segment gets its own pair.

We pop a pair (seg, tet) from $queue$ and operate on it (lines 3–13). In order to avoid processing the same pair twice, we mark the pair as visited. This visit flag is implemented as an array of tetrahedron IDs for each segment. If the points of tet have both positive and negative distance to the seg, tet may intersect with a surface component of seg. The pair then joins the output pairs P_t.

As we have checked the evolution inside tet for seg, we push adjacent tetrahedra tet_a's in $queue$ for visiting it later as long as the component continues tet_a (lines 14–18). This continuation happens when tet and tet_a's touching face intersects seg in the range.

Finally, we pass tet to its neighbors seg_n's (lines 19–23). If a control point between seg and seg_n does not move during a drag, the evolution of component is independent of seg_n. We do not process such seg_n (line 20). Otherwise, we check whether the component of (seg, tet) continues to seg_n (line 20). If so, we put (seg_n, tet) in $queue$.

After we visited all the (seg, tet) pairs, we extract the fiber surface components in them using the method by Klacakanski et al.

Algorithm 1 Follow the deformation of surface components

Input: Source control polygon P_s, target control polygon P_t, Pairs (segment, active tetrahedron) $tets_s$ of P_s

Output: Pairs (segment, active tetrahedron) $tets_t$ of P_t

```
queue ← tets_s
while queue is not empty do
    pop (seg, tet) from queue
    if (seg, tet) is visited then
        continue
    end if
    mark (seg, tet) as visited
    if seg is not dragged then
        continue
    end if
    if (seg, tet) has base fiber surface of target segment then
        put (seg, tet) in tets_t
    end if
    for tetrahedron tet_a adjacent to tet do
        if (seg, tet_a) is not visited and shared_face(tet, tet_a) intersects segment sweep in
        range then
            put (seg, tet_a) in queue
        end if
    end for
    for segment seg_n neighboring seg do
        if control point between seg and seg_n moved and (seg_n, tet) is not visited and tet
        intersects sweep border in range then
            put (seg_n, tet) in queue
        end if
    end for
end while
```

5 Outcomes

We build a proof-of-concept interface with C++. We use VT for the data structure and Qt for the GUI. The interface lets us display a pre-computed scatterplot or continuous scatterplot [1]. We selected a few typical, simple, datasets to evaluate our Reeb-free approach. We show that one can in fact achieve component-wise flexing of fiber surfaces. Our serial implementation is run on a PC with Intel Xeon CPU (3.20 GHz, 20 MB Cache) and 64 GB RAM. Each dragging completes in several seconds.

5.1 The Analysis of the Algorithm

As expected, the computation time scales linearly with the number of active tetrahedra the algorithm, Algorithm 1, visits (Fig. 5). This demonstrates the fact that the size of the features the user is interested determines the response time. If the

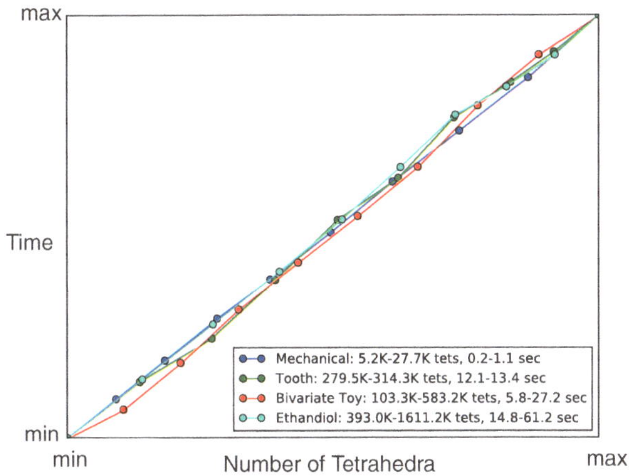

Fig. 5 The duration of dragging scale linearly with the number of tetrahedra we process. The examples share the same control polygon: it is defined as the diagonal line from the point (0, 0) to (1, 1) in the normalized range, and is dragged towards (0.6, 0.4) to produce 10 samples evenly along the trace

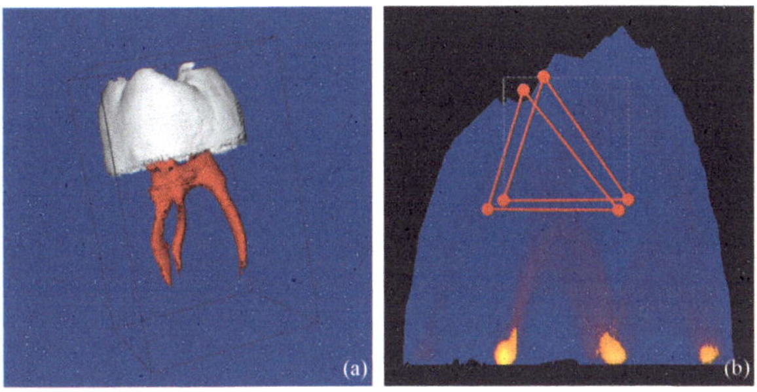

Fig. 6 Tooth dataset. The two fields are CT value and its gradient magnitude. (**a**) The domain and (**b**) the range. We start by drawing two control polygons that contain either only the crown or root. We then move them into a region that contains both. The evolution of the crown in white (or root in red) is restricted to the crown (root)

tetrahedra are distributed evenly in the range, the number of tetrahedra being swept shall scale linearly with the distance the dragged control point moves away. We can see this in the plot as the near-constant distance between two neighboring points of each line.

5.2 A Simple Proof of Concept: The Tooth Dataset

The tooth dataset (Fig. 6) gives a simple proof of concept for our flexible fiber surface. We subdivided each cube of the input regular grid into 6 tetrahedra with the Freudenthal tessellation (see [6]), so that the tetrahedral faces are consistent across neighboring cubes.

We report that the features are simple to identify without the Reeb space since we can identify the boundary of objects as hyperbolic curves [17], and their overlaps in the range resolve in the high gradient regions.

5.3 *Comparison with the Flexible Isosurface*

We take some small head CT dataset with the resolution of 50 × 50 × 50 for comparing our results with the flexible isosurface. In Fig. 7, we have visualized the dataset with the flexible isosurface and with our Reeb-free interface. Due to the overlap of features in the range, the interaction with the domain is essential for

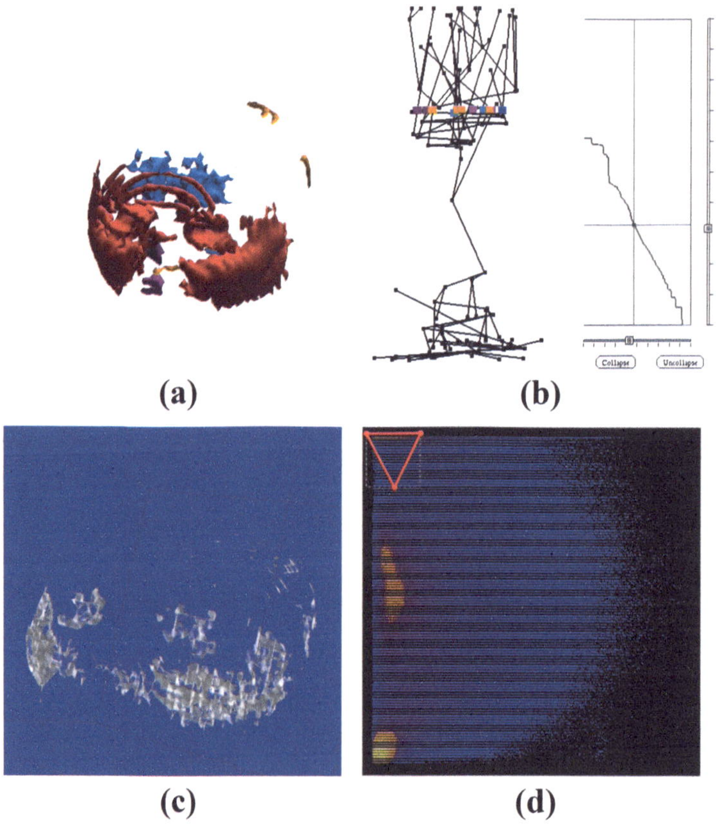

Fig. 7 Comparing the original flexible isosurface (**a**, **b**) and our Reeb-free flexible fiber surface (**c**, **d**) under severe overlaps of features in the range

our Reeb-free interface. The simplified contour tree is a significant advantage of the original flexible isosurface interface since it gives hints to an experienced user about the surface component evolution.

6 Discussion

We now discuss the consequence of our generalization of flexible isosurface to Reeb-free multi-field analysis.

Analysis of the Algorithm The number of steps required for tracking a component is $O(n_T)$, where n_T is the number of tetrahedra to be visited in our method. n_T shall be close to the number of tetrahedra necessary to extract the fiber surface partitioning the domain [24] with the Reeb space extraction [27]. Though our implementation is serial, the approach can apparently extend itself to distributed systems by locally running Algorithm 1 in each node with occasional communications between different nodes. Though this requires further research, it should be more feasible than computing the Reeb graph of such systems.

Evolution The evolution of fiber surface components lets us understand how multivariate values distribute, and especially how the features continue in the domain.

Simplification We did not simplify the topology of connected components although this was available in the flexible isosurface concept by Carr et al. thanks to the contour tree.

Global Exploration We show a scatterplot to provide the user with a global context to the analysis. The cruciality of this lack depends on the dataset to be analyzed. Datasets with similar objects tend to suffer because their image overlap in the range. The scatterplot can be peeled [27] for an effective exploration, though such an operation assumes pre-computing the Reeb space. Even if the Reeb space were available, navigating the user in the abundance of features is a challenge. This is because visualizing the Reeb space becomes rapidly challenging as it grows complex [24].

7 Conclusion and Future Work

We extended the flexible isosurface to multi-field without requiring Reeb space analysis. In particular, we generalized the semantics of component-wise pre-image evolution to multi-fields. Our approach does not require computing the pre-image topology explicitly. The algorithm identifies the connected components of fiber surfaces, and sweeps the range to track them. The lack of global pre-image topology and simplification is a downside of this approach (although rendering the Reeb space is itself an unsolved challenge). Through experiments for rather simple datasets, we demonstrated that the interaction in the domain does not necessarily require the Reeb space.

We see a few future directions: the global navigation and simplification of data that are affordable for non-experts of topological analysis; extension to different cell types and interpolants.

Acknowledgments We thank Julien Tierny at Sorbonne Universities UPMC for offering some of the datasets [28]. This work was supported by the German Federal Ministry of Education and Research (HD(CP)^2 project, grant number 01LK1501C) and the Engineering and Physical Sciences Research Council (EPSRC) project EP/J013072/1.

References

1. Bachthaler, S., Weiskopf, D.: Continuous scatterplots. IEEE Trans. Vis. Comput. Graph. **14**(6), 1428–1435 (2008)
2. Bajaj, C.L., Pascucci, V., Schikore, D.R.: The contour spectrum. In: Proceedings of IEEE Visualization '97, pp. 167–173 (1997)
3. Bloomenthal, J.: Polygonization of implicit surfaces. Comput. Aided Geom. Des. **5**(4), 341–355 (1988)
4. Carr, H., Duke, D.: Joint contour nets. IEEE Trans. Vis. Comput. Graph. **20**(8), 1100–1113 (2014)
5. Carr, H., Snoeyink, J., Axen, U.: Computing contour trees in all dimensions. Comput. Geom. **24**(2), 75–94 (2003)
6. Carr, H., Moller, T., Snoeyink, J.: Artifacts caused by simplicial subdivision. IEEE Trans. Vis. Comput. Graph. **12**(2), 231–242 (2006)
7. Carr, H., Snoeyink, J., van de Panne, M.: Flexible isosurfaces: simplifying and displaying scalar topology using the contour tree. Comput. Geom. Theory Appl. **43**(1), 42–58 (2010)
8. Carr, H., Geng, Z., Tierny, J., Chattopadhyay, A., Knoll, A.: Fiber surfaces: generalizing isosurfaces to bivariate data. Comput. Graphics Forum **34**(3), 241–250 (2015)
9. Carr, H.A., Weber, G.H., Sewell, C.M., Ahrens, J.P.: Parallel peak pruning for scalable SMP contour tree computation. In: Proceedings of 2016 IEEE 6th Symposium on Large Data Analysis and Visualization (LDAV), pp. 75–84 (2016)
10. Chiang, Y.J., Lu, X.: Progressive simplification of tetrahedral meshes preserving all isosurface topologies. Comput. Graphics Forum **22**(3), 493–504 (2003)
11. Doraiswamy, H., Natarajan, V.: Computing Reeb graphs as a union of contour trees. IEEE Trans. Vis. Comput. Graph. **19**(2), 249–262 (2013)
12. Duke, D., Carr, H., Knoll, A., Schunck, N., Nam, H.A., Staszczak, A.: Visualizing nuclear scission through a multifield extension of topological analysis. IEEE Trans. Vis. Comput. Graph. **18**(12), 2033–2040 (2012)
13. Edelsbrunner, H., Harer, J., Patel, A.K.: Reeb spaces of piecewise linear mappings. In: Proceedings of the Twenty-fourth Annual Symposium on Computational Geometry, SCG '08, pp. 242–250 (2008)
14. Howie, C., Blake, E.: The mesh propagation algorithm for isosurface construction. Comput. Graphics Forum **13**(3), 65–74 (1994)
15. Huettenberger, L., Heine, C., Carr, H., Scheuermann, G., Garth, C.: Towards multifield scalar topology based on pareto optimality. Comput. Graphics Forum **32**(3pt3), 341–350 (2013)
16. Kettner, L., Rossignac, J., Snoeyink, J.: The Safari interface for visualizing time-dependent volume data using isosurfaces and contour spectra. Comput. Geom. **25**(1), 97–116 (2003)

17. Kindlmann, G., Durkin, J.W.: Semi-automatic generation of transfer functions for direct volume rendering. In: Proceedings of the 1998 IEEE Symposium on Volume Visualization, VVS '98, pp. 79–86 (1998)
18. Klacansky, P., Tierny, J., Carr, H., Geng, Z.: Fast and exact fiber surfaces for tetrahedral meshes. IEEE Trans. Vis. Comput. Graph. **23**(7), 1782–1795 (2017)
19. Lorensen, W.E., Cline, H.E.: Marching cubes: a high resolution 3D surface construction algorithm. ACM SIGGRAPH Comput. Graph. **21**(4), 163–169 (1987)
20. Manders, E.M.M., Hoebe, R., Strackee, J., Vossepoel, A.M., Aten, J.A.: Largest contour segmentation: a tool for the localization of spots in confocal images. Cytometry **23**(1), 15–21 (1996)
21. Pascucci, V., Cole-McLaughlin, K.: Parallel computation of the topology of level sets. Algorithmica **38**(1), 249–268 (2003)
22. Pascucci, V., Scorzelli, G., Bremer, P.T., Mascarenhas, A.: Robust on-line computation of Reeb graphs: simplicity and speed. ACM Trans. Graph. **26**(3), 58 (2007)
23. Reeb, G.: Sur les points singuliers d'une forme de Pfaff complètement intégrable ou d'une fonction numérique. Comptes Rendus l'Acadèmie des Sciences de Paris **222**, 847–849 (1946)
24. Sakurai, D., Saeki, O., Carr, H., Wu, H.Y., Yamamoto, T., Duke, D., Takahashi, S.: Interactive visualization for singular fibers of functions $f : \mathbb{R}^3 \to \mathbb{R}^2$. IEEE Trans. Vis. Comput. Graph. **22**(1), 945–954 (2016)
25. Takahashi, S., Takeshima, Y., Fujishiro, I.: Topological volume skeletonization and its application to transfer function design. Graph. Model. **66**(1), 24–49 (2004)
26. Tarjan, R.E.: Efficiency of a good but not linear set union algorithm. J. ACM **22**(2), 215–225 (1975)
27. Tierny, J., Carr, H.: Jacobi fiber surfaces for bivariate Reeb space computation. IEEE Trans. Vis. Comput. Graph. **23**(1), 960–969 (2017)
28. Tierny, J., Favelier, G., Levine, J.A., Gueunet, C., Michaux, M.: The Topology ToolKit. Technicl report, CNRS/UPMC. https://topology-tool-kit.github.io/
29. van Kreveld, M., van Oostrum, R., Bajaj, C., Pascucci, V., Schikore, D.: Contour trees and small seed sets for isosurface traversal. In: Proceedings of the Thirteenth Annual Symposium on Computational Geometry, SCG '97, pp. 212–220 (1997)
30. Weber, G.H., Dillard, S.E., Carr, H., Pascucci, V., Hamann, B.: Topology-controlled volume rendering. IEEE Trans. Vis. Comput. Graph. **13**(2), 330–341 (2007)
31. Wyvill, B., McPheeters, C., Wyvill, G.: Animating soft objects. Vis. Comput. **2**(4), 235–242 (1986)

Topological Subdivision Graphs for Comparative and Multifield Visualization

Christian Heine and Christoph Garth

Abstract We propose that a topological model of a real-valued function can be employed to define a spatial subdivision of the function's domain. When multiple topologically-induced subdivisions for the same or different functions on the same domain are combined, a finer spatial subdivision arises: the topological subdivision complex. The topological subdivision graph then gives adjacency relations among the d-cells of the subdivision complex and can be used to describe similarities among topological models. We apply this idea to give new topological models for multiple real-valued functions (multifields), extending contour trees and Morse-Smale complexes to these problem settings, and we illustrate our idea for piecewise-linear functions. We also discuss how our work relates to joint contour nets.

1 Introduction

A common approach to study the behavior of physical phenomena is to simulate them and then analyze the results—often visually. To overcome the problem of occlusions in 3D and the increasingly finer resolution of simulation models, the focus has shifted to identifying specific *features*, e.g. spatial regions that locally behave in a similar manner. It has been shown that certain topological structures yield suitable candidates of such features. Since some phenomena cannot be captured by criteria addressing spatiotemporal variation of just one physical quantity, feature detection needs to consider characteristic patterns in and relationships between multiple physical quantities. For a general overview of such multifield visualizations we refer the interested reader to the surveys by Fuchs and Hauser [13] and Heine et al. [16].

C. Heine (✉)
Image and Signal Processing Group, University of Leipzig, Leipzig, Germany
e-mail: cheine@informatik.uni-leipzig.de

C. Garth
AG Computational Topology, University of Kaiserslautern, Kaiserslautern, Germany
e-mail: garth@cs.uni-kl.de

H. Carr et al. (eds.), *Topological Methods in Data Analysis and Visualization V*, Mathematics and Visualization, https://doi.org/10.1007/978-3-030-43036-8_13

This paper introduces the idea, that the well-understood topological models for single real-valued functions can be trivially extended to multiple real-valued functions on a common domain, since each topological model for single functions gives rise to a subdivision of the domain and that these subdivisions can be combined to give a fine subdivision encompassing all information present in its constituents. Thus, unlike methods that extract features directly from multifields, our method possibly omits some information spread across multiple fields. But the combination of information extracted from single fields allows further applications: It can be applied to compare or combine different topological models for the same function.

2 Related Work

There have been multiple approaches to extend topological models of single real-valued fields to multifields. Edelsbrunner and Harer [9] defined *Jacobi sets* as the critical points of one function restricted to all preimages of the remaining functions. This extends the notion of critical points to multiple functions, but can be applied only when the number of functions does not exceed the dimension of the domain. Huettenberger et al. [17] use techniques from multicriteria optimization to define Pareto sets—an extension of critical points to multifields, but preserving the signedness of the single fields. While the number of input functions is not constrained as in the Jacobi set case, the more functions there are, the more likely a point becomes Pareto. Pareto sets get less informative and detailed the bigger they become. In contrast, our method always increases in detail the more functions the input comprises.

Edelsbrunner et al. [12] defined the Reeb space of a set of functions f_i as the quotient space for the equivalence relation induced by the connected components of all of f_is' preimages, thereby extending the concept of Reeb graphs to multifields. However, while for single functions, the preimages' components can be differentiated based on their topological properties, this does not happen for multiple functions. Without such built-in differentiation, preimage components need to be organized differently. Chattopadhyay et al. [6] proposed Jacobi structures, which map Jacobi sets into the Reeb space, integrating information of both models.

Carr and Duke [4] proposed joint contour nets (JCNs): a JCN is the quotient space induced by the connected components of all of $\tilde{f}_i$s' preimages, where $\tilde{f}_i$ denotes a range-quantized version of f_i. Similarly, Singh et al. [22]'s Mapper algorithm defines a set of overlapping regions in the functions' range space, covering it, and stores spatial relations among the regions' preimages' connected components. These methods work on the basis of preimages and their connected components, but make no direct use of topological features present in the single input functions to point out distinct or idiosyncratic subsets of the functions. Compared to this preimage-focused strategy, topological subdivision graphs focus on integrating information from topology-induced subdivisions of *single* fields. Still, they can be made to yield results similar to joint contour nets as will be shown in Sect. 6.

Another approach is presented by Schneider et al. [20]. It extracts so-called largest contours from the single functions, uses the spatial overlap between largest contours of different functions to quantify similarity, and then uses clustering techniques to identify groups of strongly overlapping largest contours. However, the result does not integrate the information from these single features to multifield features. If the input functions are sufficiently similar, one can use, e.g. Günther et al. [15], who presented a method to compute confidence regions for critical points and relate them by a nesting structure similar to merge trees. Wu and Zhang [25] proposed to compute the arithmetic mean of the input functions, use single-function topological methods on the result, and annotate it by measures of variation. Our method is not limited to input functions that are pairwise similar.

3 Background

This section contains the topological concepts used in the remainder of this paper. For brevity, we assume that the reader is familiar with certain topological concepts, notably topological spaces, simplicial complexes, as well as piecewise linear functions (see, e.g., Edelsbrunner and Harer [10]).

A critical point [1] p of a function $f : \mathbb{M} \to \mathbb{R}$ defined on a d-manifold is a point where $\nabla f = 0$, i.e., where f's gradient vanishes. We assume f to be a Morse function, meaning that critical points are isolated and have distinct function values.

A *level set* $f^{-1}(v)$ of a function f for a value v is simply the preimage of v with respect to f, i.e., the set of all points p where $f(p) = v$. Each level set may consist of multiple connected components. An equivalence relation can then be defined for points of the domain: p and q are equivalent, if they belong to the same component of some level set, called *contour*. We will refer to this equivalence relation as $\sim_C$ in the remainder of this paper. The *Reeb graph* [19] is defined as the quotient topology on $\mathbb{M}$ with respect to $\sim_C$. Informally, this process can be thought of as representing each contour as a point and contour adjacency as point adjacency. If the domain is simply connected, i.e. each closed curve can be smoothly contracted to a point, the Reeb graph contains no loops and is called the *contour tree* [2]. A *superlevel set* of f for the value v is defined as f's preimage of the interval $[v, \infty)$ and a *sublevel set* is defined analogously as f's preimage of the interval $(-\infty, v]$. Similar to level sets, superlevel and sublevel sets may consist of multiple connected components, and we may define an equivalence relation among points with function value v, based on whether they lie in the same superlevel or sublevel set for v, respectively. The quotient topology using this equivalence relation gives the *merge tree*.

An *integral line* of a smooth function f is a curve of maximal length along which the tangent is collinear to the function's gradient. For each critical point, its descending cell comprises the integral lines that converge on it, and the ascending cell comprises the integral lines that originate from it. When ascending and descending cells intersect only transversally, the domain subdivision induced by this equivalence relation (two points are equivalent if each lies on an integral line that in the limit connects the same pair of critical points) is a CW complex called the *Morse-Smale complex*.

A *subdivision* of a set Ω is a collection of sets S such that (1) the union over all sets in S is Ω, and (2) the members of S are pairwise disjoint. We will only consider subdivisions where each set of S is connected. Subdivisions of a set arise naturally as the equivalence classes of equivalence relations. Given two subdivisions of Ω, S_1 and S_2, their *joint subdivision* $S_{1,2}$ is the subdivision of Ω where, whenever $s_1 \in S_1$ and $s_2 \in S_2$ are not disjoint, the connected components of their intersection are members of $S_{1,2}$. The definition extends naturally to the joint subdivision of more than two subdivisions.

4 Topology-Induced Spatial Subdivision

In this section, we use contour trees and Morse-Smale complexes to illustrate how topological models subdivide the domain and how to combine this information.

We were motivated by the question "What is the smallest structure from which the contour tree for each function of a collection can be reconstructed?" Its answer was inspired by joint contour nets, which can be viewed as solving the problem for augmented contour trees of range-quantized functions. We wished to remove these restrictions and in the process realized that joint contour nets combine the information from a preimage-based subdivision of each function. We first noted that the subdivision for each field can be constructed in other manners and later that the idea generalizes to when the subdivision for each field need not result from the same topological method for each field. Under this premise, the information from multiple topological models of the same function can be combined.

Recall that in the contour tree case, two points are called equivalent if they belong to the same connected component of a level set for their common function value, formally expressed by the equivalence relation $\sim_C$. It may now be observed that each contour C partitions a simply-connected domain into at least two regions ($\mathbb{M} \setminus C$ generally consists of multiple connected components) and that each critical point lies either on the contour itself or in exactly one of these regions. We can define an equivalence relation among contours in the following way: Two contours C_1, C_2 are equivalent, if any critical point p of f is either member of both C_1 and C_2 or it is member of some connected component X of $\mathbb{M} \setminus C_1$ and member of some connected component Y of $\mathbb{M} \setminus C_2$ such that any critical point of f is member of X iff it is also member of Y. Informally, two contours are equivalent if they partition the set of critical points in the same way. Note that the definition implies that contours that have a critical point as member can only be equivalent to themselves. We will refer to this equivalence relation by $\sim_S$, and we can combine it with $\sim_C$ to a new equivalence relation $\sim_{CT}$ that amounts to define two points as equivalent, if they lie on the same node or edge of the contour tree. We can use $\sim_{CT}$ to subdivide the domain; the result can be stored in a CW complex.

For the more general case of a Reeb graph in a non-simply-connected domain, $\sim_S$ is not discriminative enough. One anonymous reviewer suggested the following criterion, which is also simpler than ours for contour trees: Two contours c_1, c_2 of the preimage for f_1, f_2 are equivalent if they are identical or if they are both subset of one connected component X of the preimage of $[\min\{f_1, f_2\}, \max\{f_1, f_2\}]$ and X has no critical point as member. This definition is the basis for the algorithms presented by Doraiswamy and Natarajan [8] and Tierny and Carr [23].

For merge trees, we define two points p, q as equivalent, if the smallest superlevel/sublevel set component containing p is the same as the one for q. Just as with level sets we can construct a coarser equivalence relation: two superlevel/sublevel set components are equivalent, when they contain the same critical points and thus by extension, two points are equivalent, if they are mapped to the same edge of the merge tree. To ensure that the decomposition results in a proper CW complex, we furthermore need to distinguish superlevel/sublevel set components that contain a critical point on their boundary from those without.

Morse-Smale complexes are already CW complexes and their cells partition the domain. However, we would like to point out that the Morse-Smale complex results from combining two domain subdivisions: one that treats points as equivalent when they have the same critical point at the terminating end of their integral line and one similarly for the beginning of their integral line. Morse-Smale complexes already employ the principle we suggest to use for any topological model: combine the information from all single functions' topologically-driven subdivision.

We were initially also considering whether, in straight analogy, the domain subdivision arising from a contour tree is just the combined domain subdivision induced by the superlevel and the sublevel merge tree. However, this is not the case, as can be seen from the counterexample in Fig. 1. The resulting combination

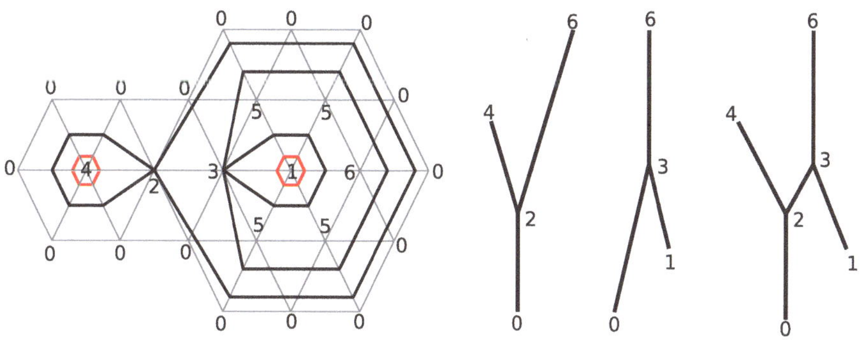

Fig. 1 Counterexample to the hypothesis that the combined domain subdivision induced by the merge tree for super- and sublevel sets is the same as the domain subdivision induced by the contour tree. From left to right: input function, merge tree for superlevel sets, merge tree for sublevel sets, contour tree. The spurious red contours arise from the merge trees' domain subdivision. The contour around 4 is part of the boundary of the sublevel set component for 3 and the contour around 1 is part of the superlevel set component for 2. The pattern in the contour tree seems to be the necessary and sufficient condition for the existence of spurious contours

would be too fine. However, one can define contour tree domain subdivisions via the equivalence of upward and downward monotone paths ending at critical points [7]. This emphasizes the difference between concepts of critical points that can be reached from a point p via monotone paths, i.e., domain paths along which the function's values are changing monotonously, and the set of critical points that are contained in the smallest superlevel/sublevel set component containing the point p.

5 Topological Subdivision Complex and Graph

The *topological subdivision complex* is not a single mathematical concept. Rather it is any CW complex that is derived in a fashion similarly to the examples in the previous section from the topological models of single functions or a joint subdivision of multiple topological subdivision complexes. Given a set of domain subdivisions, regardless of whether they result from the topological models of the same kind for different functions, topological models of different kinds for the same function, or any combination thereof, we can trivially compute a joint subdivision, or, equivalently, combine the equivalence relations giving rise to the domain subdivisions.

The resulting subdivision can be displayed directly in two or three dimensions, but for three or more dimensions it will be easier to view its dual, which we call the *topological subdivision graph*. Its nodes are the set of d-cells of the joint subdivision, and we connect two nodes by an edge if they have a common d-1-face. Furthermore, we annotate each node with its cell volume (for visualization purposes) and annotate each edge with a list of subdivisions which do not separate the two cells, i.e., in which the union of the two cells are a subset of a common cell.

The latter annotation ensures that one can quickly compute a quotient graph that equals the topological subdivision graph for a subset of domain subdivisions. E.g., imagine two subdivisions (H and V) of a square, one splitting the square horizontally and one splitting the domain vertically. The topological subdivision graph of the joint subdivision contains four nodes corresponding to the four 2-cells: upper-left, upper-right, lower-left, and lower-right. There are now two H-edges connecting the lower with the upper nodes and two V-edges connecting the left with the right nodes. To get the topological subdivision graph for the H-subdivision one just collapses all nodes connected by V-edges. The process of computing a quotient topological subdivision graph thus becomes a simple graph algorithm in strict analogy to how joint contour nets for a subset of input functions can be computed from the joint contour net for all functions [4]. For merge trees and contour trees, edges can be additionally annotated with a direction, based on the ordering of function values in the connected nodes.

Our work was originally motivated by finding a small data structure that contains all information from the individual functions' contour trees. Note that this annotation and the quotient graph operation on the topological subdivision graph will not give these contour trees, because the subdivision graph retains only information from the complex's d-cells and their adjacencies, but critical points reside in 0-cells. Instead, a joint subdivision complex can be annotated as follows: for each cell of the joint subdivision we list all cells of the base subdivision complex's cells that it is a subset of. Based on this information it is again simple to perform quotient complex operations. The contour tree can be reconstructed from the subdivision complex for one function as follows: Each cell of dimension less than d has a critical point, and two critical points are connected, when their cells are incident on a common d-cell.

6 Computation

A general algorithm for our method may only be sketched, because we do not limit ourselves to particular topological methods to drive construction of the base subdivisions. For illustration, we will give the details of our current proof-of-concept implementation that can combine information from multiple Morse-Smale complexes and contour trees. Our implementation is currently restricted to piecewise linear functions on a common simplicial complex in 2D. Before joining the information, we compute contour tree and Morse-Smale complex for each input function separately. To compute the contour trees, we use the algorithm by Carr et al. [5] and to compute Morse-Smale complexes we use the algorithm by Edelsbrunner et al. [11]. We choose this latter algorithm purely out of convenience. One may use instead the algorithm by Bremer et al. [3] that computes the geometry of the Morse-Smale complex's cells more accurately, or a combinatorial Morse-Smale complex. Algorithms for this are present, e.g., in Shivashankar and Natarajan [21] or Tierny et al. [24].

When using the algorithm by Edelsbrunner or a combinatorial Morse-Smale complex, the 1-cells of the resulting Morse-Smale complex are a subset of the input domain's edges. Therefore combining this information is trivial: two triangles of the domain belong to the same 2-cell in the joint subdivision complex if they share an edge that does not belong to any 1-cell of the Morse-Smale complexes. The triangles are further subdivided based on the functions' contour trees.

Let S_0 be the domain subdivision given by the simplicial complex and $f_1, \ldots, f_n$ be the input functions, our implementation conceptually first computes the topological subdivision complex for the subdivisions $\{S_0, \ldots, S_n\}$ and then uses the quotient operation defined in the previous section to remove S_0. This is in strict analogy to the algorithm by Carr and Duke [4]. This approach allows us to work on a triangle-by-triangle basis, simplifying implementation: We never explicitly compute any S_i, rather we compute the needed convex polygonal patches on the fly, and compute line intersections in local triangle coordinates, which is numerically preferable.

In particular, we iterate over all triangles. For each triangle we start with a set P of convex patches that is initially just the triangle's geometry in local coordinates. Then we iterate over fields. The triangle's endpoints, by construction of the contour tree, must lie along a monotone path in the tree. Thus, for each saddle that we pass along this path, the triangle contains a part of its contour which is a line segment. We extend this line segment to a line and split all patches it intersects, updating an incidence graph between points, line segments, and patches. Since all line segments from the same field are parallel withing the same triangle, we can perform the test and split efficiently by sweeping along the incidence graph. This sweeping also ensures consistency with the information in the contour trees. After computing all patches in this fashion, we sweep the incidence graphs of adjacent triangles and merge points and line segments along the shared edge. We use a simple connected components algorithm on the patches' adjacency graph and compute the quotient graph to obtain the topological subdivision complex's 2-cells and their adjacencies.

Since the patches are split iteratively with each field, one would like to manage the numerical error. However, since each point of each patch is at the intersection of two line segments which are parts of contours from different fields, their position in local coordinates numerically only depends on the saddles' function values and the three triangle endpoints' function values for two fields. We thus store geometry at the edges only, namely using the coefficients α_i for the linear equation $\alpha_0 + \alpha_1\beta_1 + \alpha_2\beta_2 = 0$ (β_i refer to local coordinates), and compute points' positions by intersecting adjacent faces of a patch. Since lines are only added, but their geometry is never changed, the numerical error is independent of the number of input fields.

The runtime of our algorithm is dominated by patch splitting and can be bounded from above by a function linear in the number of triangles N and linear in the number of saddles s_i for each field: $O(N \prod s_i)$. Since each saddle's contour typically runs through few triangles and not all saddle contours of one field intersect all saddle contours of all other fields in all triangles, the runtime is much lower in practice. A stricter upper bound on the runtime is the number of patches. The computation of connected components and the quotient graph to remove the subdivision given by the triangulation are linear in the number of patches.

We use a simple straight-line graph drawing to show the topological subdivision graph: we center nodes on the cell they represent and use their sizes to indicate their cell's volume. This enables comparison of topological models for different functions. If, say, the functions yield very similar contour-tree-based domain subdivisions, we can expect the topological subdivision graph to be mostly tree-like as well, with only minor additional loops, but whose node volumes are notably smaller. Note that the topological subdivision graph for each field does not contain loops, but the topological subdivision graph for the joint subdivision might. In our implementation, we currently use topological simplification for each field separately, but more appropriate would be a method to simplify the graph directly. We leave other visualizations of topological subdivision graphs and their simplification for future work.

7 Results

We used the cross-section of a fluid flow simulation inside a cylinder. The flow is steady and rotationally symmetric. We use a cross-section along the axis of

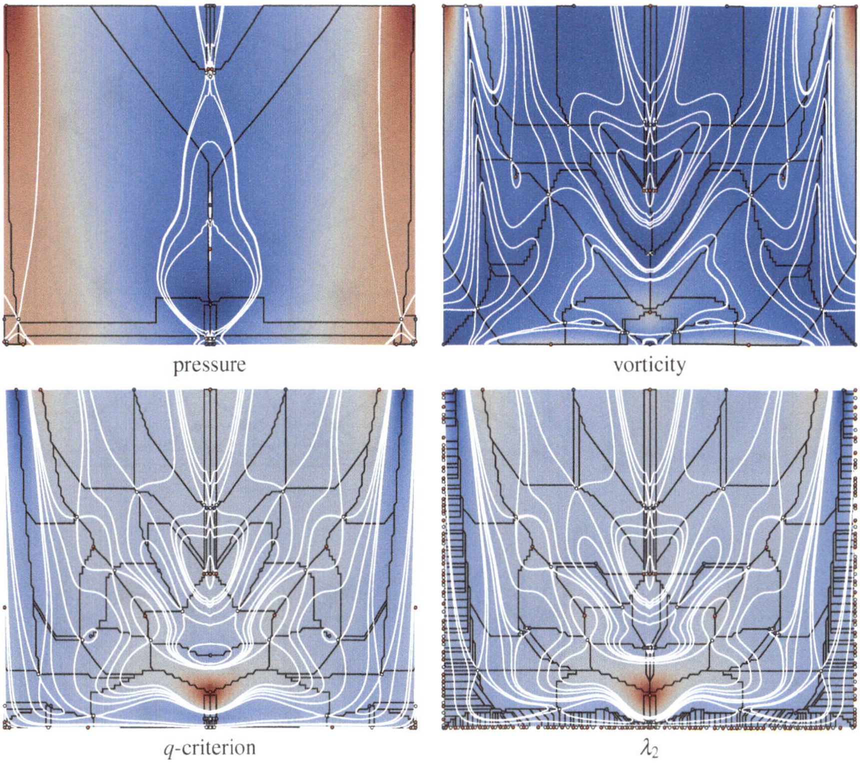

Fig. 2 Cross-section through a flow simulation inside a cylinder. Four quantities are shown. White lines show all contours that contain a critical point (as a set-theoretic member). Black lines indicate Morse cells (separatrices)

rotation. The data set contains 29704 triangles on 15105 points. Of particular interest is the identification of vortical structures, which tend to occur in areas of low pressure and high vorticity. Furthermore, the quantities q-criterion and λ_2 [18] are typically associated with vortices. Figure 2 shows color maps overlayed by contour trees subdivision and Morse-Smale complexes for these four quantities. Note that asymmetries arise from simulation of simplicity that is used to compute the topological models. The computation of the single fields' contour trees, Morse-Smale complexes, their subdivision, and the topological subdivision graph took less than 0.4 s. Our proof-of-concept implementation has not yet been optimized for speed.

Figure 3 illustrates and showcases the topological subdivision graph of a cross-section through a cylindrical flow. From the color maps alone, the fields look similarly complex, but the depiction of the Morse-Smale complex illustrate that pressure has fewer cells in its Morse-Smale complex. When we combine both Morse-Smale complexes, we can see that the resulting subdivision graph closely

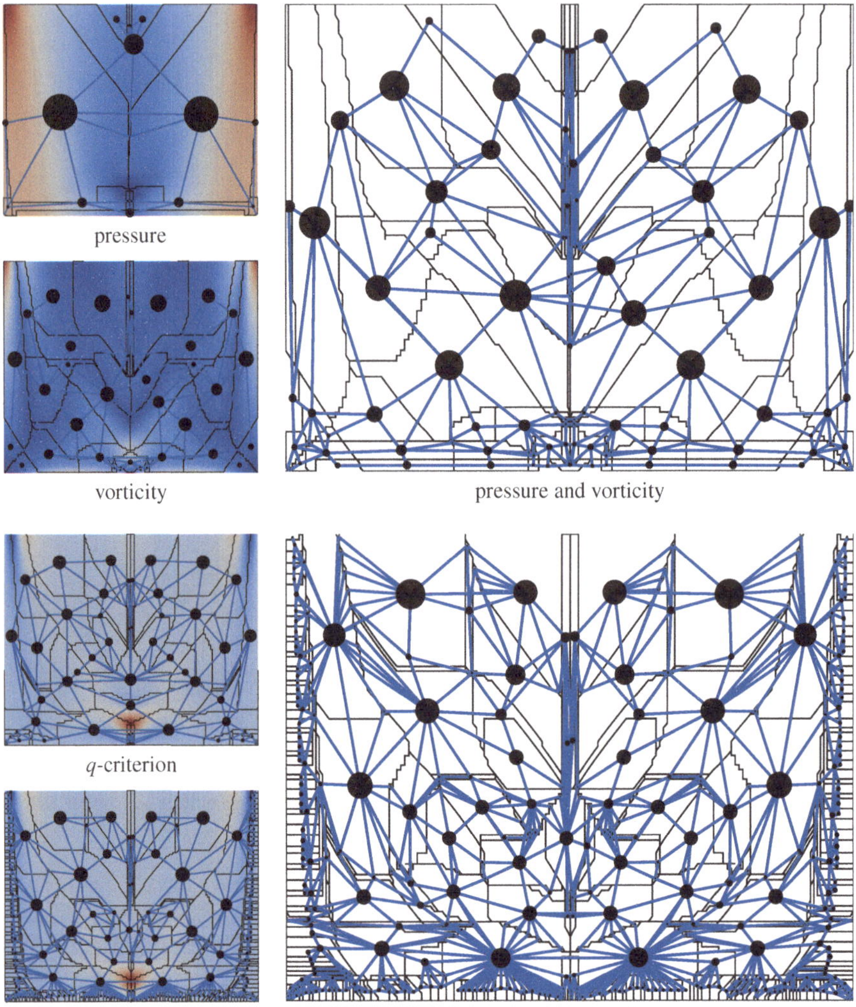

Fig. 3 Combined topological subdivision graphs from the subdivision graphs induced by the Morse-Smale complex of different quantities in the cross-section of a cylindrical fluid flow. Black circles show cell centers, circle size is proportional to cell size, blue lines show cell adjacency

resembles the subdivision graph for vorticity, so vorticity already contained most information and pressure added little more. We can also see that the subdivision graph for q-criterion and λ_2 are highly similar, with the notable exception that the Morse-Smale complex for λ_2 contains many spurious cells at the boundary. We suppose this is because our λ_2 computation has less data near the boundary to compute smooth gradients. A next step would be to identify and remove such small features meaningfully (c.f. Sect. 8). We leave such considerations for future work.

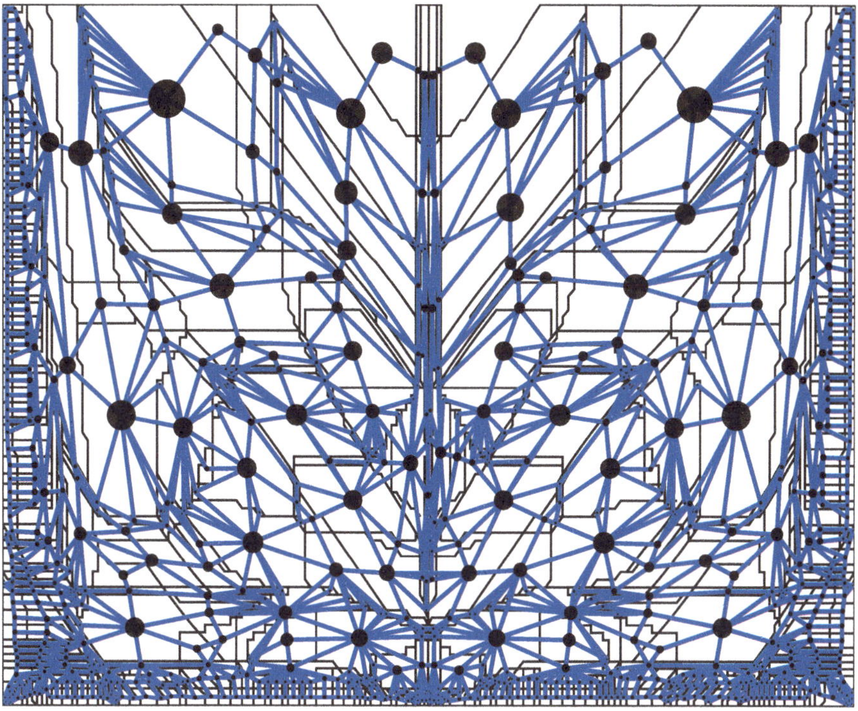

Fig. 4 Cross-section through a flow simulation inside a cylinder. Topological subdivision graph for four quantities are shown

The joint subdivision graph also demonstrates that both subdivision are very similar, but we noticed the fan-like structures in the upper outer parts of the image. These result from the 1-cells of the two Morse-Smale complexes not being fully congruent. This could be an artifact of using the algorithm by Edelsbrunner et al. [11], where the gradient path can switch from one mesh edge to another for minute changes in function value at the mesh nodes. The problem could be alleviated by using a different algorithm. However, the graphs allow to study the systematic difference of these two models, most easily visible in the middle and lower middle part of the image. The joint subdivision graph integrates the information from both effectively. Figure 4 shows the topological subdivision graph for all four quantities. Apart from the mentioned spurious cells, the cells are roughly the same size and shape, indicating that the information from the different functions could be integrated quite well.

Figure 5 gives an example for the topological subdivision graph of the contour tree, the Morse-Smale complex, and their combination for the pressure field of the same dataset. Since gradient paths are always perpendicular to contours, the information of a contour tree and a Morse-Smale complex typically integrates well.

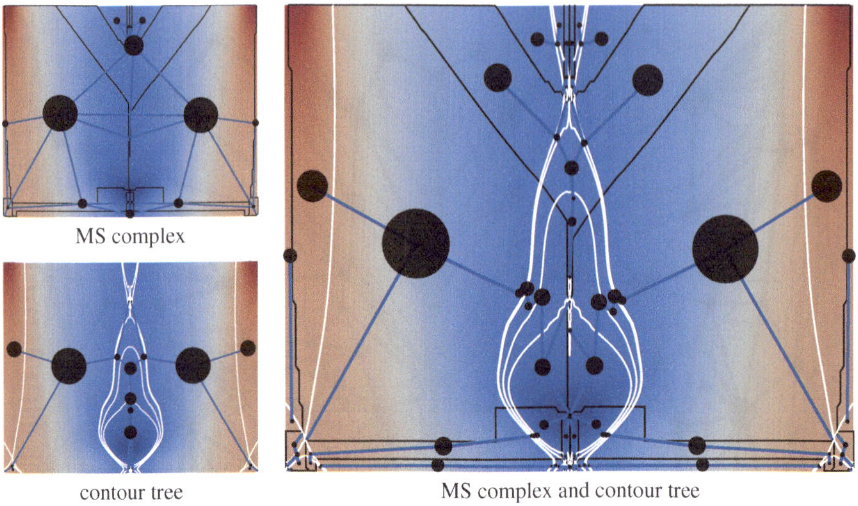

Fig. 5 Topological subdivision graph for the MS complex, contour tree, and their combination, of the pressure field of a flow inside a cylinder

8 Discussion

The presented method was largely inspired by joint contour nets [4]. One may view our method as a generalization: the domain subdivisions implicit in joint contour nets are based on a simple equivalence relation: connected components of the preimages $f_i^{-1}([k\Delta_i, (k+1)\Delta_i))$ for all $k \in \mathbb{Z}$. We basically replaced this by preimages of topological cells from a topological model for each field, removing the parameters Δ_i in the process. Although not fully equivalent, topological subdivision models can be used to compute joint contour nets. For each input function's range define the following intervals: $[k\Delta_i]$, $(k\Delta_i, (k+1)\Delta_i)$, for all $k \in \mathbb{Z}$, and define two points equivalent, if their functions' values fall all inside the same interval. This slightly different construction is necessary to ensure the topological subdivision complex is indeed a complex. Similarly, the Mapper algorithm [22] uses a set of overlapping intervals $(k\Delta_i - \epsilon, (k+1)\Delta_i + \epsilon)$ for each function. When $\epsilon > 0$ approaches 0, the above variation of our method can be considered equivalent. However, we would like to stress that such methods that only consider the functions' range, fail to integrate important topological information present in the single functions.

The method currently can only be considered to be a building block. Although it uses all information present in the topological models of single functions, it only generates a set of possible feature candidates. Which of these correspond to features meaningful to a particular application depends both on the distribution of function values inside the regions, and ultimately also on the goal of the application. An open challenge here is to find typical patterns of value distributions.

The biggest challenge for the method is to find a suitable method of topological simplification. One option is to simplify the joint topological subdivision complex, but this can only consider the size of cells and may not always be consistent: it is not clear whether there are still subdivision complexes that will result in the joint complex when combined. Of course, one can always simplify the topological models for each field separately. But this will not help in a situation like in Fig. 3, where the 1-cells of two Morse-Smale complexes ran close but were not exactly identical. Also, when simplifying different topological models of the same field, it is not clear which simplification should take precedence when it comes to updating the underlying real-valued function, since the changes needed to make one topological model of a function simpler might conflict with the requirements of making a different topological model of the same function simpler. In particular if one wants the resulting function to be geometrically simple as well (c.f. Günther et al. [14]).

9 Conclusion

We presented a new method suitable for integrating the information from different topological models of the same scalar function, topological models of different scalar functions, and combinations thereof. The basic idea is to use domain subdivisions, which can serve as a "common language", and integrate the information from all subdivisions in a finer subdivision. Our initial experiments on a real-world simulation dataset of a fluid flow inside a cylinder showed the feasibility of the approach. The proposed method needs an algorithm for removing small features that arise from noise or are insignificant. But it is challenging to find a method that is general enough to work with any kind and combination of topological model.

By construction, the topological subdivision complex for a set of contour trees contains information from them; the single contour trees can be extracted using a suitable quotient space operation. Can we prove that this is the smallest structure with this property? Finally, although we illustrated topological subdivision graphs only for Morse-Smale complexes and contour trees, the method can easily be adapted to other topological models—not necessarily models for single fields. For instance, Pareto sets [17] can also be interpreted as a domain subdivision. Under certain circumstances Jacobi sets [9] can also subdivide a domain. It would be interesting to apply our approach to these methods and extend it to vector and tensor field topology as well. We leave such considerations for the future.

Acknowledgments We would like to thank the anonymous reviewers for in-depth and helpful remarks and suggestions to improve the paper, in particular the reviewer that pointed out the equivalence relation for contours of Reeb graphs.

References

1. Banchoff, T.F.: Critical points and curvature for embedded polyhedral surfaces. Am. Math. Mon. **77**(5), 475–485 (1970)
2. Boyell, R.L., Ruston, H.: Hybrid techniques for real-time radar simulation. In: Proceedings of the Fall Joint Computer Conference (AFIPS '63) (Fall), pp. 445–458. ACM, New York (1963)
3. Bremer, P.T., Edelsbrunner, H., Hamann, B., Pascucci, V.: A topological hierarchy for functions on triangulated surfaces. IEEE Trans. Vis. Comput. Graph. **10**(4), 385–396 (2004)
4. Carr, H., Duke, D.: Joint contour nets: computation and properties. In: Proceedings of the IEEE Pacific Visualization Symposium (PacificVis), pp. 161–168 (2013)
5. Carr, H., Snoeyink, J., Axen, U.: Computing contour trees in all dimensions. Comput. Geom. **24**(2), 75–94 (2003)
6. Chattopadhyay, A., Carr, H., Duke, D., Geng, Z.: Extracting Jacobi structures in Reeb spaces. In: Elmqvist, N., Hlawitschka, M., Kennedy, J. (eds.) EuroVis—Short Papers. The Eurographics Association, Switzerland (2014)
7. Chiang, Y.J., Lenz, T., Lu, X., Rote, G.: Simple and optimal output-sensitive construction of contour trees using monotone paths. Comput. Geom. **30**(2), 165–195 (2005)
8. Doraiswamy, H., Natarajan, V.: Output-sensitive construction of Reeb graphs. IEEE Trans. Visual. Comput. Graph. **18**, 146–159 (2011)
9. Edelsbrunner, H., Harer, J.: Jacobi sets. In: Cucker, F., DeVore, R., Olver, P., Süli, E. (eds.) Foundations of Computational Mathematics, pp. 37–57. Cambridge University Press, Cambridge (2004)
10. Edelsbrunner, H., Harer, J.: Computational Topology: An Introduction. American Mathematical Society, New York (2010)
11. Edelsbrunner, H., Harer, J., Zomorodian, A.: Hierarchical Morse complexes for piecewise linear 2-manifolds. In: Proceedings of the Seventeenth Annual Symposium on Computational Geometry (SCG '01), pp. 70–79. ACM, New York (2001)
12. Edelsbrunner, H., Harer, J., Patel, A.K.: Reeb spaces of piecewise linear mappings. In: Proceedings of the Twenty-fourth Annual Symposium on Computational Geometry (SCG '08), pp. 242–250. ACM, New York (2008)
13. Fuchs, R., Hauser, H.: Visualization of multi-variate scientific data. Comput. Graphics Forum **28**(6), 1670–1690 (2009)
14. Günther, D., Jacobson, A., Reininghaus, J., Seidel, H.P., Sorkine-Hornung, O., Weinkauf, T.: Fast and memory-efficienty topological denoising of 2D and 3D scalar fields. IEEE Trans. Vis. Comput. Graph. **20**(12), 2585–2594 (2014)
15. Günther, D., Salmon, J., Tierny, J.: Mandatory critical points of 2D uncertain scalar fields. Comput. Graphics Forum **33**(3), 31–40 (2014)
16. Heine, C., Leitte, H., Hlawitschka, M., Iuricich, F., De Floriani, L., Scheuermann, G., Hagen, H., Garth, C.: A survey of topology-based methods in visualization. Comput. Graphics Forum **35**(3), 643–667 (2016)
17. Huettenberger, L., Heine, C., Carr, H., Scheuermann, G., Garth, C.: Towards multifield scalar topology based on Pareto optimality. Comput. Graphics Forum **32**(3), 341–350 (2013)
18. Jeong, J., Hussain, F.: On the identification of a vortex. J. Fluid Mech. **285**, 69–94 (1995)
19. Reeb, G.: Sur les points singuliers d'une forme de Pfaff completement intégrable ou d'une fonction numérique. Comput. Rendus de L'Académie des Séances, Paris **222**(847–849) (1946)
20. Schneider, D., Heine, C., Carr, H., Scheuermann, G.: Interactive comparison of multifield scalar data based on largest contours. Comput. Aided Geom. Des. **30**(6), 521–528 (2013)
21. Shivashankar, N., Natarajan, V.: Efficient software for programmable visual analysis using Morse-Smale complexes. In: Carr, H., Garth, C., Weinkauf, T. (eds.) Topological Methods in Data Analysis and Visualization IV, pp. 317–331. Springer, Cham (2017)

22. Singh, G., Mèmoli, F., Carlsson, G.: Topological methods for the analysis of high dimensional data sets and 3D object recognition. In: Botsch, M., Pajarola, R., Chen, B., Zwicker, M. (eds.) Eurographics Symposium on Point-Based Graphics, pp. 91–100. The Eurographics Association, Switzerland (2007)
23. Tierny, J., Carr, H.: Jacobi fiber surfaces for bivariate Reeb space computation. IEEE Trans. Vis. Comput. Graph. **23**(1), 960–969 (2017)
24. Tierny, J., Favelier, G., Levine, J.A., Gueunet, C., Michaux, M.: The topology toolkit. IEEE Trans. Vis. Comput. Graph. **24**(1), 832–842 (2018)
25. Wu, K., Zhang, S.: A contour tree based visualization for exploring data with uncertainty. Int. J. Uncertain. Quantif. **3**(3), 203–223 (2013)

Part V
Other Forms of Topology

Interpreting Galilean Invariant Vector Field Analysis via Extended Robustness

Bei Wang, Roxana Bujack, Paul Rosen, Primoz Skraba, Harsh Bhatia, and Hans Hagen

Abstract The topological notion of robustness introduces mathematically rigorous approaches to interpret vector field data. Robustness quantifies the structural stability of critical points with respect to perturbations and has been shown to be useful for increasing the visual interpretability of vector fields. However, critical points, which are essential components of vector field topology, are defined with respect to a chosen frame of reference. The classical definition of robustness, therefore, depends also on the chosen frame of reference. We define a new Galilean invariant robustness framework that enables the simultaneous visualization of robust critical points across the dominating reference frames in different regions of the data. We also demonstrate a strong connection between such a robustness-based framework with the one recently proposed by Bujack et al., which is based on the determinant of the Jacobian. Our results include notable observations regarding the definition of stable features within the vector field data.

B. Wang (✉)
University of Utah, Salt Lake, Utah
e-mail: beiwang@sci.utah.edu

R. Bujack
Los Alamos National Laboratory, Nuevo México, NM, USA
e-mail: bujack@lanl.gov

P. Rosen
University of South Florida, Tampa, FL, USA
e-mail: prosen@usf.edu

P. Skraba
Jozef Stefan Institute, Ljubljana, Slovenia
e-mail: primoz.skraba@ijs.si

H. Bhatia
Lawrence Livermore National Laboratory, Livermore, CA, USA
e-mail: hbhatia@llnl.gov

H. Hagen
Technical University Kaiserlautern, Kaiserslautern, Germany
e-mail: hagen@informatik.uni-kl.de

H. Carr et al. (eds.), *Topological Methods in Data Analysis and Visualization V*, Mathematics and Visualization, https://doi.org/10.1007/978-3-030-43036-8_14

1 Introduction

Motivation Understanding vector fields is integral to many scientific applications ranging from combustion to global oceanic eddy simulations. Critical points of a vector field (i.e., zeros of the field) are essential features of the data and play an important role in describing and interpreting the flow behavior. However, vector field analysis based on critical points suffers a major drawback: the interpretation of critical points depends upon the chosen frame of reference. Just like the velocity field itself, they are not Galilean invariant. Figure 1 highlights this limitation, where the critical points in a simulated flow (the von Kármán vortex street) are visible only when the velocity of the incoming flow is subtracted.

The extraction of *meaningful* features in the data therefore depends on a *good* choice of a reference frame. Oftentimes, there exists no single frame of reference

Fig. 1 Visualization of the flow behind a cylinder without (**a**) and with (**b**) the background flow removed, where the colormap encodes the speed of the flow. For comparison, (**c**) shows the corresponding Galilean invariant vector field introduced by Bujack et al. which is constructed from the extrema of the determinant of the Jacobian. The Galilean invariant critical points are marked with red nodes for vortices/sources/sinks and with blue nodes for saddles. Image courtesy of Bujack et al. [3]. (**d**) Galilean invariant vector field constructed from the extended robustness. The local maxima of the extended robustness field are marked with red nodes

that enables simultaneous visualization of *all* relevant features. For example, it is not possible to find one single frame that simultaneously shows the von Kármán vortex street from Fig. 1b and the first vortex formed directly behind the obstacle in Fig. 1a. To overcome such a drawback, a framework recently introduced by Bujack et al. [3] considers every point as critical and locally adjusts the frame of reference to enable simultaneous visualization of dominating frames that highlight features of interest. Such a framework selects a subset of critical points based on Galilean invariant criteria, and visualizes their frame of reference in their local neighborhood. Galilean invariance refers to the principle that Newton's laws hold in all frames moving at a uniform relative velocity. Thus, a Galilean invariant property is one that does not change when observed in different frames with uniform motion relative to each other. The extrema of the determinant of the Jacobian are particular examples of Galilean invariant critical points [3], and they simultaneously capture all relevant features in the data, as illustrated in Fig. 1c. The intuition is that the determinant of the Jacobian determines the type of critical point, and since the Jacobian is Galilean invariant, its extrema (with a magnitude away from zero) correspond to *stable* critical point locations where small perturbations in the field do not change their types. Such Galilean invariant critical points, in general, do not overlap with the classical zeros of the vector fields; however, each has a frame of reference in which it is a zero of the field. Such a perspective is useful in revealing features beyond those obtainable with a single frame of reference (e.g., Fig. 1c).

The topological notion of robustness, on the other hand, considers the *stability* of critical points with respect to perturbations. Robustness, a concept closely related to *topological persistence* [10], quantifies the stability of critical points, and, therefore, assesses their significance with respect to perturbations to the field. Intuitively, the robustness of a critical point is the minimum amount of perturbation necessary to cancel it within a local neighborhood. Robustness, therefore, helps in interpreting a vector field in terms of its structural stability. Several studies have shown it to be useful for increasing the visual interpretability of vector fields [29] in terms of feature extraction, tracking [24], and simplification [25–27].

Contributions In this paper, we present new and intriguing observations connecting the Jacobian based and robustness based notions in quantifying stable critical points in vector fields. In particular, we address the following questions: *Can we interpret Galilean invariant vector field analysis based on the determinant of the Jacobian via the notion of robustness? What are the relations between these two seemingly different notions?* Our contributions are:

- We extend the definition of robustness by considering every point as a critical point and introduce the notion of the *extended robustness* field by assigning each point in the domain its robustness when it is made critical with a proper frame of reference.
- We prove that the extended robustness satisfies the criterion of Galilean invariance, where the local maxima of the extended robustness field are the Galilean invariant critical points.

- We prove, theoretically, that the determinant of the Jacobian is a lower bound for the extended robustness at the same point.
- We demonstrate, visually, that the extended robustness helps to interpret the Jacobian-based Galilean invariant vector field analysis, in particular, that the extrema of the determinant of the Jacobian coincide with the local maxima of the extended robustness (Fig. 1c–d).

2 Related Work

Vector Field Analysis and Reference Frames The analysis of vector fields depends upon the chosen frame of reference [20–22], as the observed vector field changes with changes in frames. In particular, for any one given point, it is always possible to create a frame of reference where this point becomes critical. Therefore, it is important to carefully choose a physically meaningful frame for analysis. In this regard, uniformly moving frames are of particular importance as they preserve many properties of interest, thus providing a *Galilean invariant* analysis.

Because of the physical importance of a feature descriptor to be independent from a Galilean change of frame of reference, many popular vector field feature detectors are Galilean invariant. In particular, a number of vortex detection techniques, such as the λ_2-[18], Q-[17], and Δ-[8] criterion, compute the Jacobian of the field, which, being a spatial derivative, discards uniform motion.

Simpler solutions to guarantee Galilean invariance in vector field analysis involve subtracting the mean vector to highlight the fluctuations in the field. In recent literature, more advanced techniques have been presented to derive vectors for subtraction to determine an expressive frame of reference, e.g., from the Helmholtz-Hodge decomposition [1, 30] or the boundary-induced flow [9]. In general, Galilean invariant frames have been employed extensively for vector field analysis [3, 7, 8, 19, 23].

Nevertheless, since Galilean invariance is limited to compensating for uniform motion, there exist techniques to perform the analysis in more sophisticated frames. For example, Haller [15] extracted vortices using time-dependent translations and time-dependent rotations; Günther et al. [14] described computation of vortices in rotational frames; Fuchs et al. [13] used time-varying frames built upon the notion of "unsteadiness"; and Bhatia et al. [1] proposed using new frames to represent harmonic background flows.

In this work, we consider Galilean invariance to be the key property for defining robustness for critical points across reference frames and extend the framework by Bujack et al. [3].

Robustness The topological notion of robustness is closely related to the topological persistence [10]. Unlike persistence, which is used extensively for the analysis and visualization of scalar field data, robustness, first introduced by Edelsbrunner et al. [11], can be employed for vector field data [6, 12]. Recent work by Wang et al. [29] assigned robustness to critical points in both stationary and time-varying vector fields and obtained a hierarchical structural description of the data. Such a hierarchical description implies simplification strategies that perform critical point cancellations in both 2D [25, 26] and 3D [27]. The robustness framework also gives a fresh interpretation of the notion of feature tracking, in particular, critical point tracking, where robust critical points could provably be tracked more easily and more accurately in the time-varying setting [24].

Since robustness of critical points is not invariant to reference frames, in our work, we aim to define a new robustness framework that addresses such a challenge and enables the simultaneous visualization of robustness across local, dominating reference frames.

3 Technical Background

We revisit some technical background before describing our results, namely, the notions of Galilean invariance, reference frame adjustment, Jacobian-based Galilean invariant vector fields, and robustness.

Galilean Invariance Let $v : \mathbb{R}^2 \to \mathbb{R}^2$ denote a 2D vector field describing the instantaneous velocity of a flow. A *Galilean transformation* of a point $x \in \mathbb{R}^2$ is the composition of a translation $b : \mathbb{R} \to \mathbb{R}^2$ with $\dot{b} = const$, and a rigid body rotation $A \in SO(2)$ [3]. A point whose position in the original frame is x, then has the coordinate in the transformed frame [28] as

$$x' = Ax + b. \tag{1}$$

A vector field $v(x)$ is *Galilean invariant* (GI) if it transforms under a *Galilean transformation*, according to the rule $v'(x') = Av(x)$ [28]. Similarly, a scalar field $s(x)$ and a matrix field $M(x)$ are called GI if $s'(x') = s(x)$ and $M'(x') = AM(x)A^{-1}$, respectively.

Reference Frame Adjustment Every point in a vector field can be transformed into a critical point by the addition of a constant vector. For a vector field $v : \mathbb{R}^2 \to \mathbb{R}^2$ and a point $x_0 \in \mathbb{R}^2$, we define the associated vector field $v_{x_0} : \mathbb{R}^2 \to \mathbb{R}^2$ with its frame of reference based on x_0 by

$$v_{x_0}(x) := v(x) - v(x_0). \tag{2}$$

Such a vector field v_{x_0} has a permanent critical point at x_0, because $v_{x_0}(x_0) = v(x_0) - v(x_0) = 0$. For a given position $x_0 \in \mathbb{R}^2$, the vector field v_{x_0} is GI, because from $v'(x') = dx'/dt \overset{(1)}{=} d(Ax+b)/dt = Av(x) + \dot{b}$ follows $v'_{x'_0}(x') \overset{(2)}{=} v'(x') - v'(x'_0) = Av(x) + \dot{b} - Av(x_0) - \dot{b} = A(v(x) - v(x_0)) \overset{(2)}{=} Av_{x_0}(x)$.

Jacobian-Based Galilean Invariant Vector Fields Recall $v : \mathbb{R}^2 \to \mathbb{R}^2$ is a 2D vector field, where $v(x) = \dot{x} = dx/dt = (v_1(x), v_2(x))^T$. Let J denote the Jacobian of a velocity field,

$$J = \nabla v(x) = \begin{pmatrix} \partial v_1(x)/\partial x_1 & \partial v_1(x)/\partial x_2 \\ \partial v_2(x)/\partial x_1 & \partial v_2(x)/\partial x_2 \end{pmatrix}.$$

The determinant of the Jacobian, $\det(J)$, is shown to be a GI scalar field [3], that is, $\det J'(x') = \det J(x)$. Such a determinant can be used to categorize first-order critical points, that is, a negative determinant corresponds to a saddle, whereas a positive determinant corresponds to a source, a sink, or a vortex.

A point $(x_0) \in \mathbb{R}^2$ is a Jacobian-based *Galilean invariant critical point* (GICP) of a vector field $v : \mathbb{R}^2 \to \mathbb{R}^2$ if it is a critical point of the determinant of the Jacobian, i.e., $\nabla \det(J) := \nabla \det(\nabla v(x_0)) = 0$ [3]. Bujack et al. [3] restrict this definition to the negative minima and the positive maxima of the determinant field. The former form saddles, whereas the latter form sources/sinks/vortices in the velocity field in some specific frame of reference. Each GICP comes with its own frame of reference in which it becomes a classical critical point.

To visualize the GICPs simultaneously, Bujack et al. [3] introduced the notion of *Galilean invariant vector field* (GIVF) that is applicable beyond Jacobian-based GICPs. The basic idea is to construct a derived vector field that locally assumes the inherent frames of references of each GICP. Such a derived vector field is constructed by subtracting a weighted average of the velocities of the GICPs, $x_1, \ldots, x_n$, of the vector field v.

Formally, let $v : \mathbb{R}^2 \to \mathbb{R}^2$ be a vector field, $x_1, \ldots, x_n \in \mathbb{R}^2$ a set of GICPs, and w_i the weights of a linear interpolation problem $\sum_{i=1}^{n} w_i(x)v(x_i)$ with weights $w_i : \mathbb{R}^2 \to \mathbb{R}$ (and a mapping $x \mapsto w_i(x)$) that are invariant under Galilean transformation, that is,

$$w_i'(x') = w_i(x),$$

and the weights add up to one, $\forall x \in \mathbb{R}^2 : \sum_{i=1}^{n} w_i(x) = 1$. Then, the GIVF $\bar{v} : \mathbb{R}^2 \to \mathbb{R}^2$ is defined by

$$\bar{v}(x) := v(x) - \sum_{i=1}^{n} w_i(x)v(x_i).$$

In this paper, we use inverse distance weighting with exponent 2. Most commonly used weights satisfy such a condition are the ones from constant, barycentric, bilinear, and inverse distance interpolations [3].

Remark Locally the transformation in defining a GIVF is a Galilean change of reference. However depending on the chosen interpolation scheme, the points between the GICPs are transformed by a mixture of the transformations of their neighbors. This mixture does not generally result in a Galilean transformation globally. As a result, the Jacobian of the GIVF and the original field are not identical.

Although the suggested method does not transform the field through a Galilean transformation itself, it does not contradict the fact that the GIVF defined above is invariant with respect to the Galilean transformation [3]. Such a transformed vector field is GI, because any vector field that differs from the original one through a Galilean transformation will result in the same GIVF, which means that the GIVF and the original field would generally not produce the same output. In a nutshell, the method is GI, but not idempotent.

Robustness Let $f, h : \mathbb{R}^2 \to \mathbb{R}^2$ be two continuous 2D vector fields. We define the distance between the two mappings as $\mathrm{d}(f, h) = \sup_{x \in \mathbb{R}^2} ||f(x) - h(x)||_2$. The field h is an *r-perturbation* of f, if $\mathrm{d}(f, h) \leq r$. Given $f : \mathbb{R}^2 \to \mathbb{R}^2$, the *robustness* of a critical point of f quantifies its stability with respect to perturbations of the vector fields [29]. Intuitively, if a critical point has a robustness value of r, then an $(r + \delta)$-perturbation h of f exists to eliminate x (via critical point cancellation); and any $(r - \delta)$-perturbation is not enough to eliminate x.

Mathematically, the robustness of critical points in our setting arises from the well group theory [12]. Given a mapping $f : \mathbb{X} \to \mathbb{Y}$ between two manifolds and a point $a \in \mathbb{Y}$, the well group theory [12] studies the robustness of the homology of the pre-image of a, $f^{-1}(a)$ with respect to perturbations of the mapping f. Roughly speaking, the homology of a topological space $\mathbb{X}$, $\mathsf{H}_*(\mathbb{X})$, measures its topological features, where the rank of the 0-, 1- and 2-dimensional homology groups corresponds to the number of connected components, tunnels, and voids, respectively. Let a be a point in $\mathbb{Y}$, and let $B_a(r)$ be a ball of radius r surrounding a. Let h be an r-perturbation of f (under some metric). The inclusion map between subspaces $h^{-1}(a) \to f^{-1}(B_a(r))$ induces a linear map $i_h : \mathsf{H}_*(h^{-1}(a)) \to \mathsf{H}_*(f^{-1}(B_a(r)))$ between their homology groups. The *well group* $W_a(r)$ is defined as $W_a(r) = \bigcap_h \mathrm{image}\, i_h$, whose elements belong to the image of each j_h for *all* r-perturbation h of f. Intuitively, its elements are *stable* under r-perturbations of the map.

When $a = 0$, $\mathbb{X} = \mathbb{Y} = \mathbb{R}^2$, $f^{-1}(0)$ are the critical points of vector fields on the plane. Chazal et al. [6] showed that in the case of vector fields, the well group could be computed from the *merge tree* of the magnitude of a vector field (i.e., $f_0 = ||f||_2$, which is a scalar function). We use the correspondences between critical points and the elements in the well groups to assign robustness values to the critical points. The merge tree of f_0 is constructed by tracking the connected components of its sublevel sets $f^{-1}(-\infty, r]$ together with their degree information as they appear and merge by increasing r from 0. Each leaf node in the tree is assigned the degree of its corresponding critical point (a saddle has a degree of -1, and a source/since has a degree of $+1$). Each internal node has a degree the sum of its subtree. The robustness of a critical point is the height of its lowest degree zero ancestor in the merge tree, see Wang et al. [29] for details.

4 Theoretical Results

We extend the definition of robustness by considering every point as a critical point. Formally, let $x_0 \in \mathbb{R}^2$ be an arbitrary point in a vector field $v : \mathbb{R}^2 \to \mathbb{R}^2$ and $R(x_0)$ be its robustness in the vector field v_{x_0}, which is associated with the frame of reference of x_0. We define the *extended robustness* $R : \mathbb{R}^2 \to \mathbb{R}$ of the point x_0 as the robustness of the critical point $x_0 \in \mathbb{R}^2$ in the vector field v_{x_0}. For a vector field $v : \mathbb{R}^2 \to \mathbb{R}^2$, we call a point a *locally robust critical point* (LRCP) if it is a local maximum in the extended robustness field, i.e.,

$$\nabla R(x_0) = 0, \quad H_R(x_0) < 0,$$

with the vector ∇R denoting the first derivative and the Hessian matrix H_R consisting of the second partial derivatives.

The following two theorems are the key theoretical contributions of the paper.

Theorem 1 *The extended robustness is a Galilean invariant scalar field. The locally robust critical points defined above are Galilean invariant.*

Proof We prove the theorem by showing that for the extended robustness $R : \mathbb{R}^2 \to \mathbb{R}$, we have $R'(x') = R(x)$. The extended robustness assigns a scalar to every point $x_0 \in \mathbb{R}^2$. Let $v'(x')$ differ from a vector field $v : \mathbb{R}^2 \to \mathbb{R}^2$ by the transformation $v'(x') = Av(x)$. The magnitude $\|v_{x_0}\|_2$ of the GI field v_{x_0} from (2) is GI. From $A \in SO(2)$, it follows that

$$\|v'_{x'_0}(x')\|_2 = \|Av_{x_0}(x)\|_2 = \|v_{x_0}(x)\|_2. \tag{3}$$

As a result, the merge trees of $v_{x_0}(x_0)$ and $v'_{x'_0}(x'_0)$ are isomorphic. Together with the invariance of the degree of a critical point with respect to orthogonal transformations, that the extended robustness is GI follows. Since the extrema of the scalar field are GI and the extended robustness field is GI, it follows that LRCPs are GI. □

Theorem 2 *At any point $x_0 \in \mathbb{R}^2$, suppose: (i) $v_{x_0} : \mathbb{R}^2 \to \mathbb{R}^2$ is generic and C^2-smooth; (ii) the directional derivative of v_{x_0} is upper bounded by a constant μ; (iii) the second (partial) derivative of v_{x_0} is upper bounded by a constant δ; and (iv) the absolute value of the determinant of the Jacobian is at least c. Then the extended robustness at x_0 is at least $\frac{c^2}{2\mu^2\delta}$.*

Proof For any point $x_0 \in \mathbb{R}^2$, let $f := v_{x_0} : \mathbb{R}^2 \to \mathbb{R}^2$. Genericity from assumption (i) of f implies that for the critical point x_0 of f, there exists a small neighborhood that contains only x_0. First, we show that a lower bound on the absolute value of the determinant of the Jacobian translates into a lower bound on the magnitude of the directional derivative of f. Let J be the Jacobian at $x_0 \in \mathbb{R}^2$ and $\det(J)$ be the determinant of the Jacobian. Assumption (iv) means that $|\det(J)| \geq c$. Let λ_1 and

λ_2 ($|\lambda_1| \geq |\lambda_2|$) be the eigenvalues of J. We have,

$$|\det(J)| = |\lambda_1 \lambda_2| \geq c.$$

Assumption (ii) means that the directional derivatives of f are upper bounded in any direction by μ, i.e., $||\frac{\partial f}{\partial u}|| \leq \mu$ for all directions $u \in \mathbb{S}^2$, which implies that the absolute values of all eigenvalues are upper bounded by μ, i.e., $|\lambda_2| \leq |\lambda_1| \leq \mu$. Hence, $|\lambda_2| \geq c' = \frac{c}{\mu}$.

Now, we show that the upper bound on the second derivative implies a lower bound on robustness. We consider the direction $u \in \mathbb{S}^2$ to be along the eigenvector associated with λ_2. At x_0, $|f(x_0)| = 0$. Since the magnitude of the directional derivative at x_0 is lower bounded by c' and there is an upper bound on the second derivative, we can bound the neighborhood size where the directional derivative becomes 0, i.e., how far from x_0 we must go in order for $\|f\|$ to stop growing. Let y be a point on the boundary of the isolating neighborhood of x_0, such that $d(y, x_0) = \varepsilon$. Then the magnitude of the directional derivative is lower bounded by $c' - \varepsilon\delta$ based on assumption (iii). The change $c' - \varepsilon\delta$ is positive for all $\varepsilon \leq \frac{c'}{\delta}$. We obtain a lower bound on the magnitude of the vector field on the boundary of the ε-neighborhood at x via integration. That is, for any y on the boundary of the isolating neighorhood of x_0,

$$|f(y)| \geq \int_0^{\varepsilon} (c' - x\delta) dx = c'\varepsilon - \frac{\delta\varepsilon^2}{2}. \tag{4}$$

For $\varepsilon \leq \frac{c'}{\delta}$, $|f(\varepsilon)|$ is an increasing function in ε; hence x_0 is the only zero in the neighborhood. To obtain a lower bound on robustness, we lower bound the magnitude of the function on the boundary of the ε-neighborhood (i.e. the neighborhood where we know that x_0 is an isolated zero). Substituting $\varepsilon = \frac{c'}{\delta} = \frac{c}{\mu\delta}$ into Eq. (4) yields the desired lower bound, i.e.,

$$|f(y)| \geq c'\varepsilon - \frac{\delta\varepsilon^2}{2} = \frac{c^2}{\mu^2\delta} - \frac{\delta c^2}{2\mu^2\delta^2} = \frac{c^2}{2\mu^2\delta}.$$

□

5 Visualization Results

We demonstrate visually that the extended robustness helps to interpret the Jacobian-based GIVF analysis. In particular, the extrema of the determinant of the Jacobian (the Jacobian-based GICPs) often coincide with the local maxima of the extended robustness (the LRCPs).

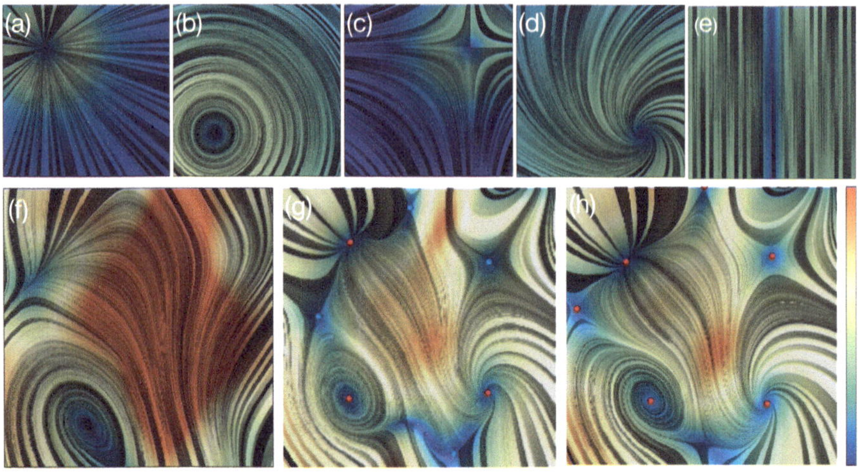

Fig. 2 Visualization of an analytic data set (**f**), which is created by superimposing five analytic fields (**a**)–(**e**). The colormap encodes the speed of the flow. For comparison, (**g**) shows the corresponding Galilean invariant vector field introduced by Bujack et al. and constructed from the extrema of the determinant of the Jacobian. The Galilean invariant critical points are marked with red nodes for vortices/sinks/sources and with blue nodes for saddles. Image courtesy of Bujack et al. [3]. (**h**) The Galilean invariant vector field introduced in this paper is constructed from the extended robustness. The local maxima of the extended robustness field are marked with red nodes

Case Study I: An Analytic Vector Field For the first case study illustrated in Fig. 2, we use an analytic vector field in (f) which contains four standard flow features, sink (a), center (b), saddle (c) and spiral source (d); each showing a different common velocity profile overlaid with a sheer flow (e) that makes it impossible to view all the flow features simultaneously. As illustrated, the GIVF based on the determinant of the Jacobian (g) simultaneously highlights the Jacobian-based GICPs, which correspond to the standard flow features described in (a)–(d). On the other hand, these flow features in (g) coincide with the features surrounding the LRCPs of the GIVF based on the extended robustness in (h).

Case Study II: A Sequence of Double Gyre We use a formula describing a double gyre vector field [2] with parameters $A = 0.25$, $\omega = 1/10$, and an extended domain $[0, 6] \times [0, 1]$. Such a dataset is smooth and requires no topological simplification (see Sect. 5.1). As shown in Fig. 3a, one vortex is visible at position (3, 0.5) within the standard frame of reference, and the Jacobian-based GIVF highlights two vortices within the same region in Fig. 3b, as shown previously [2]. These Jacobian-based critical points coincide with the LRCPs obtained via extended robustness in Fig. 3c. The separators from the robustness-based GIVF coincide with the separators from the standard frame of reference, but those from the Jacobian-based GIVF

Fig. 3 Visualization of a sequence of double gyre. (**a**) The original flow; the colormap encodes the speed of the flow. (**b**) Jacobian-based Galilean invariant vector field with highlighted critical points; the flow is color-coded by the value of the determinant. (**c**) Robustness-based Galilean invariant vector field with highlighted critical points; the flow is color-coded by extended robustness values. (**d**) Robustness-based Galilean invariant vector field without contour tree pruning

do not. This observation gives an indication that the two vortices detected by both robustness-based and Jacobian-based GIVF are likely true features, whereas the separators detected by the Jacobian-based GIVF are not (therefore partially addressing an open question in [2]). Furthermore, we illustrate the robustness-based GIVFs in Fig. 3d without topological simplification (see Sect. 5.1 for details).

Case Study III: Swirly Jet Our last case study, illustrated in Fig. 4, focuses on a flow simulation of a swirling jet entering a fluid at rest. Such a dataset has been previously studied in the work of Bujack et al. [3]. We demonstrate visually an interpretation of its corresponding Jacobian-based GIVF with the extended robustness. As shown in Fig. 4c, some but not all of the LRCPs are shown to coincide with the critical points extracted from Jacobian-based GIVF in Fig. 4b. Such a discrepancy could be due to numerical issues in computing extended robustness, discretization resolution and the noisy, non-smooth data domain. How to choose the optimal parameters for topological simplification (as discussed in Sect. 5.1), remains an open question for both Jacobian-based and robustness-based GIVFs.

Fig. 4 Visualization of the swirling jet entering a fluid at rest. (**a**) The original flow; the colormap encodes the speed of the flow. (**b**) Jacobian-based Galilean invariant vector field with highlighted critical points; the flow is color-coded by the value of the determinant. (**c**) Robustness-based Galilean invariant vector field with highlighted locally robust critical points; the flow is color-coded by extended robustness values

5.1 Topological Simplification

In our case studies, the extended robustness fields are often noisy, resulting in many insignificant local maxima. Analogously to Bujack et al. [3], we make use of the topological simplification tools for scalar fields to reduce the number of local maxima to the significant ones. For an introduction to scalar topological simplification, we recommend the work of Carr et al. [5] and Heine et al. [4, 16].

For a scalar field, a contour is a connected component of a level set, which is the set of points that all have the same value in the scalar field. If we increase this value, contours can be created at local minima, join or split at saddles, and be destroyed at local maxima of the scalar field. The contour tree is an abstraction of the scalar field that is formed from shrinking each contour to a node in the tree, where each branch starts and ends at an extremum or a saddle and corresponds to a connected component in the domain. Each branch of the contour tree comes with three popular measures: persistence, volume, and hypervolume [5, 16]. Persistence is the maximal difference of the scalar values of the components of a branch, the volume is the integral over its affiliated points, and the hypervolume is the integral over the scalar values. These measures can be used to simplify the contour tree by pruning branches that do not exceed given thresholds (see Carr et al. [5]).

We compute the contour tree of the extended robustness field and prune it with respect to persistence. The result for case study I can be found in Fig. 5.

Remark We have demonstrated that the Jacobian-based GICPs in some smooth, synthetic cases coincide with the LRCPs, whereas in noisy, real-world datasets, unambiguous equivalence among these points is difficult to find due to the resolution of the data and the different range of scalar values for topological simplification. In

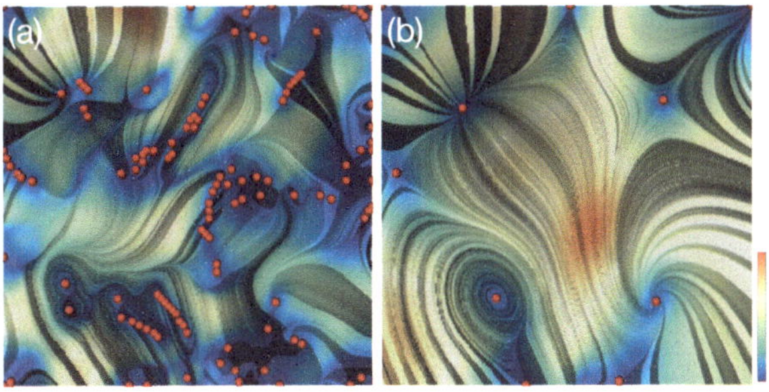

Fig. 5 Topological simplification of case study I, where the colormap encodes the speed of the flow. The robustness-based Galilean invariant vector fields before (**a**) and after (**b**) simplification are illustrated, where the locations of extended robustness local maxima are marked in red

addition, we conjecture that the determinant of Jacobian could be considered as a first-order approximation that captures the stabilities of critical points, whereas the extended robustness captures higher order information; therefore the LRCPs do not always coincide with the Jacobian-based GICPs.

The best way to select the pruning parameters for simultaneous visualization of robust critical points in different regions of the data, remains an open question. We currently use an exploratory process to choose pruning parameters so that the LRCPs are at a level comparable to the Jacobian-based GICPs.

6 Discussion

Robustness and Jacobian The Jacobian carries important information about the local behavior of a vector field, while robustness quantifies their global stability. In this work, we demonstrate their relations theoretically and visually. Furthermore, our results inspire discussions regarding different quantifiers of stable features within the vector field data.

Extended Robustness: Degeneracies and Continuity In our current framework, some critical points do not have any cancellation partner, and so have large robustness values beyond the range of the maximum vector norm in the domain. This can cause boundary effects in our visualization as some critical points are detected on the boundary. Furthermore, robustness computation also assumes that each critical point is isolated within its local neighborhood. Our datasets, however, contain regions with degenerate critical points where such isolation conditions are violated (i.e., regions where the determinant of Jacobian switches sign). For the purpose of visualization, such degeneracies are handled separately.

Small changes to the vector field may introduce partner switches in the merge tree, which lead to some discontinuities in the current computation of extended robustness (see Fig. 4c). However this does not impact our visualization results significantly. Ensuring the continuity of the extended robustness remains an open question.

Other Perturbation Metrics for Robustness The robustness framework also allows a certain flexibility in defining perturbation metrics, in the sense that the L_∞ metric defined in Sect. 3 could be replaced by other metrics such as the L_2 metric, which incorporates both the magnitude of the vectors and the area to capture a quantity closer to the energy of a perturbation. We will investigate the effect of different perturbation metrics on the computation of extended robustness and its connection with the determinant of the Jacobian.

Acknowledgments This work was partially supported by NSF IIS-1513616, National Research Agency ARRS Project TopRep N1-0058, the NNSA ASC Program and the LDRD program of Los Alamos National Laboratory under project number 20190143ER. It is published under LA-UR-17-28344.

References

1. Bhatia, H., Pascucci, V., Kirby, R.M., Bremer, P.: Extracting features from time-dependent vector fields using internal reference frames. Comput. Graphics Forum **33**(3), 21–30 (2014)
2. Bujack, R., Joy, K.I.: Lagrangian representations of flow fields with parameter curves. In: Proceedings of the IEEE Symposium on Large Data Analysis and Visualization, pp. 41–48 (2015)
3. Bujack, R., Hlawitschka, M., Joy, K.I.: Topology-inspired Galilean invariant vector field analysis. In: Proceedings of the IEEE Pacific Visualization Symposium, pp. 72–79. IEEE, New York (2016)
4. Carr, H., Snoeyink, J., Axen, U.: Computing contour trees in all dimensions. Comput. Geom. **24**(2), 75–94 (2003)
5. Carr, H., Snoeyink, J., van de Panne, M.: Simplifying flexible isosurfaces using local geometric measures. In: IEEE Visualization, pp. 497–504. IEEE, New York (2004)
6. Chazal, F., Patel, A., Skraba, P.: Computing well diagrams for vector fields on $\mathbb{R}^n$. Appl. Math. Lett. **25**(11), 1725–1728 (2012)
7. Chen, G., Palke, D., Lin, Z., Yeh, H., Vincent, P., Laramee, R.S., Zhang, E.: Asymmetric tensor field visualization for surfaces. IEEE Trans. Visual. Comput. Graph. **17**(12), 1979–1988 (2011)
8. Chong, M.S., Perry, A.E., Cantwell, B.J.: A general classification of three-dimensional flow fields. Phys. Fluids A **2**(5), 765–777 (1990)
9. Ebling, J., Wiebel, A., Garth, C., Scheuermann, G.: Topology based flow analysis and superposition effects. In: Hauser, H., Hagen, H., Theisel, H. (eds.) Topology-based Methods in Visualization. Mathematics and Visualization, pp. 91–103. Springer, Berlin (2007)
10. Edelsbrunner, H., Letscher, D., Zomorodian, A.J.: Topological persistence and simplification. Discrete Comput. Geom. **28**, 511–533 (2002)
11. Edelsbrunner, H., Morozov, D., Patel, A.: The stability of the apparent contour of an orientable 2-manifold. In: an dXavier Tricoche, V.P., Hagen, H., Tierny, J. (eds.) Topological Methods in Data Analysis and Visualization. Mathematics and Visualization, pp. 27–42. Springer, Berlin (2010)
12. Edelsbrunner, H., Morozov, D., Patel, A.: Quantifying transversality by measuring the robustness of intersections. Found. Comput. Math. **11**, 345–361 (2011)

13. Fuchs, R., Kemmler, J., Schindler, B., Waser, J., Sadlo, F., Hauser, H., Peikert, R.: Toward a Lagrangian Vector Field Topology. Comput. Graphics Forum **29**(3), 1163–1172 (2010)
14. Günther, T., Schulze, M., Theisel, H.: Rotation invariant vortices for flow visualization. IEEE Trans. Visual. Comput. Graph. **22**(1), 817–826 (2016)
15. Haller, G.: An objective definition of a vortex. J. Fluid Mech. **525**, 1–26 (2005)
16. Heine, C., Schneider, D., Carr, H., Scheuermann, G.: Drawing contour trees in the plane. IEEE Trans. Visual. Comput. Graph. **17**(11), 1599–1611 (2011)
17. Hunt, J.C.R.: Vorticity and vortex dynamics in complex turbulent flows. Trans. Can. Soc. Mech. Eng. **11**(1), 21–35 (1987)
18. Jeong, J., Hussain, F.: On the identification of a vortex. J. Fluid Mech. **285**, 69–94 (1995)
19. Kasten, J., Reininghaus, J., Hotz, I., Hege, H.C.: Two-dimensional time-dependent vortex regions based on the acceleration magnitude. IEEE Trans. Visual. Comput. Graph. **17**(12), 2080–2087 (2011)
20. Lugt, H.J.: The dilemma of defining a vortex. In: Müller, U., Roesner, K.G., Schmidt, B. (eds.) Recent Developments in Theoretical and Experimental Fluids Mechanics, pp. 309–321. Springer, Berlin (1979)
21. Perry, A.E., Chong, M.S.: Topology of flow pattern in vortex motions and turbulence. Appl. Sci. Res. **53**(3–4), 357–374 (1994)
22. Pobitzer, A., Peikert, R., Fuchs, R., Schindler, B., Kuhn, A., Theisel, H., Matkovic, K., Hauser, H.: The state of the art in topology-based visualization of unsteady flow. Comput. Graphics Forum **30**(6), 1789–1811 (2011)
23. Sahner, J., Weinkauf, T., Hege, H.C.: Galilean invariant extraction and iconic representation of vortex core lines. In: Proceedings of the IEEE VGTC Symposium on Visualization, pp. 151–160 (2005)
24. Skraba, P., Wang, B.: Interpreting feature tracking through the lens of robustness. In: Bremer, P.T., Hotz, I., Pascucci, V., Peikert, R. (eds.) Topological Methods in Data Analysis and Visualization III. Mathematics and Visualization, pp. 19–38. Springer, Berlin (2014)
25. Skraba, P., Wang, B., Chen, G., Rosen, P.: 2D vector field simplification based on robustness. In: Proceedings of the IEEE Pacific Visualization Symposium, pp. 49–56 (2014)
26. Skraba, P., Wang, B., Chen, G., Rosen, P.: Robustness-based simplification of 2D steady and unsteady vector fields. IEEE Trans. Visual. Comput. Graph. **21**(8), 930–944 (2015)
27. Skraba, P., Rosen, P., Wang, B., Chen, G., Bhatia, H., Pascucci, V.: Critical point cancellation in 3D vector fields: Robustness and discussion. IEEE Trans. Visual. Comput. Graph. **22**(6), 1683–1693 (2016)
28. Song, Y.: A note on Galilean invariants in semi-relativistic electromagnetism. arXiv:1304.6804 (2013)
29. Wang, B., Rosen, P., Skraba, P., Bhatia, H., Pascucci, V.: Visualizing robustness of critical points for 2D time-varying vector fields. In: Computer Graphics Forum, vol. 32(3pt2), pp. 221-230. Blackwell Publishing, Oxford (2013). Computational Graphics Forum, pp. 221–230
30. Wiebel, A., Garth, C., Scheuermann, G.: Localized flow analysis of 2D and 3D vector fields. In: Proceedings of the IEEE VGTC Symposium on Visualization, pp. 143–150 (2005)

Maximum Number of Transition Points in 3D Linear Symmetric Tensor Fields

Yue Zhang, Lawrence Roy, Ritesh Sharma, and Eugene Zhang

Abstract Transition points are well defined topological features in 3D tensor fields, which are important for the study of other prominent topological singularities such as wedges and trisectors. In this paper, we study the maximum number of transition points in a linear tensor field, which is important to process wedge and trisector classification along degenerate curves.

1 Introduction

3D symmetric tensor field topology consists of degenerate curves and neutral surfaces [7]. Transition points, which are degenerate points that separate wedge and trisector segments along a degenerate curve, are important topological features that are key to the study of 3D symmetric tensor field topology [12].

To the best of our knowledge, existing degenerate curve extraction methods do not explicitly extract transition points. We believe that this is largely due to the fact that it is not known how many transition points can exist in a tensor field and how to algebraically characterize them. In this paper, we attempt to address this cause by studying the minimum and maximum numbers of transition points in a 3D linear symmetric tensor field.

Y. Zhang (✉)
School of Electrical Engineering and Computer Science, 3117 Kelley Engineering Center, Oregon State University, Corvallis, OR, USA
e-mail: zhangyue@eecs.oregonstate.edu; zhangyue@oregonstate.edu

L. Roy · R. Sharma
School of Electrical Engineering and Computer Science, 1148 Kelley Engineering Center, Oregon State University, Corvallis, OR, USA
e-mail: royl@eecs.oregonstate.edu; sharmrit@eecs.oregonstate.edu

E. Zhang
School of Electrical Engineering and Computer Science, 2111 Kelley Engineering Center, Oregon State University, Corvallis, OR, USA
e-mail: zhange@eecs.oregonstate.edu

H. Carr et al. (eds.), *Topological Methods in Data Analysis and Visualization V*, Mathematics and Visualization, https://doi.org/10.1007/978-3-030-43036-8_15

2 Previous Work

There has been much work on the analysis and visualization of 2D and 3D tensor fields. We refer the readers to the recent survey by Kratz et al. [5]. Here we only refer to the research most relevant to this chapter.

Delmarcelle and Hesselink [1, 2] introduce the topology of 2D symmetric tensor fields. They point out that there are two *fundamental* types of degenerate points in a 2D symmetric tensor field, i.e., wedges and trisectors, which have a tensor index of $\frac{1}{2}$ and $-\frac{1}{2}$, respectively. Hesselink et al. later extend this work to 3D symmetric tensor fields [4] and study triple degenerate points, i.e., all eigenvalues are the same. Zheng et al. [11] point out that triple degeneracies are not structurally stable features. They further show that double degeneracies, i.e., tensors with only two equal eigenvalues, form lines in the domain. In this work and subsequent research [13], they provide a number of degenerate curve extraction methods based on the analysis of the discriminant function of the tensor field. Furthermore, Zheng et al. [12] point out that near degenerate curves the tensor field exhibits 2D degenerate patterns and define separating surfaces which are extensions of separatrices from 2D symmetric tensor field topology. Zhang et al. [10] show that there are at least two and at most four degenerate curves in a 3D *linear* symmetric tensor field under structurally stable conditions. Palacios et al. [6] introduce a design system for 3D tensor fields that allows the editing of the structure of degenerate curves in the field. Roy et al. [8] describe a method to extract degenerate curves with a high quality at an interactive speed.

However, most of the aforementioned work except [6, 12] only focuses on the linearity and planarity of the degenerate points, but not wedges, trisectors, and transition points. In this paper, we explore the minimum and maximum number of transition points in a 3D linear tensor field.

3 Background on Symmetric Tensors and Tensor Fields

We review some pertinent technical concepts in this section on tensors and tensor fields.

A K-dimensional (symmetric) tensor T has K real-valued *eigenvalues*: $\lambda_1 \geq \lambda_2 \geq \cdots \geq \lambda_K$. The largest and smallest eigenvalues are referred to as the *major eigenvalue* and *minor eigenvalue*, respectively. When $K = 3$, the middle eigenvalue is referred to as the *medium eigenvalue*. An eigenvector belonging to the major eigenvalue is referred to as a *major eigenvector*. Medium and minor eigenvectors can be defined similarly. Eigenvectors belonging to different eigenvalues are mutually perpendicular. A tensor is *degenerate* if there are repeating eigenvalues. In this case, there exists at least one eigenvalue whose corresponding eigenvectors form a higher-dimensional space than a line. When $K = 2$, a degenerate tensor must be a multiple of the identity matrix. When $K = 3$, there are two types of degenerate tensors, corresponding to three repeating eigenvalues (*triple degenerate*) and two repeating eigenvalues (*double degenerate*), respectively. There are two types of double degenerate tensors: (1) linear ($\lambda_1 > \lambda_2 = \lambda_3$) and (2) planar ($\lambda_1 = \lambda_2 > \lambda_3$). The trace of a tensor $T = (t_{ij})$ is $trace(T) = \sum_{i=1}^{3} \lambda_i$. T can be uniquely decomposed as $D + A$ where $D = \frac{trace(T)}{3}\mathbb{I}$ ($\mathbb{I}$ is the three-dimensional identity matrix) and $A = T - D$. The *deviator* A is a *traceless* tensor, i.e., $trace(A) = 0$. Note that T is degenerate if and only if A is degenerate. Consequently, it is sufficient to study the set of traceless tensors, which is closed under matrix addition and scalar multiplication.

A *tensor field* is a tensor-valued function over some domain $\Omega \subset \mathbb{R}^3$. The topology of a tensor field is defined as the set of *degenerate points*, i.e., points in the domain where the tensor field becomes degenerate.

In a 2D tensor field, there are two fundamental types of degenerate points, wedges and trisectors. They can be classified based on an invariant $\delta = \left|\begin{pmatrix} \frac{a_{11}-a_{22}}{2} & a_{12} \\ \frac{b_{11}-b_{22}}{2} & b_{12} \end{pmatrix}\right|$, where $a_{ij} = \frac{\partial t_{ij}(x,y)}{\partial x}$ and $b_{ij} = \frac{\partial t_{ij}(x,y)}{\partial y}$, i.e., the partial derivatives of the ij-th entry of the tensor field. A degenerate point $\mathbf{p}_0$ is a wedge when $\delta(\mathbf{p}_0) > 0$ and a trisector when $\delta(\mathbf{p}_0) < 0$. When $\delta(\mathbf{p}_0) = 0$, $\mathbf{p}_0$ is a higher-order degenerate point, which is structurally unstable.

In 3D symmetric tensor fields, a degenerate point can be classified by the linear-planar classification and the wedge-trisector classification. In the former, a degenerate point is either *triple degenerate*, *linear degenerate*, or *planar degenerate*. While triple degeneracies can exist, they are structurally unstable, i.e., they can disappear under arbitrarily small perturbations. In contrast, linear and planar degenerate points are structurally stable, i.e., they persist under sufficiently small perturbations in the tensor field. Moreover, under structurally stable conditions such points form curves, along which the tensor field is either always linear degenerate or always planar degenerate. While it is possible that linear and planar degenerate points are isolated points or form surfaces and volumes, these three scenarios do not persist under arbitrarily small perturbation in the field, i.e., are structurally unstable.

A degenerate point can also be classified based on the so-called wedge-trisector classification. Given a degenerate point $\mathbf{p}_0$, let $n = (\alpha, \beta, \gamma)$ be the non-repeating eigenvector at $\mathbf{p}_0$. The plane P that passes through $\mathbf{p}_0$ whose normal is n is referred to the *repeating plane* at $\mathbf{p}_0$. When projecting the 3D tensor field onto P, one obtains a 2D symmetric tensor field which, under structurally stable conditions, has exactly one degenerate point, $\mathbf{p}_0$. In the 2D tensor field, $\mathbf{p}_0$ can be either a wedge, a trisector, or a higher-order and thus structurally unstable degenerate point. In these cases, $\mathbf{p}_0$ will be referred to respectively as a wedge, a trisector, and a transition point in the 3D tensor field. Figure 1 provides examples of such degenerate points and their non-repeating planes with a 3D tensor field. Here and in the remaining figures in the paper, we use the following color scheme for degenerate points: yellow (planar wedge), green (linear wedge), red (planar trisector), and blue (linear trisector).

Note that while a higher-order degenerate point is structurally unstable, a transition point is structurally stable in 3D tensor fields. Moreover, a transition point

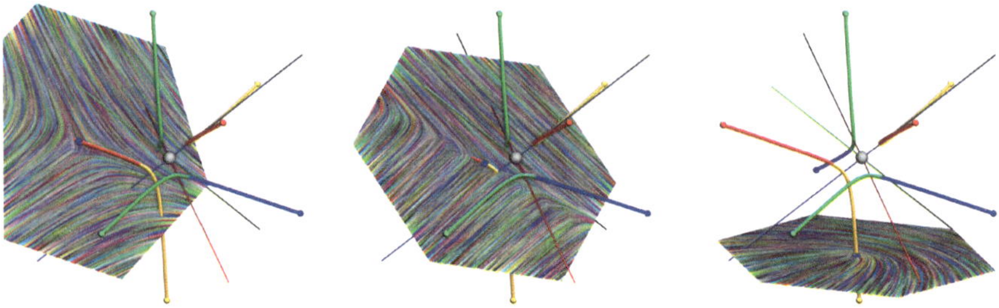

(a) (b) (c)

Fig. 1 Along a degenerate curve, the projection of the tensor field onto the repeating planes can exhibit 2D degenerate patterns such as a trisector (**a**) or a wedge (**c**). Between segments of wedges (yellow) and trisectors (red), transition points can appear (**b**)

is *not* the same as triple degenerate points. At the transition point, the repeating plane is tangent to the degenerate curve.

4 Transition Points in 3D Symmetric, Traceless Tensor Fields

We first note that the set of all traceless and symmetric tensors with configuration $\begin{pmatrix} a & b & c \\ b & d & e \\ c & e & -a-d \end{pmatrix}$ form a five dimensional linear space $\mathbb{T}$ spanned by the basis $T_a = \begin{pmatrix} 1 & 0 & 0 \\ 0 & 0 & 0 \\ 0 & 0 & -1 \end{pmatrix}$, $T_d = \begin{pmatrix} 0 & 0 & 0 \\ 0 & 1 & 0 \\ 0 & 0 & -1 \end{pmatrix}$, $T_b = \begin{pmatrix} 0 & 1 & 0 \\ 1 & 0 & 0 \\ 0 & 0 & 0 \end{pmatrix}$, $T_c = \begin{pmatrix} 0 & 0 & 1 \\ 0 & 0 & 0 \\ 1 & 0 & 0 \end{pmatrix}$, and $T_e = \begin{pmatrix} 0 & 0 & 0 \\ 0 & 0 & 1 \\ 0 & 1 & 0 \end{pmatrix}$. Any tensor in this space can be expressed as $t_a T_a + t_b T_b + t_c T_c + t_d T_d + t_e T_e$ for some $t_a, t_b, t_c, t_d, t_e \in \mathbb{R}$. For convenience, it can be written in the vector form $(t_a, t_d, t_b, t_c, t_e)$.

A 3D symmetric, traceless linear tensor field has the following form

$$LT(x, y, z) = T_0 + xT_x + yT_y + zT_z \tag{1}$$

where $T_0 = \begin{pmatrix} a_0 & b_0 & c_0 \\ b_0 & d_0 & e_0 \\ c_0 & e_0 & -a_0 - d_0 \end{pmatrix}$, $T_x = \begin{pmatrix} a_x & b_x & c_x \\ b_x & d_x & e_x \\ c_x & e_x & -a_x - d_x \end{pmatrix}$, $T_y = \begin{pmatrix} a_y & b_y & c_y \\ b_y & d_y & e_y \\ c_y & e_y & -a_y - d_y \end{pmatrix}$, and $T_z = \begin{pmatrix} a_z & b_z & c_z \\ b_z & d_z & e_z \\ c_z & e_z & -a_z - d_z \end{pmatrix}$ are symmetric, traceless matrices. Under structurally stable conditions, T_0, T_x, T_y, and T_z are linearly independent. In this section we study the number of transition points in such a tensor field.

Zhang et al. [10] show that the degenerate points in a 3D linear tensor field satisfy the following system of equations

$$h(\alpha, \beta, \gamma) = 0 \quad (2)$$

$$\alpha^2 + \beta^2 + \gamma^2 = 1 \quad (3)$$

where (α, β, γ) is a unit non-repeating eigenvector and $h(\alpha, \beta, \gamma)$ is a homogeneous quadratic polynomial.

A transition point, as a degenerate point, must satisfy Eqs. (2) and (3). However, while degenerate points form curves under structurally stable conditions, transition points are isolated points. This indicates that one more condition is needed in terms of α, β, and γ.

Given a linear symmetric tensor field $LT(x, y, z) = T_0 + xT_x + yT_y + zT_z$, its projection onto any plane is also a 2D linear tensor field [9]. Consequently, the discriminant function δ is constant for the plane. We define a plane to be a *wedge plane* if $\delta > 0$, a *trisector plane* if $\delta < 0$, and more relevantly a *transition plane* if $\delta = 0$. A transition point must have its repeating plane as a transition plane. Therefore, characterizing transition planes gives us the additional condition to characterize transition points.

The following result from [9] is important to our analysis of transition planes.

Theorem 1 *Given a 3D linear tensor field $LT = T_0 + xT_x + yT_y + zT_z$ and a plane P, the discriminant function δ of the projection of LT onto P is a function of only T_x, T_y, and T_z.*

This leads to the following results:

Corollary 1 *Given a 3D linear tensor field $LT = T_0 + xT_x + yT_y + zT_z$ and two parallel planes P_1 and P_2, P_1 is a transition plane if and only if P_2 is a transition plane.*

Corollary 2 *Given two 3D linear tensor field $LT = T_0 + xT_x + yT_y + zT_z$ and $LT' = T_0' + xT_x + yT_y + zT_z$, a plane P is a transition plane for LT if and only if P is also a transition plane for LT'.*

Corollary 1 states that whether a plane is a transition plane depends only on the normal of the plane. Therefore, it is sufficient to only consider planes $P : \alpha x + \beta y + \gamma z = 0$, where $(\alpha, \beta, \gamma)^t$ is a unit vector and can be modelled by $\mathbb{RP}^2$, the two-dimensional real projective space.

Corollary 2 states that adding a constant tensor to the field will not change whether a plane is a transition plane. We can therefore set $T_0 = 0$ while finding the transition planes. Under these simplification conditions, we have the following result:

Lemma 1 *Given a linear symmetric tensor field $LT(x, y, z) = xT_x + yT_y + zT_z$, its projection onto the plane $P : \alpha x+\beta y+\gamma z = n\cdot p = 0$ has either one degenerate point or a line of degenerate points. The former occurs when $\delta \neq 0$ while the latter occurs when $\delta = 0$, i.e., transition plane.*

Proof Select a coordinate system (O, X', Y') for the plane P where O is the origin. Then the projection tensor field has the following form in this coordinate system:

$$\begin{pmatrix} a_1x' + a_2y' & b_1x' + b_2y' \\ b_1x' + b_2y' & c_1x' + c_2y' \end{pmatrix} \tag{4}$$

Consequently, a degenerate point in T' satisfies:

$$a_1x' + a_2y' = c_1x' + c_2y' \tag{5}$$

$$b_1x' + b_2y' = 0 \tag{6}$$

The above system corresponds to the intersection of two lines in P. Either the two lines intersect at one point, i.e., the degenerate point, or they are the same line, i.e., every point on the line is a degenerate point. These two cases correspond precisely to the conditions $\delta \neq 0$ and $\delta = 0$, respectively. □

This characterization of transition planes is essential in our analysis of the maximum number of transition points.

We now consider $Q_n \subset \mathbb{T}$, the set of traceless, symmetric tensors whose projection onto the plane $P : n \cdot p = 0$ are 2D degenerate tensors (not necessarily traceless). Here, n is the normal to the plane. Note that the set of 2D degenerate tensors is a codimension-two linear subspace of the set of 2D symmetric tensors. Therefore, since the projection is clearly surjective, Q_n is also a codimension-two linear subspace of $\mathbb{T}$. That is, Q_n is a three-dimensional linear subspace of $\mathbb{T}$. Q_n can be parameterized by a three-dimensional linear subspace W as follows.

Let q_n be the linear map from W which is isomorphic to $\mathbb{R}^3 = \{r = (u, v, w)|u, v, w \in \mathbb{R}\}$ to the set of 3×3 symmetric matrices defined as

$$q_n(r) = \frac{1}{2}(n\,r^t + r\,n^t) - \frac{1}{3}(r \cdot n)\mathbb{I} \tag{7}$$

Note that $trace(q_n(r)) = trace(\frac{1}{2}(n\,r^t + r\,n^t) - \frac{1}{3}(r \cdot n)\mathbb{I}) = \frac{1}{2}(trace(n\,r^t) + trace(r\,n^t)) - \frac{1}{3}r \cdot ntrace(\mathbb{I}) = \frac{1}{2}(trace(r^t\,n) + trace(n^t\,t)) - r \cdot n = 0$. Consequently, q_n is a map from $\mathbb{R}^3$ to $\mathbb{T}$, the set of traceless tensors.

Furthermore, we show that $q_n(r) \in Q_n$ for any $r \in \mathbb{R}^3$. To see this, we return to the domain of the linear tensor field (which is not W) and consider the plane P_n. We can choose a coordinate system X', Y' for the plane. Let $M = \begin{pmatrix} X' & Y' \end{pmatrix}$, which is a 3×2 matrix. The projection of a 3×3 tensor K onto the plane P_n is therefore $M^t K M$. In particular, $M^t q_n(r) M = M^t(\frac{1}{2}(n\,r^t + r\,n^t) - \frac{1}{3}(r \cdot n)\mathbb{I})M = \frac{1}{2}M^t n\,r^t M + \frac{1}{2}M^t r\,n^t M - \frac{1}{3}(r \cdot n)M^t M$. Since X' and Y' are both perpendicular to n, we have $M^t n = n^t M = 0$. Furthermore, $M^t M$ is the 2D identity matrix, i.e., degenerate. Consequently, $q_n(r) \in Q_n$ for any $r \in \mathbb{R}^3$. This means q_n is a map from $\mathbb{R}^3$ to Q_n.

Furthermore, the map is an injection. This can be verified by studying the kernel of the map, i.e., for what $r = (u, v, w) \in W$, $q_n(r)$ is the zero tensor. Note that $q_n(u, v, w) = \begin{pmatrix} \frac{2\alpha u-\beta v-\gamma w}{3} & \frac{\beta u+\alpha v}{2} & \frac{\gamma u+\alpha w}{2} \\ \frac{\beta u+\alpha v}{2} & \frac{-\alpha u+2\beta v-\gamma w}{3} & \frac{\beta w+\gamma v}{2} \\ \frac{\gamma u+\alpha w}{2} & \frac{\beta w+\gamma v}{2} & \frac{-\alpha u-\beta v+2\gamma w}{3} \end{pmatrix}$. Given any non-zero $n = (\alpha, \beta, \gamma)$, the matrix $\begin{pmatrix} \frac{2\alpha}{3} & -\frac{\beta}{3} & -\frac{\gamma}{3} \\ -\frac{\alpha}{3} & \frac{2\beta}{3} & -\frac{\gamma}{3} \\ \frac{\beta}{2} & \frac{\alpha}{2} & 0 \\ \frac{\gamma}{2} & 0 & \frac{\alpha}{2} \\ 0 & \frac{\gamma}{2} & \frac{\beta}{2} \end{pmatrix}$ is rank 3. Therefore, for the following equations to hold,

$$\frac{2\alpha u - \beta v - \gamma w}{3} = 0 \tag{8}$$

$$\frac{-\alpha u - \beta v + 2\gamma w}{3} = 0 \tag{9}$$

$$\frac{\beta u + \alpha v}{2} = 0 \tag{10}$$

$$\frac{\gamma u + \alpha w}{2} = 0 \tag{11}$$

$$\frac{\beta w + \gamma v}{2} = 0 \tag{12}$$

we must have $u = v = w = 0$. That is, q_n is an injection. This means that q_n has rank 3, which is also the dimension of Q_n, so the map must be a surjection. Therefore, q_n is an isomorphism between W and Q_n, i.e., W is a parameterization of Q_n.

Thus far, we have identified a parameterization for Q_n, the set of symmetric, traceless tensors whose projection onto the plane P_n is degenerate. Given the tensor field $LT(x, y, z) = xT_x + yT_y + zT_z$, it maps $\mathbb{R}^3$ isomorphically to $U \subset \mathbb{T}$. When restricted on the plane P_n, LT maps P_n (isometric to $\mathbb{R}^2$) to a two-dimensional linear subspace $N \subset \mathbb{T}$. Under structurally stable conditions, $N_0 = N \bigcap Q_n$ is a zero-dimensional linear subspace, i.e., the zero tensor. This happens when P_n is *not* a transition plane. On the other hand, when N_0 is a one-dimensional linear space, i.e., there exists a point $p_0 = (x_0, y_0, z_0) \in P_n$ (p_0 is not the origin in $\mathbb{R}^3$) such that $x_0T_x + y_0T_y + z_0T_z \in Q_n$, then P_n is a transition plane.

In this case, $x_0 T_x + y_0 T_y + z_0 T_z = q_n(r_0)$ for some non-zero $r_0 \in W$. Recall that U is a codimension-two subset of $\mathbb{T}$, so there exist two linear functions such that

$$f_0(q_n(r_0)) = 0 \tag{13}$$

$$g_0(q_n(r_0)) = 0 \tag{14}$$

Note that q_n is linear in terms of r_0. Consequently, both $f_0'(r_0) = f_0(q_n(r_0))$ and $g_0'(r_0) = g_0(q_n(r_0))$ are linear in terms of r_0.

Furthermore, p_0 is in P_n and so is perpendicular to n, i.e., $(x_0, y_0, z_0) \cdot n = 0$. Because T_x, T_y, and T_z are linearly independent, the linear map $LT : \mathbb{R}^3 \to \mathbb{T}$ given by the field will have a left inverse $TL : \mathbb{T} \to \mathbb{R}^3$, such that $TL(LT(\mathbf{p})) = \mathbf{p}$, for any $\mathbf{p} \in \mathbb{R}^3$ (e.g. TL could be LT's pseudoinverse). Since $q_n(r_0) = LT(p_0)$, $TL(q_n(r_0)) = p_0$. Therefore, we have

$$0 = n \cdot p_0 = n \cdot TL(q_n(r_0)) \tag{15}$$

Let $d_0(r_0) = n \cdot TL(q_n(r_0))$. This function is also a linear function of r_0, since both TL and q_n are linear functions with respect to their arguments. Therefore, r_0 must satisfy the following system of linear equations:

$$f_0'(r_0) = 0 \tag{16}$$

$$g_0'(r_0) = 0 \tag{17}$$

$$d_0(r_0) = 0 \tag{18}$$

The above system can be rewritten as

$$y_f \cdot r_0 = 0 \tag{19}$$

$$y_g \cdot r_0 = 0 \tag{20}$$

$$y_d \cdot r_0 = 0 \tag{21}$$

where y_f and y_g are vector-valued linear functions of n. To understand y_d, we consider $n \cdot TL(q_n(r_0)) = n \cdot TL(\frac{1}{2}(n\, r_0^t + r_0\, n^t) - \frac{1}{3}(r_0 \cdot n)\mathbb{I})$. This is a homogeneous quadratic function of n and a homogeneous linear function of r_0. Consequently, it can be written as $y_d \cdot r_0$ where y_d is a vector-valued quadratic function of n.

Given our assumption that $r_0 \neq 0$, the above linear system is under-determined. Therefore, the determinant of the matrix formed by y_f, y_g, and y_d must be zero. This determinant is a quartic polynomial of n, which we refer to as $j(n)$. Consequently, when P_n is a transition plane, we have

$$j(n) = 0 \tag{22}$$

We now return to the characterization of a transition point p_0 for $LT(x, y, z) = T_0 + xT_X + yT_y + zT_z$. Its unit non-repeating eigenvector n must satisfy

$$h(n) = 0 \tag{23}$$

$$j(n) = 0 \tag{24}$$

on $\mathbb{RP}^2$, the two-dimensional real-projective space. To study the maximum number of solutions to the system, we borrow Bézout Theorem from Algebraic Geometry [3]:

Theorem 2 *Let f_0 and g_0 be two homogeneous polynomials in three variables of degree d and e, respectively. Let C_f and C_g be the curves defined by $f_0 = 0$ and $g_0 = 0$ in the complex projective space $\mathbb{CP}^2$. Assume that C_f and C_g do not have any common component, then they intersect at exactly $d * e$ points in $\mathbb{CP}^2$, counted with multiplicity.*

By Bézout's Theorem, there can be at most $8 = 2 \times 4$ solutions as h and j are quadratic and quartic, respectively. This leads to the following result:

Theorem 3 *Under structurally stable conditions, a 3D linear tensor field has at most eight transition points.*

In addition, we have the following result:

Theorem 4 *Under structurally stable conditions, a 3D linear tensor field has an even number of transition points, counting multiplicity.*

A degenerate curve is divided into wedge segments and trisector segments by transition points. Under structurally stable conditions, a degenerate curve extends to infinity on both ends. Consequently, we classify a degenerate curve as either a WW curve (the two end segments are both wedge segments), a WT curve (one end segment is a wedge segment and the other end segment is a trisector segment), and a TT curve (the two end segments are both trisector segments).

Along a WW or TT degenerate curve, there must be an even number of transition points. In contrast, along a WT degenerate curve, there must be an odd number of transition points. In particular, the following is true.

Theorem 5 *Under structurally stable conditions, a WW or TT degenerate curve can have at least zero transition points and at most eight transition points. In addition, a WT degenerate curve can have at least one transition point and at most seven transition points.*

This leads to the following result regarding the lower-bound of the number of transition points:

Theorem 6 *Under structurally stable conditions, a 3D linear tensor field can have as few transition points as the number of WT degenerate curves in the field.*

Knowing that there are either two or four degenerate curves in a linear tensor field, we have the following nine scenarios: (1) two WW curves, (2) two WT curves,

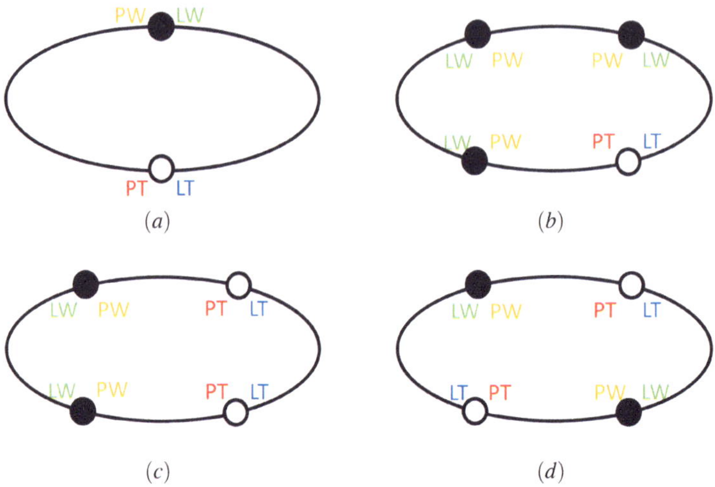

Fig. 2 This figure shows that for each of the nine configurations, it is possible to have as few transition points as the theoretical lower bound (the number of WT curves) in the field

(3) two TT curves, (4) four WW curves, (5) two WW curves and two WT curves, (6) one WW curve, two WT curves, and one TT curve, (7) four WT curves, (8) two WT curves and two TT curves, and (9) four TT curves.

Note that there are four scenarios which are theoretically possible but structurally unstable: (1) one WW curve and one TT curve, (2) three WW curves and one TT curve, (3) two WW curves and two TT curves, and (4) one WW curve and three TT curves. To see this, we examine Fig. 2. In (a), there are two degenerate points at ∞: one wedge (black dot) and one trisector (white dot). Notice that the wedge ∞ point is end to two degenerate curves, one with a linear wedge at the end and one with planar wedge at the end. Similarly, the trisector ∞ point also corresponds to a linear trisector point at the end of a degenerate curve and a planar trisector at the end of some degenerate curve. Recall that all degenerate points including the ∞ points form an ellipse (shown in black). Consequently, the two segments divided by the wedge ∞ point and the trisector ∞ point form two segments, both of which are WT curves.

For four degenerate curves, there are four ∞ degenerate points (Fig. 2b–d). Assume that there are three wedge ∞ points and one trisector ∞ point, then the three wedges ∞ points must be consecutive along the ellipse (b), leading to two WW curves and two WT curves. That is, it is not possible to have three WW curves and one TT curve. Similarly, if there are three trisector ∞ points and one wedge ∞ point (not shown in the figure), they must divide the degenerate curves into two TT types and two WT types. Consequently, it is not possible to have three TT curves and one WW curve.

Finally, when there are two wedge ∞ points and two trisector ∞ points, there are two scenarios. In the first case, the two wedge ∞ points are next to each other along the ellipse, and so are the two trisector ∞ points (Fig. 2c). Consequently, there are

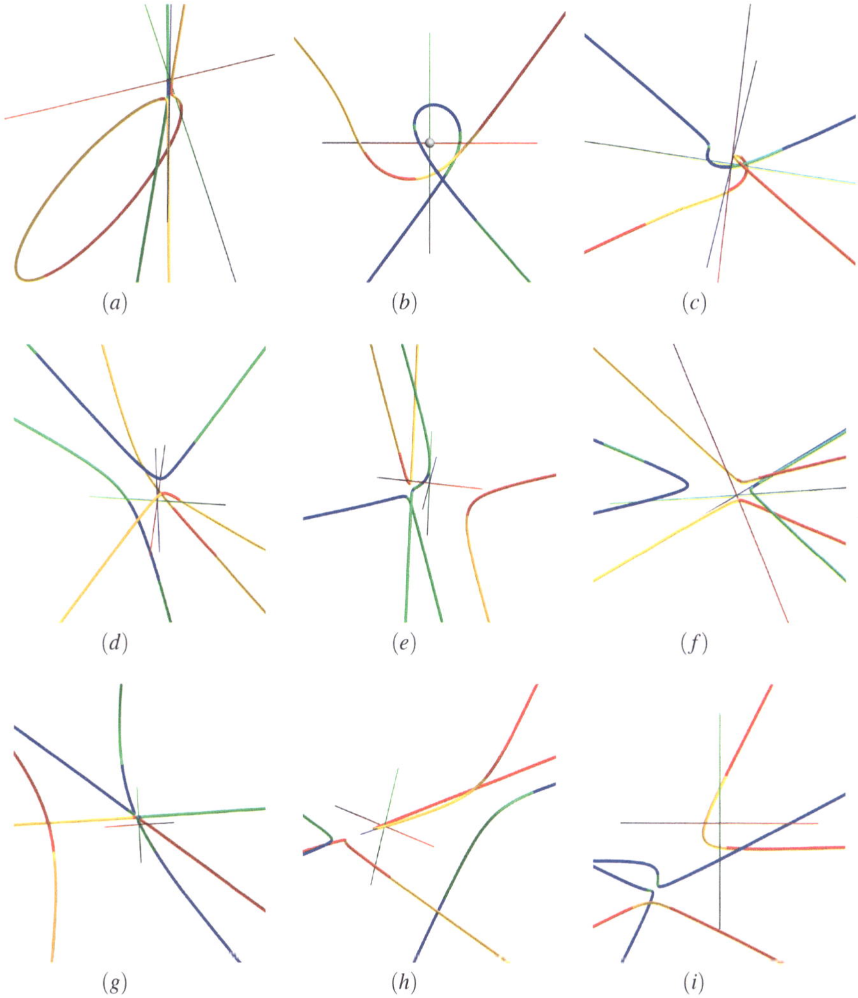

Fig. 3 This figure shows that for each of the nine configurations, it is possible to have eight transition points (theoretical upper bound) in the field. From left to right, the numbers in each triple are the number of WW curves, WT curves, and TT curves, respectively. (**a**) 2/0/0. (**b**) 0/2/0. (**c**) 0/0/2. (**d**) 4/0/0. (**e**) 2/2/0. (**f**) 1/2/1. (**g**) 0/4/0. (**h**) 0/2/2. (**i**) 0/0/4

one WW curve, one TT curve, and two WT curves. In the second case, the wedge and trisector ∞ points appear alternatingly along the ellipse, leading to four WT curves. Consequently, it is not possible to have two WW curves and two TT curves.

Each of nine legal scenarios can be encoded as (p, q, r) where p, q and r are the number of WW curves, WT curves, and TT curves, respectively. For example, the scenario of two WW curves is encoded as $2/0/0$.

Our experiments have shown that both the upper-bound and the lower-bound can be reached for each of the nine scenarios (upper-bound: Fig. 3; lower-bound: Fig. 4).

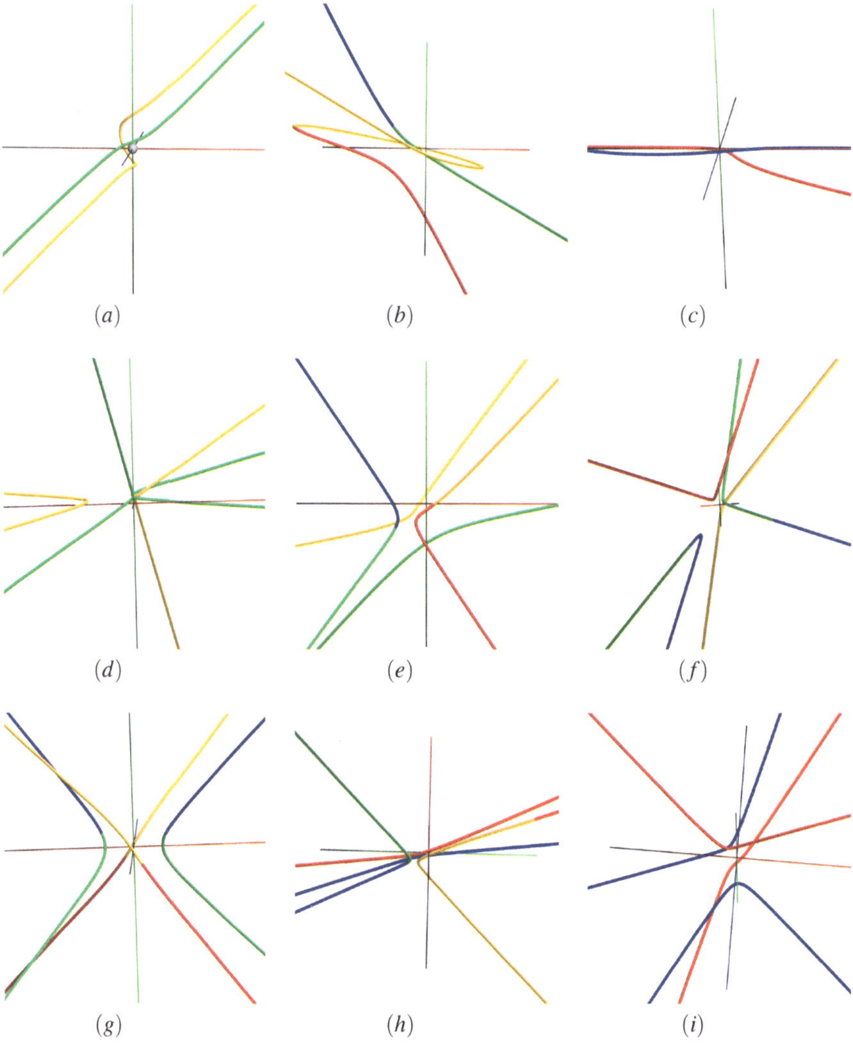

Fig. 4 This figure shows that for each of the nine configurations, it is possible to have as few transition points as the theoretical lower bound (the number of WT curves) in the field. (**a**) 2/0/0. (**b**) 0/2/0. (**c**) 0/0/2. (**d**) 4/0/0. (**e**) 2/2/0. (**f**) 1/2/1. (**g**) 0/4/0. (**h**) 0/2/2. (**i**) 0/0/4

Therefore, these bounds are not only tight in general, there are tight for each of the nine scenarios. In particular, there can be zero transition points in the field (Fig. 4a, c, d, and i).

We have also observed that a linear tensor field can have zero, two, four, six, and eight transition points (respectively Figs. 4a, b, g, 5a, and 3a). Moreover, along a single WW or TT curve, there can be zero, two, four, and six degenerate points (respectively Figs. 4a, 3d, b a). Along a WT degenerate curve, there can be one,

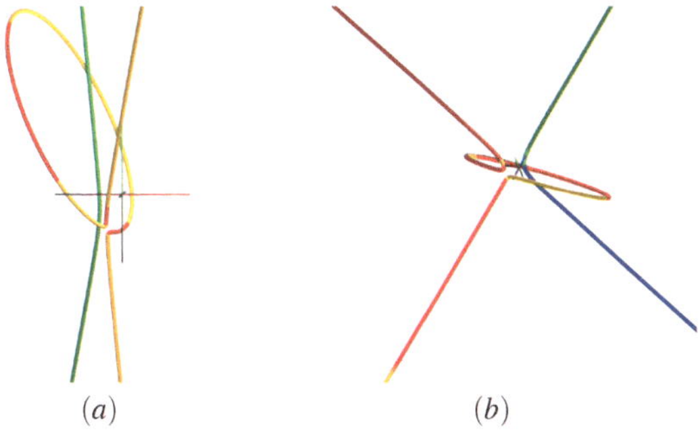

Fig. 5 This figure shows that a 3D linear tensor field can have a total of six transition points 6/0 (**a**) or seven transition points along one degenerate curve 7/1 (**b**)

three, five, and seven degenerate points (respectively Figs. 3e, f, b, and 5b). We therefore conjecture that along a degenerate curve, it is not possible to have eight degenerate points.

5 Conclusion

In this paper, we study the number of transition points in a 3D linear tensor field. We show that under structurally stable conditions there are at most 8 transition points. Moreover, we show that the minimum number of transition points is the same as the number WT degenerate curves in the field. Both of these bounds are tight for the nine scenarios in a linear tensor field.

In addition, we have established the theoretical lower-bound and upper-bound on the number of transition points that can occur on a single degenerate curve, which are zero (min) and eight (max) for WW and TT curves and one (min) and seven (max) for WT curves.

In practice, we have the lower-bounds for WW, WT, and TT degenerate curves to be tight, i.e., there are tensor fields which have degenerate curves with the given lower-bound transition points. Furthermore, we have also identified 7 to be a tight upper-bound for WT degenerate curves. On the other hand, the observed upper-bound for a WW or TT degenerate curve is *six*, which is a conjecture that we plan to investigate.

In the future, we plan to strive for a tight upper bound on the number of transition points on one WW or TT degenerate curve in a 3D linear tensor field. In addition, we plan to study the bifurcations in a 3D tensor field.

Acknowledgments We are also grateful for the constructive comments from our anonymous reviewers. Yue Zhang was partially supported by National Science Foundation Award 1566236. Eugene Zhang was partially supported by National Science Foundation Award 1619383.

References

1. Delmarcelle, T., Hesselink, L.: Visualizing second-order tensor fields with hyperstream lines. IEEE Comput. Graph. Appl. **13**(4), 25–33 (1993)
2. Delmarcelle, T., Hesselink, L.: The topology of symmetric, second-order tensor fields. In: Proceedings IEEE Visualization '94 (1994)
3. Fulton, W.: Algebraic Curves. Mathematics Lecture Note Series. W.A. Benjamin, New York (1974)
4. Hesselink, L., Levy, Y., Lavin, Y.: The topology of symmetric, second-order 3D tensor fields. IEEE Trans. Vis. Comput. Graph. **3**(1), 1–11 (1997)
5. Kratz, A., Auer, C., Stommel, M., Hotz, I.: Visualization and analysis of second-order tensors: moving beyond the symmetric positive-definite case. Comput. Graph. Forum **32**(1), 49–74 (2013). http://dblp.uni-trier.de/db/journals/cgf/cgf32.html#KratzASH13
6. Palacios, J., Roy, L., Kumar, P., Hsu, C.Y., Chen, W., Ma, C., Wei, L.Y., Zhang, E.: Tensor field design in volumes. ACM Trans. Graph. **36**(6), 188:1–188:15 (2017). http://doi.acm.org/10.1145/3130800.3130844
7. Palacios, J., Yeh, H., Wang, W., Zhang, Y., Laramee, R.S., Sharma, R., Schultz, T., Zhang, E.: Feature surfaces in symmetric tensor fields based on eigenvalue manifold. IEEE Trans. Vis. Comput. Graph. **22**(3), 1248–1260 (2016). http://dx.doi.org/10.1109/TVCG.2015.2484343
8. Roy, L., Kumar, P., Zhang, Y., Zhang, E.: Robust and fast extraction of 3d symmetric tensor field topology. IEEE Trans. Vis. Comput. Graph. **25**(1), 1102–1111 (2019). https://doi.org/10.1109/TVCG.2018.2864768
9. Zhang, Y., Palacios, J., Zhang, E.: Topology of 3D Linear Symmetric Tensor Fields, pp. 73–91. Springer International Publishing, Cham (2015). http://dx.doi.org/10.1007/978-3-319-15090-1_4
10. Zhang, Y., Tzeng, Y.J., Zhang, E.: Maximum number of degenerate curves in 3d linear tensor fields. In: Carr, H., Garth, C., Weinkauf, T. (eds.) Topological Methods in Data Analysis and Visualization IV, pp. 221–234. Springer International Publishing, Cham (2017)
11. Zheng, X., Pang, A.: Topological lines in 3d tensor fields. In: Proceedings IEEE Visualization 2004, VIS '04, pp. 313–320. IEEE Computer Society, Washington, DC (2004) http://dx.doi.org/10.1109/VISUAL.2004.105
12. Zheng, X., Parlett, B., Pang, A.: Topological structures of 3D tensor fields. In: Proceedings IEEE Visualization 2005, pp. 551–558 (2005)
13. Zheng, X., Parlett, B.N., Pang, A.: Topological lines in 3d tensor fields and discriminant hessian factorization. IEEE Trans. Vis. Comput. Graph. **11**(4), 395–407 (2005)

Discrete Poincaré Duality Angles as Shape Signatures on Simplicial Surfaces with Boundary

Konstantin Poelke and Konrad Polthier

Abstract We introduce and explore the concept of *discrete Poincaré duality angles* as an intrinsic measure that quantifies the metric-topological influence of boundary components to compact surfaces with boundary. Based on a discrete Hodge-Morrey-Friedrichs decomposition for piecewise constant vector fields on simplicial surfaces with boundary, the discrete Poincaré duality angles reflect a deep linkage between metric properties of the spaces of discrete harmonic Dirichlet and Neumann fields and the topology of the underlying surface, and may act as a new kind of shape signature. We provide an algorithm for the computation of these angles and discuss them on several exemplary surface models.

1 Introduction

Hodge-type decomposition statements form an indispensable tool for the analysis and structural understanding of vector fields and more generally differential forms on manifolds. Dating back at least to Helmholtz' classical result [18] on the decomposition of a vector field into a divergence-free and a rotation-free component, there has been a remarkable evolution of extensions and generalizations. Nowadays there is a well-developed theory for Hodge decompositions of differential forms of Sobolev class (see [15] for an overview), which is of central importance e.g. for finite element Galerkin methods for problems involving vector fields such as Maxwell's equations or Navier-Stokes systems. A surprising property is the strong linkage of certain spaces of harmonic forms to the topology of the underlying manifold, whose first encounter is given by de Rham's theorem, stating that on a closed oriented Riemannian manifold the space of harmonic k-forms is isomorphic to the k-th cohomology with real coefficients. On a surface with non-empty boundary, the corresponding statement applies to the spaces of harmonic Dirichlet fields $\mathcal{H}_D^1$

K. Poelke (✉) · K. Polthier
Freie Universität Berlin, Berlin, Germany
e-mail: konstantin.poelke@fu-berlin.de; konrad.polthier@fu-berlin.de

H. Carr et al. (eds.), *Topological Methods in Data Analysis and Visualization V*, Mathematics and Visualization, https://doi.org/10.1007/978-3-030-43036-8_16

and Neumann fields $\mathscr{H}_N^1$, which are subspaces of all harmonic fields with certain boundary conditions imposed. However, there are now two decompositions—one including $\mathscr{H}_D^1$, the other one including $\mathscr{H}_N^1$—and in general there is no single L^2-orthogonal decomposition including both these spaces at the same time. A recent result by Shonkwiler [16, 17] identifies the reason for this non-orthogonality as the existence of non-empty subspaces representing the *interior cohomology* of the manifold (in contrast to the *cohomology induced by the boundary components*), which establishes another astonishing linkage between metric properties and the topology. In particular, the principal angles between $\mathscr{H}_N^1$ and $\mathscr{H}_D^1$ seem to act as an indicator for the influence of boundary components on the overall geometry and therefore as a theoretical shape signature.

Contributions The main contribution of this article is the introduction of discrete Poincaré duality angles in Sect. 4, based on a discretization of harmonic Neumann and Dirichlet fields by piecewise constant vector fields on simplicial surfaces with boundary. Furthermore, we provide an algorithm for their numerical computation, using a singular value decomposition of a matrix whose size only depends on the topological complexity of the surface. Finally, we compute these angles for a few exemplary models and discuss their interpretation as shape signatures.

Related Work The literature on Hodge-type decomposition statements is vast and has a long history. For a modern treatment and a good overview on smooth and Sobolev-class Hodge-type decomposition statements see [15] and the literature referenced therein. The recent work by Shonkwiler [16, 17] introduces the concept of Poincaré duality angles and the splitting of Neumann and Dirichlet fields into interior cohomology- and boundary cohomology-representing subspaces on smooth oriented Riemannian manifolds. A general background on differential forms, Hodge decompositions and homology theory of manifolds can be found in standard textbooks such as [5, 9] and [10].

For the numerical treatment of vector fields there is a variety of discretization strategies available, cf. the overview article [3]. For instance, the *finite element exterior calculus* [2] by Arnold et al. introduces families of spaces of polynomial differential forms of arbitrary degree and generalizes classical ansatz spaces such as the Raviart-Thomas elements or Nédélec's elements. The *discrete exterior calculus* [7] by Hirani defines discrete differential forms as synonyms for simplicial cochains and relies on a dual grid to derive metric-dependent properties. Here we focus on a discretization by *piecewise constant vector fields* (PCVFs). Their usage and analysis in geometry processing tasks goes back at least to the work by Polthier and Preuss [13] and Wardetzky [19]. The interplay of linear Lagrange and Crouzeix-Raviart elements as ansatz spaces for gradients and cogradients which is central to these works can already be found in [1] for the special case of a simply-connected domain in $\mathbb{R}^2$. Since then, PCVFs have become a main ingredient for frame field modelling [20], remeshing [6, 14] or surface parameterization [8], just to name a few examples.

A complete, structurally consistent set of Hodge-type decompositions for PCVFs on simplicial surfaces and solids with boundary has been recently developed by

Poelke and Polthier [11, 12], and we refer the reader to these works for all details concerning theory, discretization, implementation and numerical solving left out in this article.

Outline Section 2 reviews the necessary background of smooth Hodge-type decompositions and Poincaré duality angles on smooth manifolds. In Sect. 3 we state the most important results from a discretization by piecewise constant vector fields as developed in [11, 12]. Furthermore, some exemplary angles between discrete harmonic Neumann and Dirichlet fields are computed, serving as a motivation for the introduction of discrete Poincaré duality angles, which are then introduced, computed and discussed in Sect. 4.

2 Hodge-Type Decompositions, Topology and Duality Angles

In its modern formulation, the Hodge decomposition theorem states that on a closed oriented Riemannian manifold M the space Ω^k of smooth k-forms can be decomposed L^2-orthogonally as

$$\Omega^k = \mathrm{d}\Omega^{k-1} \oplus \delta\Omega^{k+1} \oplus \mathcal{H}^k \tag{1}$$

where $\mathcal{H}^k$ is the space of harmonic k-forms satisfying $\mathrm{d}\omega = \delta\omega = 0$. Here and in the following, $\oplus$ always denotes an L^2-orthogonal direct sum. A remarkable result is de Rham's theorem which provides an isomorphism $\mathcal{H}^k \cong H^k(M)$ between the space of harmonic k-forms and the k-th cohomology with real coefficients on M, and therefore identifies the dimension of $\mathcal{H}^k$ as a *topological* invariant.

As soon as the manifold M has a non-empty boundary $\partial M \neq \emptyset$, Eq. (1) is no longer valid. Instead, the analogous splitting is now given by *two* decomposition statements known as the *Hodge-Morrey-Friedrichs decomposition* (see [15]):

$$\begin{aligned}\Omega^k &= \mathrm{d}\Omega_D^{k-1} \oplus \delta\Omega_N^{k+1} \oplus (\mathcal{H}^k \cap \mathrm{d}\Omega^{k-1}) \oplus \mathcal{H}_N^k \\ &= \mathrm{d}\Omega_D^{k-1} \oplus \delta\Omega_N^{k+1} \oplus (\mathcal{H}^k \cap \delta\Omega^{k+1}) \oplus \mathcal{H}_D^k.\end{aligned}$$

Here, the subscript $_D$ denotes *Dirichlet boundary conditions* (i.e. the tangential part $\mathbf{t}(\omega)$ of a differential form ω has to vanish along ∂M) and $_N$ denotes *Neumann boundary conditions* (i.e. the normal part $\omega\mid_{\partial M} -\mathbf{t}(\omega)$ has to vanish along ∂M) which are imposed on the corresponding spaces. Again, there are isomorphisms $\mathcal{H}_N^k \cong H^k(M)$ and $\mathcal{H}_D^k \cong H^k(M, \partial M)$, respectively, with the latter space $H^k(M, \partial M)$ denoting the k-th *relative* cohomology of M.

With respect to the characterization of vector fields on surfaces with boundary, a natural question is whether there is a single orthogonal decomposition including $\mathcal{H}_N^1$ and $\mathcal{H}_D^1$ at the same time. To this end, we say that a surface M *is of type*

$\Sigma_{g,m}$, if M is a compact orientable surface of genus $g \geq 0$ with $m \geq 0$ boundary components. We have the following result [11, Lemma 2.4.5]:

Lemma 1 *Let M be a surface of type $\Sigma_{0,m}$. Then there is an L^2-orthogonal decomposition*

$$\Omega^1 = \mathrm{d}\Omega^0_D \oplus \delta\Omega^2_N \oplus (\mathrm{d}\Omega^0 \cap \delta\Omega^2) \oplus \mathscr{H}^1_D \oplus \mathscr{H}^1_N.$$

Lemma 1 includes the common case of two-dimensional flat domains embedded in $\mathbb{R}^2$. On the other hand, if $g \geq 1$ the sum $\mathscr{H}^1_D \oplus \mathscr{H}^1_N$ is not L^2-orthogonal any more. A recent result by Shonkwiler [16, 17] identifies subspaces of $\mathscr{H}^1_D$ and $\mathscr{H}^1_N$, or more generally $\mathscr{H}^k_D$ and $\mathscr{H}^k_N$, representing the cohomology corresponding to the *inner topology* of M as the reason for this non-orthogonality. Shonkwiler introduces the spaces

$$\begin{aligned}
\mathscr{H}^k_{N,\mathrm{co}} &:= \mathscr{H}^k_N \cap \delta\Omega^{k+1} \\
\mathscr{H}^k_{N,\partial\mathrm{ex}} &:= \{\omega \in \mathscr{H}^k_N : \iota^*\omega \in \mathrm{d}\Omega^{k-1}(\partial M)\} \\
\mathscr{H}^k_{D,\mathrm{ex}} &:= \mathscr{H}^k_D \cap \mathrm{d}\Omega^{k-1} = \star\mathscr{H}^{n-k}_{N,\mathrm{co}} \\
\mathscr{H}^k_{D,\partial\mathrm{co}} &:= \{\omega \in \mathscr{H}^k_D : \iota^*(\star\omega) \in \mathrm{d}\Omega^{n-k-1}(\partial M)\} = \star\mathscr{H}^{n-k}_{N,\partial\mathrm{ex}}
\end{aligned} \tag{2}$$

of *coexact Neumann*, *boundary-exact Neumann*, *exact Dirichlet* and *boundary-coexact Dirichlet k-forms*, respectively. Here, $\iota : \partial M \hookrightarrow M$ denotes the inclusion, $\star$ is the Hodge star and n is the dimension of M. It is then shown [16, Thm. 2.1.3] that always $\mathscr{H}^k_{D,\mathrm{ex}} \perp \mathscr{H}^k_N$ and $\mathscr{H}^k_{N,\mathrm{co}} \perp \mathscr{H}^k_D$, but in general $\mathscr{H}^k_{N,\partial\mathrm{ex}} \not\perp \mathscr{H}^k_{D,\partial\mathrm{co}}$. Furthermore, the subspaces $\mathscr{H}^k_{D,\mathrm{ex}}$ and $\mathscr{H}^k_{N,\mathrm{co}}$ can be directly related to cohomology information that is induced by the boundary components, whereas the critical subspaces $\mathscr{H}^k_{N,\partial\mathrm{ex}}$ and $\mathscr{H}^k_{D,\partial\mathrm{co}}$ are related to the "interior topology" of the manifold, and it is this presence of interior topology that causes the orthogonality to fail. The amount of failure can be measured by the *Poincaré duality angles*, which are the principal angles between $\mathscr{H}^k_D$ and $\mathscr{H}^k_N$. Since the boundary-representing subspaces are always orthogonal, the interesting, non-trivial principal angles therefore arise between the spaces $\mathscr{H}^k_{N,\partial\mathrm{ex}}$ and $\mathscr{H}^k_{D,\partial\mathrm{co}}$. Shonkwiler computes these angles analytically for the complex projective space with an open ball removed, and the Grassmannian with a tubular neighbourhood of a sub-Grassmannian removed [16, Chap. 3 & 4]. In both cases the Poincaré duality angles quantify how far the manifolds are from being closed. In particular, shrinking the size of the boundary component lets the angles tend to zero, whereas increasing the size lets them tend to $\pi/2$. Furthermore, the order of convergence seems to encode the codimension of the removed submanifold, which is posed as a conjecture [16, Conj. 3]. For general manifolds, though, the analytic computation of these angles seems difficult or even intractable.

3 Discrete Neumann and Dirichlet Fields

Now, let $M_h \subset \mathbb{R}^3$ be a compact, orientable simplicial surface with (possibly empty) boundary, triangulated by affine triangles, equipped with the locally Euclidean metric. We use the subscript h to distinguish between the simplicial surface and discrete function spaces, and their smooth counterparts. It can be thought of as a discretization parameter and commonly refers to the maximum diameter of the affine triangles in the triangulation of M_h.

Let $\mathscr{X}_h$ denote the space of PCVFs on M_h, which are given by one tangent vector X_T per affine triangle T of M_h, and let $\mathscr{L}$ and $\mathscr{F}$ denote the finite element spaces of linear Lagrange and Crouzeix-Raviart elements on M_h. To be precise, $\mathscr{L}$ denotes the space of all continuous functions on M_h that are linear when restricted to an individual triangle T, and $\mathscr{F}$ is the space of all L^2-functions on M_h that are represented by a linear function when restricted to an individual triangle T such that the values of these linear representatives agree at the edge midpoint between any two adjacent triangles. Therefore, an element in $\mathscr{L}$ is uniquely defined by its values at vertices, whereas an element in $\mathscr{F}$ is uniquely defined by its values at edge midpoints. We denote by $\mathscr{L}_0 \subset \mathscr{L}$ the subspace of all linear Lagrange functions that vanish at the boundary, and by $\mathscr{F}_0 \subset \mathscr{F}$ the subspace of all Crouzeix-Raviart functions that vanish at all midpoints of boundary edges. Furthermore, we denote by

$$\begin{aligned} \nabla\mathscr{L} &:= \{\nabla\varphi : \varphi \in \mathscr{L}\} & \nabla\mathscr{L}_0 &:= \{\nabla\varphi : \varphi \in \mathscr{L}_0\} \\ J\nabla\mathscr{F} &:= \{J\nabla\psi : \psi \in \mathscr{F}\} & J\nabla\mathscr{F}_0 &:= \{J\nabla\psi : \psi \in \mathscr{F}_0\}, \end{aligned}$$

the gradient and cogradient spaces, respectively, formed by these ansatz functions. Here, ∇ is the element-wise surface gradient and J denotes a counter-clockwise (with respect to a fixed unit normal field) rotation by $\pi/2$ in the tangent plane of each triangle. In other words, if X_T is a vector in the tangent plane of a triangle T, and N_T is a unit normal on T, then $JX_T := N_T \times X_T$ acts as the cross-product with N_T.

It is not hard to prove that the spaces $\nabla\mathscr{L}_0$ and $J\nabla\mathscr{F}_0$ are L^2-orthogonal to each other [12, Sec. 3.1], even if $\partial M_h = \emptyset$ in which case we have $\mathscr{L}_0 = \mathscr{L}$ and $\mathscr{F}_0 = \mathscr{F}$. Let $\mathscr{H}_h$ denote the space of *discrete harmonic PCVFs*, defined as the L^2-orthogonal complement of the sum $\nabla\mathscr{L}_0 \oplus J\nabla\mathscr{F}_0$. If $\partial M_h = \emptyset$, i.e. if M_h is a closed surface of genus g, then there is a single, orthogonal decomposition of the space of PCVFs, given by

$$\mathscr{X}_h = \nabla\mathscr{L} \oplus J\nabla\mathscr{F} \oplus \mathscr{H}_h \tag{3}$$

and furthermore $\dim \mathscr{H}_h = 2g$ [19, Thm. 2.5.2]. On the other hand, if $\partial M_h \neq \emptyset$, i.e. $m \geq 1$, there are now two discrete Hodge-Morrey-Friedrichs decompositions [12, Cor. 3.3]:

Definition and Lemma 2 *If $M_h \neq \emptyset$, there are two L^2-orthogonal decompositions*

$$\mathscr{X}_h = \nabla \mathscr{L}_0 \oplus J\nabla \mathscr{F}_0 \oplus (\mathscr{H}_h \cap \nabla \mathscr{L}) \oplus \mathscr{H}_{h,N} \tag{4}$$

$$= \nabla \mathscr{L}_0 \oplus J\nabla \mathscr{F}_0 \oplus (\mathscr{H}_h \cap J\nabla \mathscr{F}) \oplus \mathscr{H}_{h,D} \tag{5}$$

where $\mathscr{H}_{h,N}$ and $\mathscr{H}_{h,D}$ are the spaces of discrete Neumann and Dirichlet fields, *defined as the L^2-orthogonal complement of the sum of the first three subspaces in Eqs.* (4) *and* (5)*, respectively.*

Furthermore, we have discrete de Rham isomorphisms $\mathscr{H}_{h,N} \cong H^1(M_h)$ and $\mathscr{H}_{h,D} \cong H^1(M_h, \partial M_h)$, and by Poincaré-Lefschetz duality it follows $\dim \mathscr{H}_{h,D} = \dim \mathscr{H}_{h,N} = 2g + m - 1$ for a surface of type $\Sigma_{g,m}$ with $m \geq 1$. Bases for the spaces $\mathscr{H}_{h,N}$ and $\mathscr{H}_{h,D}$ are shown in Fig. 1 for a surface of type $\Sigma_{0,3}$ and in Fig. 2 on a hand model with four boundary components, which is of type $\Sigma_{1,4}$.

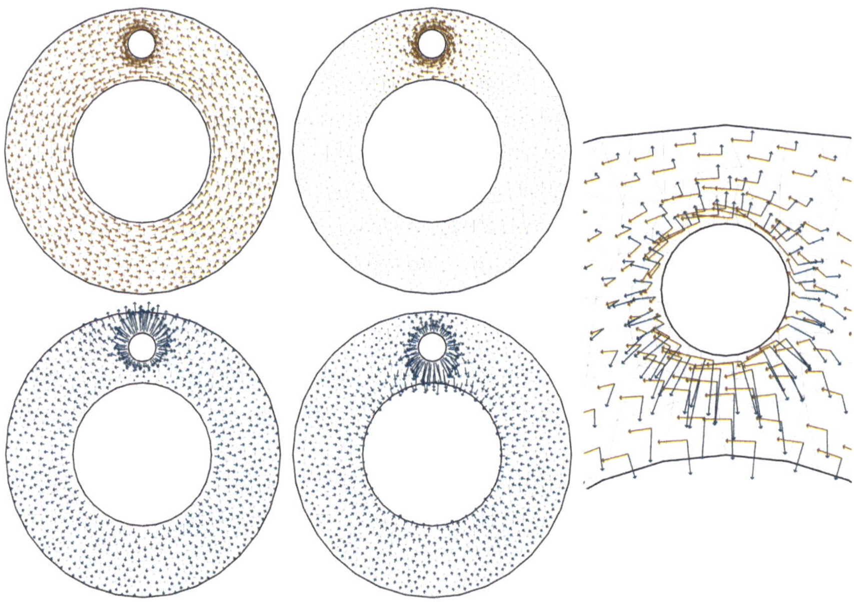

Fig. 1 Basis fields $X_{N,1}, X_{N,2}$ for $\mathscr{H}_{h,N}$ (top row) and $X_{D,1}, X_{D,2}$ for $\mathscr{H}_{h,D}$ (bottom row) on an annulus with a hole ("AwH"), which is a surface of type $\Sigma_{0,3}$. As in the smooth case, discrete Dirichlet fields are characterized by having a vanishing tangential component along the boundary, whereas discrete Neumann fields have an almost vanishing normal component (for details on why the normal component is not necessarily strictly vanishing in this discretization, see [12, Section 3.1]). Each discrete Neumann field is L^2-orthogonal to each discrete Dirichlet field on this model. Locally, though, these fields need not be orthogonal, as can be seen in the rightmost image, where $X_{N,1}$ and $X_{D,2}$ are shown in a close-up

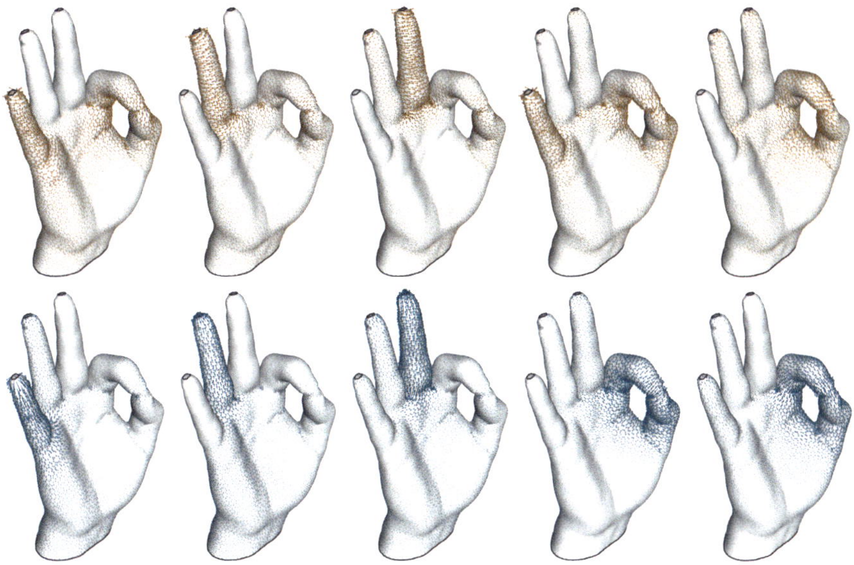

Fig. 2 Bases for the spaces $\mathscr{H}_{h,N}$ (top row) and $\mathscr{H}_{h,D}$ (bottom row) on a hand model with four boundary components (one hole cut out at each finger tip and the fourth one at the wrist)

Table 1 Angles between the basis fields for $\mathscr{H}_{h,N}$ and $\mathscr{H}_{h,D}$ on the flat AwH-model from Fig. 1 and the TwC-model from Fig. 3 in radians. The bold values in the TwC table belong to the two pairs in the close-up Fig. 4

AwH	$X_{D,1}$	$X_{D,2}$
$X_{N,1}$	1.57	1.57
$X_{N,2}$	1.57	1.57

TwC	$X_{D,1}$	$X_{D,2}$	$X_{D,3}$
$X_{N,1}$	2.30	1.57	**0.74**
$X_{N,2}$	1.62	1.57	1.55
$X_{N,3}$	2.41	1.58	**2.31**

As in the smooth case in Lemma 1, for simplicial surfaces of type $\Sigma_{0,m}$ both spaces are always L^2-orthogonal to each other and consequently there is a single complete discrete decomposition [12, Lemma 3.10]

$$\mathscr{X}_h = \nabla \mathscr{L}_0 \oplus J\nabla \mathscr{F}_0 \oplus (J\nabla \mathscr{F} \cap \nabla \mathscr{L}) \oplus \mathscr{H}_{h,D} \oplus \mathscr{H}_{h,N}. \tag{6}$$

The numerical angles in Table 1 confirm this result for the surface of type $\Sigma_{0,3}$ from Fig. 1. Each angle $\alpha = \measuredangle(X, Y)$ is computed as usual by

$$\cos \alpha = \frac{\langle X, Y \rangle_{L^2}}{\|X\|_{L^2} \, \|Y\|_{L^2}} \quad \text{for} \quad X \in \mathscr{H}_{h,N}, \; Y \in \mathscr{H}_{h,D}.$$

Note that the orthogonality of the shown vector fields is always meant with respect to the L^2-product on $\mathscr{X}_h$. Locally, these fields are in general not orthogonal, see the rightmost image in Fig. 1.

In contrast, the example in Fig. 3 shows bases for the three-dimensional spaces $\mathscr{H}_{h,N}$ and $\mathscr{H}_{h,D}$ on a torus with a cylinder attached, which is of type $\Sigma_{1,2}$. Whereas both the second Neumann and Dirichlet fields form an angle of almost $\pi/2$ to all other fields, this is not true for the those fields, whose masses concentrate on the toroidal region. Figure 4 shows a close-up of two pairs of fields on the toroidal region, one forming locally acute angles, the other forming locally obtuse angles. As their mass on the cylindrical region is negligible, the local situation here dominates the L^2-angle, and indeed the first pairing forms an acute L^2-angle of 0.74 radians,

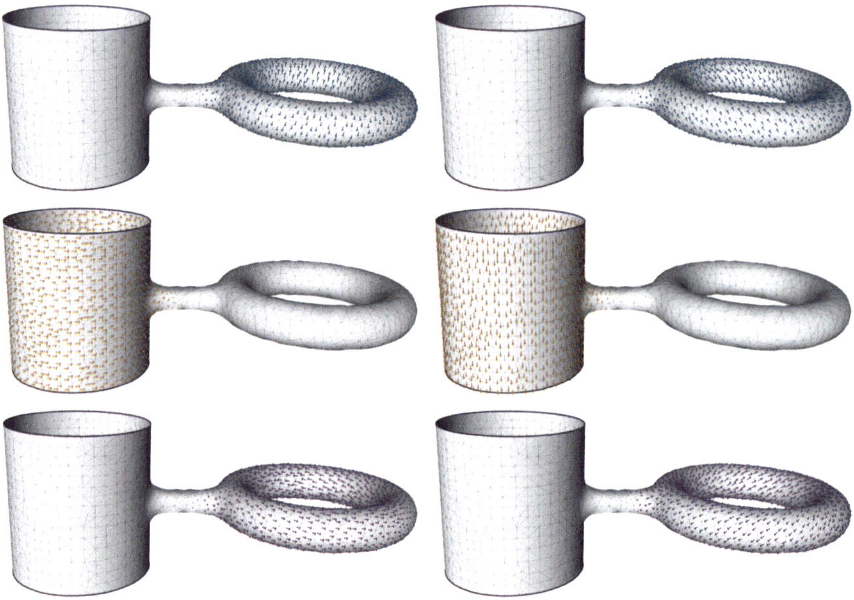

Fig. 3 Basis fields $X_{N,1}$, $X_{N,2}$, $X_{N,3}$ for $\mathscr{H}_{h,N}$ (left column, top to bottom) and $X_{D,1}$, $X_{D,2}$, $X_{D,3}$ for $\mathscr{H}_{h,D}$ (right column, top to bottom) on a torus with a cylinder attached ("TwC"), which is topologically a surface of type $\Sigma_{1,2}$. The fields in the first and third row all concentrate their mass in the same fashion along the longitudinal and latitudinal cycles that reflect homology generated by the torus

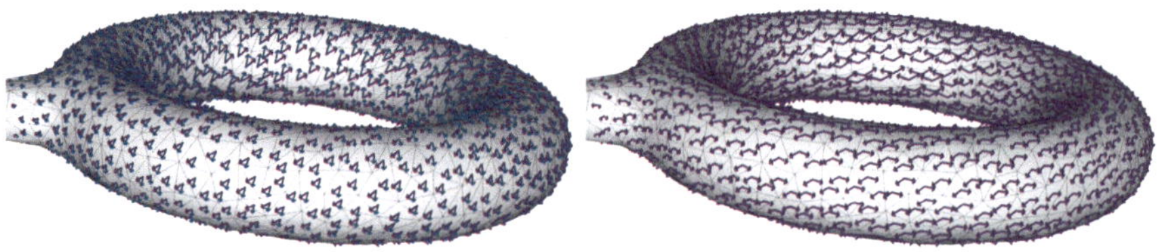

Fig. 4 Two pairings of Neumann and Dirichlet fields from the bases shown in Fig. 3. The left image shows the first Neumann field $X_{N,1}$ and the third Dirichlet field $X_{D,3}$, forming locally acute angles on each triangle on the torus region. The right image shows the third Neumann field $X_{N,3}$ and the third Dirichlet field $X_{D,3}$, forming obtuse angles

whereas the second pairing forms an obtuse L^2-angle of 2.31 radians, see Table 1. Consequently, the spaces $\mathscr{H}_{h,N}$ and $\mathscr{H}_{h,D}$ cannot appear simultaneously in a single orthogonal decomposition on this model.

4 Discrete Poincaré Duality Angles as Shape Signatures

The previous examples depend on the particular choice of basis vector fields whose pairwise angles are measured. In order to get rid of this dependence, we consider instead the principal angles between the vector spaces $\mathscr{H}_{h,N}$ and $\mathscr{H}_{h,D}$, which are independent of any concrete choice of basis. In accordance with the smooth situation we call these principal angles *discrete Poincaré duality angles* and define them as follows:

Definition 3 (Discrete Poincaré Duality Angles) The *discrete Poincaré duality angles* on M_h are the principal angles $0 \leq \theta_1 \leq \theta_2 \leq \cdots \leq \theta_{2g+m-1} \leq \pi/2$ between the spaces $\mathscr{H}_{h,N}$ and $\mathscr{H}_{h,D}$, defined recursively by

$$\theta_1 := \measuredangle(u_1, v_1) := \min\{\measuredangle(u, v) : u \in \mathscr{H}_{h,N}, v \in \mathscr{H}_{h,D}\}$$

$$\theta_k := \measuredangle(u_k, v_k) := \min\left\{\measuredangle(u, v) : \begin{array}{l} u \in \mathscr{H}_{h,N}, v \in \mathscr{H}_{h,D} \text{ with } u \perp u_i, v \perp v_i \\ \text{for all } i = 1, \ldots, k-1 \end{array}\right\}.$$

We stress again that the sequence of discrete Poincaré duality angles, being defined as principal angles between linear subspaces of $\mathscr{X}_h$, only depends on $\mathscr{H}_{h,N}$ and $\mathscr{H}_{h,D}$, but not on a concrete choice of vectors. In particular, whereas the sequence $\theta_1, \theta_2, \ldots$ is uniquely determined for a given simplicial surface M_h, the vectors u_k, v_k realizing an angle θ_k are not (cf. [4]).

A trivial consequence of Definition 3 is that if $\theta_i = \pi/2$ for all i, then clearly $\mathscr{H}_{h,N} \perp \mathscr{H}_{h,D}$. In general, though, these angles measure the deviation of the spaces $\mathscr{H}_{h,N}$ and $\mathscr{H}_{h,D}$ from being orthogonal to each other. In this sense they can be thought of as a quantitative value that measures how far away the two discrete Hodge-Morrey-Friedrichs-decompositions Eqs. (4) and (5) are from being either the single decomposition Eq. (1) on a closed surface of type $\Sigma_{g,0}$ or the complete orthogonal decomposition Eq. (6) on a domain coming from a sphere, i.e. a surface of type $\Sigma_{0,m}$.

This is illustrated by the sequence in Fig. 5, which shows a torus surface of type $\Sigma_{1,2}$. In the beginning the boundary components are almost negligible and start to grow until the final surface in the sequence is ultimately dominated by the boundary components. This behaviour is numerically reflected by the sequence of the corresponding Poincaré duality angles given in Table 2.

Figure 6 shows a similar sequence of a genus-2-surface with three boundary components, so it is of type $\Sigma_{2,3}$. Here, the leftmost growing boundary component has a dramatic influence on the pair (θ_3, θ_4), corresponding to the left torus region. The other duality angles remain mostly unaffected.

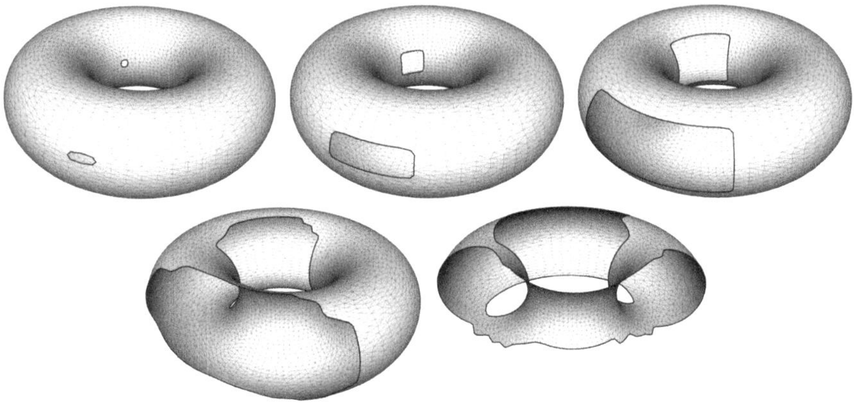

Fig. 5 A torus sequence (denoted "g1h2(a) – g1h2(e)") with two growing boundary components. Although all five models are topologically equivalent, the influence of the boundary components is drastically increasing and turns the initially almost closed surface into a surface dominated by its boundary. This is reflected by the numerical values in Table 2

Table 2 Poincaré duality angles for all experiments

Model	Type	θ_1	θ_2	θ_3	θ_4	θ_5	θ_6
TwC	$\Sigma_{1,2}$	$1.73 \cdot 10^{-4}$	$1.76 \cdot 10^{-4}$	1.57	–	–	–
Hand	$\Sigma_{1,4}$	0.03	0.03	1.57	1.57	1.57	–
g1h2(a)	$\Sigma_{1,2}$	0.08	0.09	1.57	–	–	–
g1h2(b)	$\Sigma_{1,2}$	0.28	0.28	1.57	–	–	–
g1h2(c)	$\Sigma_{1,2}$	0.62	0.62	1.57	–	–	–
g1h2(d)	$\Sigma_{1,2}$	0.95	0.95	1.57	–	–	–
g1h2(e)	$\Sigma_{1,2}$	1.32	1.32	1.57	–	–	–
g2h3(a)	$\Sigma_{2,3}$	0.13	0.14	0.26	0.27	1.57	1.57
g2h3(b)	$\Sigma_{2,3}$	0.14	0.15	0.70	0.70	1.57	1.57
g2h3(c)	$\Sigma_{2,3}$	0.21	0.21	1.14	1.14	1.57	1.57

A dash means that these angles do not exist for the respective surface, as there are always only $2g + m - 1$ duality angles

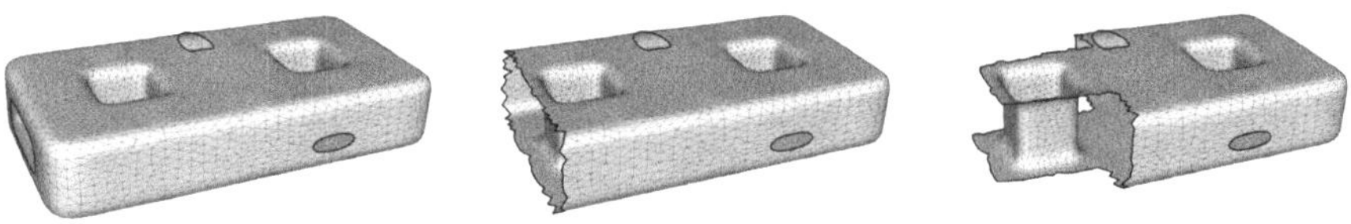

Fig. 6 A sequence of a surface (denoted "g2h3(a) – g2h3(c)") of type $\Sigma_{2,3}$. The leftmost boundary is growing and dominates the left torus component. This is reflected by the increasing angles θ_3 and θ_4 in Table 2, corresponding to the left toroidal region. Note that the other toroidal region remains mostly unaffected by the growing boundary component—the angles θ_1 and θ_2 merely increase from 0.13 to 0.21

The small angles θ_1 and θ_2 on the TwC-model suggest that the torus region is almost decoupled from the boundary components, and indeed this was the intention behind this rather artificial model. Here, $\mathscr{H}_{h,N}$ and $\mathscr{H}_{h,D}$ almost collapse to a single harmonic space of dimension 2 which would exist on a perfect torus. Moreover, by comparing with the values in Table 1 we see that the vector fields $X_{N,1}$, $X_{N,3}$ and $X_{D,1}$, $X_{D,3}$ in Fig. 3 are far from realizing the discrete Poincaré duality angles.

To a lesser extent, this situation is also prevalent for the hand model, where the torus region formed by thumb and index finger appears more integrated in the remaining part of the surface. Consequently, the angles θ_1 and θ_2, albeit small, are of magnitudes larger than in the TwC-model.

Note that in all examples in Table 2, there exist principal angles of value 1.57. This can be explained as follows. In analogy to the smooth case Eq. (2) there are splittings

$$\mathscr{H}_{h,N,\mathrm{co}} := \mathscr{H}_{h,N} \cap J\nabla\mathscr{F}$$

$$\mathscr{H}_{h,N,\partial\mathrm{ex}} := (\mathscr{H}_{h,N,\mathrm{co}})^{\perp} \cap \mathscr{H}_{h,N}$$

$$\mathscr{H}_{h,D,\mathrm{ex}} := \mathscr{H}_{h,D} \cap \nabla\mathscr{L}$$

$$\mathscr{H}_{h,D,\partial\mathrm{co}} := (\mathscr{H}_{h,D,\mathrm{ex}})^{\perp} \cap \mathscr{H}_{h,D}$$

of *discrete coexact Neumann fields, discrete boundary-exact Neumann fields* and so on, and it holds $\mathscr{H}_{h,N,\mathrm{co}} \perp \mathscr{H}_{h,D}$ and $\mathscr{H}_{h,D,\mathrm{ex}} \perp \mathscr{H}_{h,N}$ [12, Section 3.4]. Furthermore, if M_h is of type $\Sigma_{g,m}$ with $m \geq 1$, then we have [12, Lemmas 3.8/3.9]

$$\dim \mathscr{H}_{h,N,\mathrm{co}} = \mathscr{H}_{h,D,\mathrm{ex}} = m - 1.$$

This explains the $(m-1)$-many right angles in each of the experiments in Table 2.

Finally, it should be noted that there is a subtlety in the discrete theory that may arise in pathological examples and depends only on the grid combinatorics, i.e. the connectivity on the triangulation: whereas in the smooth case it is always $\mathscr{H}^k_N \cap \mathscr{H}^k_D = \{0\}$ [15, Thm. 3.4.4], the corresponding intersection of discrete spaces $\mathscr{H}_{h,N} \cap \mathscr{H}_{h,D}$ need not be trivial, resulting in invalid duality angles of value zero. This may happen if the discretization of the boundary is too low in comparison to the topological complexity (i.e. the numbers m and g of $\Sigma_{g,m}$) of the surface, or if there are very coarsely triangulated regions that form the only connections of the boundaries to the rest of the surface. A precise treatment as well as a criterion that guarantees the validity of the statement $\mathscr{H}_{h,N} \cap \mathscr{H}_{h,D} = \{0\}$ is given in [11, Section 3.4]. While it is important to keep this in mind, most models that arise in practice are not affected by this pathology.

Numerical Computation of Poincaré Duality Angles The numerical computation of principal angles between subspaces of a vector space is classical [4] and can be easily computed by means of the singular value decomposition (SVD). To this end we first compute orthonormal bases (ONB) $\mathscr{B}_N$ and $\mathscr{B}_D$ for the subspaces

$\mathcal{H}_{h,N}$ and $\mathcal{H}_{h,D}$, respectively. We refer the reader to [12, Sec. 4.1] for details on how to obtain these bases. By [4, Thm. 1] the principal angles are then the inverse cosines of the singular values of the matrix $\left(\langle u, v\rangle_{L^2}\right)_{u\in\mathcal{B}_N, v\in\mathcal{B}_D}$, which is of dimension $(2g+m-1)\times(2g+m-1)$. As $2g+m-1$ is typically small, the computation of this SVD poses no problems in terms of memory consumption or computation time. Listing 1 summarizes this procedure.

Listing 1 Numerical computation of discrete Poincaré duality angles

```
Input: Simplicial surface mesh M_h
  Compute ONB B_N for H_{h,N}
  Compute ONB B_D for H_{h,D}
  Assemble the matrix M := (<u, v>_{L^2})_{u∈B_N, v∈B_D}
  Compute the SVD M = Y · Σ · Z^t with Σ = diag(ζ_1, ..., ζ_{2g+m-1}), ζ_i ≥ ζ_{i+1}
  return {θ_i := arccos ζ_i}_{i=1,...,2g+m-1}
```

5 Conclusion and Outlook

We have introduced the notion of *discrete Poincaré duality angles* on simplicial surfaces with boundary as an intrinsic quantity that sets the geometry, i.e. the metric properties of the surface, in relation to its topology, which is determined by its boundary components and the genus of the corresponding closed surface. In particular, these angles measure the influence of the boundary components on the overall geometry and quantify how far the surface is from being a closed surface. Rephrasing this fact algebraically, they quantify how far the two discrete Hodge-Morrey-Friedrichs decompositions Eqs. (4) and (5) differ from either the single decomposition Eq. (3) on closed surfaces of type $\Sigma_{g,0}$ or the complete orthogonal decomposition Eq. (6) on surfaces of type $\Sigma_{0,m}$ in which both the spaces $\mathcal{H}_{h,N}$ and $\mathcal{H}_{h,D}$ appear simultaneously as orthogonal subspaces. In this sense, the vector $(\theta_1, \ldots, \theta_{2g+m-1})$ of discrete Poincaré duality angles can be considered as a new intrinsic shape signature for a geometric-topological classification of simplicial surfaces with boundary.

On the other hand, it is still not clear in which way *precisely* the angles between Dirichlet and Neumann fields are related to the boundary components of M_h. The examples explicitly computed in [16] are a first starting point for the search of a relation that could be even quantitatively described, but as mentioned in [17], in general they still seem to be poorly understood. In this respect, the numerical computation of discrete Poincaré duality angles complements the smooth theory and may provide further insights. Moreover, for arbitrarily complex surfaces that arise from CAD-modelling or 3D scanning, an analytic computation seems out of reach.

Therefore, a better understanding of this correlation is very promising with regard to applications including metric-topological shape classification, extraction of certain vector field components with controlled characteristics and parametrization tasks of surfaces with boundary, and remains for future work.

Acknowledgments The authors thank the anonymous reviewers for their detailed and valuable feedback. This work was supported by the Einstein Center for Mathematics Berlin. The Laurent's hand model on which the hand model in Fig. 2 is based on is provided courtesy of INRIA by the AIM@SHAPE-VISIONAIR Shape Repository.

References

1. Arnold, D.N., Falk, R.S.: A uniformly accurate finite element method for the Reissner–Mindlin plate. SIAM J. Numer. Anal. **26**(6), 1276–1290 (1989)
2. Arnold, D.N., Falk, R.S., Winther, R.: Finite element exterior calculus, homological techniques, and applications. Acta Numer. **15**, 1–155 (2006)
3. Bhatia, H., Norgard, G., Pascucci, V., Bremer, P.-T.: The Helmholtz-Hodge decomposition - a survey. IEEE Trans. Vis. Comput. Graph. **19**(8), 1386–1404 (2013)
4. Björck, Å., Golub, G.H.: Numerical methods for computing angles between linear subspaces. Math. Comput. **27**(123), 579–594 (1973)
5. Bredon, G.E.: Topology and Geometry. Graduate Texts in Mathematics. Springer, New York (1993)
6. Dong, S., Kircher, S., Garland, M.: Harmonic functions for quadrilateral remeshing of arbitrary manifolds. Comput. Aided Geom. Des. **22**(5), 392–423 (2005)
7. Hirani, A.N.: Discrete exterior calculus. PhD thesis, California Institute of Technology (2003)
8. Kälberer, F., Nieser, M., Polthier, K.: QuadCover – surface parameterization using branched coverings. Comput. Graph. Forum **26**(3), 375–384 (2007)
9. Lee, J.M.: Introduction to Smooth Manifolds. Graduate Texts in Mathematics. Springer, New York (2003)
10. Munkres, J.R.: Elements of algebraic topology. Advanced Book Classics. Perseus Books, Cambridge, MA (1984)
11. Poelke, K.: Hodge-Type Decompositions for Piecewise Constant Vector Fields on Simplicial Surfaces and Solids with Boundary. PhD thesis, Freie Universität Berlin (2017)
12. Poelke, K., Polthier, K.: Boundary-aware Hodge decompositions for piecewise constant vector fields. Comput.-Aided Des. **78**, 126–136 (2016)
13. Polthier, K., Preuss, E.: Identifying vector field singularities using a discrete Hodge decomposition. In: Hege, H.-C., Polthier, K. (eds.) Visualization and Mathematics III, pp. 113–134. Springer, New York (2003)
14. Schall, O., Zayer, R., Seidel, H.-P.: Controlled field generation for quad-remeshing. In: Proceedings of the 2008 ACM Symposium on Solid and Physical Modeling, SPM '08, pp. 295–300. ACM, New York (2008)
15. Schwarz, G.: Hodge Decomposition: A Method for Solving Boundary Value Problems. Lecture Notes in Mathematics. Springer, Berlin, Heidelberg (1995)
16. Shonkwiler, C.: Poincaré Duality Angles on Riemannian Manifolds with Boundary. PhD thesis, University of Pennsylvania (2009)
17. Shonkwiler, C.: Poincaré duality angles and the Dirichlet-to-Neumann operator. Inverse Prob. **29**(4), 045007 (2013)
18. von Helmholtz, H.: Über Integrale der hydrodynamischen Gleichungen, welche den Wirbelbewegungen entsprechen. Journal für die reine und angewandte Mathematik **55**, 25–55 (1858)
19. Wardetzky, M.: Discrete Differential Operators on Polyhedral Surfaces - Convergence and Approximation. PhD thesis, Freie Universität Berlin (2006)
20. Xu, K., Zhang, H., Cohen-Or, D., Xiong, Y.: Dynamic harmonic fields for surface processing. Comput. Graph. **33**(3), 391–398 (2009)

Index

H. Carr et al. (eds.), *Topological Methods in Data Analysis and Visualization V*, Mathematics and Visualization, https://doi.org/10.1007/978-3-030-43036-8

R

S

T

The manufacturer's authorised representative in the EU is Springer Nature Customer Service Centre GmbH, Europaplatz 3, 69115 Heidelberg, Germany. If you have any concerns regarding our products, please contact ProductSafety@springernature.com

Printed and bound by CPI Group (UK) Ltd, Croydon, CR0 4YY
30/09/2026
02225456-0018